Technician mathematics 1
Third edition

Series Editor:

D. R. Browning, B.Sc., F.R.S.C., C.Chem., A.R.T.C.S.
Formerly Principal Lecturer and Head of Chemistry, Bristol Polytechnic

Other titles in this series:

Technician mathematics 2 Third edition *J O Bird and A J C May*
Technician mathematics 3 Second edition *J O Bird and A J C May*
Technician mathematics 4/5 Second edition *J O Bird and A J C May*
Mathematics for electrical technicians 4/5 Second edition *J O Bird and A J C May*

Other mathematics titles:

Mathematics for scientific and technical students *H G Davies and G A Hicks*
Mathematical formulae Second edition *J O Bird and A J C May*

Technician mathematics 1

Third edition

J. O. Bird
B.Sc.(Hons), C.Math., F.I.M.A., C.Eng., M.I.E.E., F.Coll.P., F.I.E.I.E.

A. J. C. May
B.A., C.Eng., M.I.Mech.E., F.I.E.I.E., M.B.I.M.

Longman Scientific & Technical

Longman Scientific & Technical
Longman Group UK Limited
Longman House, Burnt Mill, Harlow
Essex CM20 2JE, England
and Associated Companies throughout the world

Copublished in the United States with
John Wiley & Sons, Inc., 605 Third Avenue, New York
NY 10158

First published 1977
Second edition 1982
Third edition 1994

British Library Cataloguing in Publication Data
A catalogue entry for this title is available from the British Library.

ISBN 0-582-23426-3

Set by 16SS in 10/12 Times
Printed in Malaysia by CL

Contents

Acknowledgements

We are indebted to the City of Portsmouth, Passenger Transport Department for our Table 3.5 from *Leigh Park Area Timetable* and also to British Rail 'Intercity guide to services' for our Tables 3.3 and 3.4.

Preface

Technician mathematics 1 is the first of a series which deal simply and carefully with the fundamental mathematics essential in the development of technicians and engineers.

The material for the third edition of this successful textbook has been updated and expanded to cover all the likely requirements of Mathematics modules contained in the Business and Technology Education Council First Certificate and Diploma courses and to provide valuable revision for students studying Mathematics modules required for National Certificate or National Diploma courses. Four-figure tables are assumed to be redundant; the electronic calculator is now an accepted essential tool in technician and engineering studies. The modifications in this third edition reflect this change.

The aim of the book is to consolidate basic mathematical principles and establish a common base for further progress, taking into account the wide variety of approaches to mathematics previously encountered at earlier stages of education.

Each topic considered in the text is presented in a way that assumes in the reader little previous knowledge of that topic. This practical mathematics book contains nearly 400 detailed worked problems followed by some 750 further problems, all with answers. Although specifically written for the BTEC first level Mathematics module, the book will also be suitable for NVQ and gNVQ studies and for students studying for their GCSE.

I would like to express my appreciation for the friendly co-operation and helpful advice given by the publishers, and to Mrs. Elaine Woolley for the excellent typing of the manuscript.

Finally, I would like to pay tribute to my colleague, friend and co-author Tony May who sadly died in 1991 after a short illness.

John O. Bird
Highbury College, Portsmouth

Chapter 1

Revision of basic arithmetic

1.1 Arithmetic operations

Whole numbers such as +5, +16, and +7 are called **positive integers**, while whole numbers such as −3, −8, and −29, are called **negative integers**. The four basic arithmetic operations associated with these numbers are shown in Table 1.1.

Table 1.1

Operation	Symbol	Example
Addition	+	$12 + 4 = 16$
Subtraction	−	$12 - 4 = 8$
Multiplication	×	$12 \times 4 = 48$
Division	÷	$12 \div 4 = 3$

When positive and negative integers occur in the same calculation the following rules apply.

Unlike signs give a negative overall sign.

Like signs give a positive overall sign.

The addition of +4 to −2, is written $4 + -2$ since, if no sign is shown in front of an integer, a positive sign is assumed. The + and − are unlike signs and give a negative overall sign. Thus $4 + -2 = 4 - 2 = 2$.

Similarly, subtracting −2 from 4 is written $4 - -2$. The − and − are like signs and give a positive overall sign, hence $4 - -2 = 4 + 2 = 6$

In multiplication + × + gives + and + × − gives −,
− × − gives + and − × + gives −.

Thus $4 \times -2 = -8$, whereas $-4 \times -2 = +8$, which is just written as 8.

For division $4 \div -2 = -2$ (unlike signs) and
$-4 \div -2 = 2$ (like signs).

Positive and negative integers can be represented by points marked on a line. The number 0 is the centre of the lines and 1, 2, 3, . . . are represented by spaces to the right of 0. To find the value of $3 + 2$ involves moving 3 places along the line to the right and then a further two places to the right since addition of positive integers is obtained by movement to the right. Figure 1.1 shows that doing this results in finishing by the number 5.

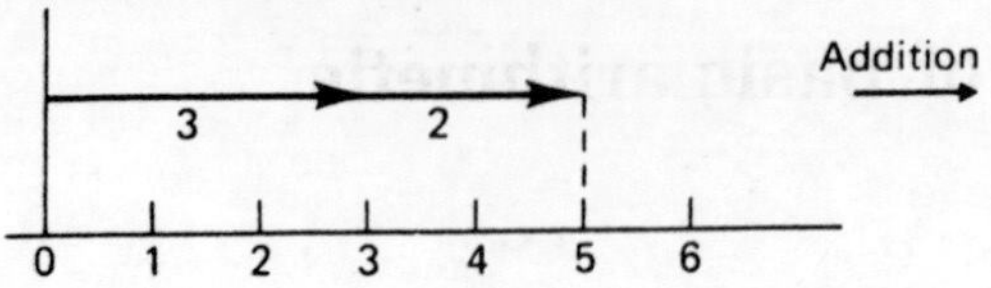

Fig. 1.1

The value of 3 − 2 is found by moving 3 places to the right and then 2 places to the left, since subtraction of positive integers is obtained by movement to the left. Figure 1.2 shows that the value of 3 − 2 is 1.

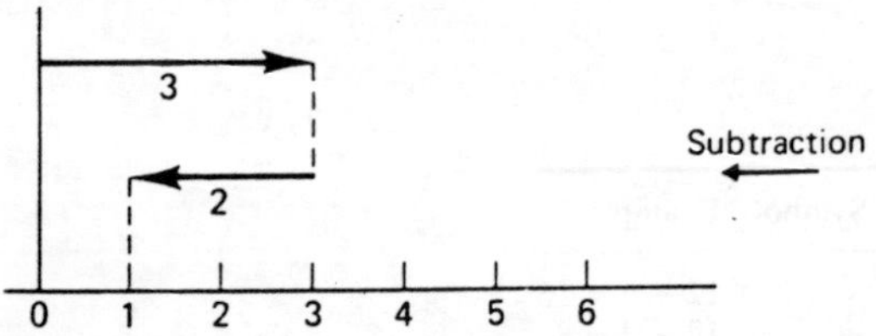

Fig. 1.2

Negative integers −1, −2, −3, . . . are represented by spaces to the left of 0. The value of 3 − 5 can be found by starting at 0 and moving 3 places to the right and then 5 places to the left. Figure 1.3 shows that 3 − 5 is −2.

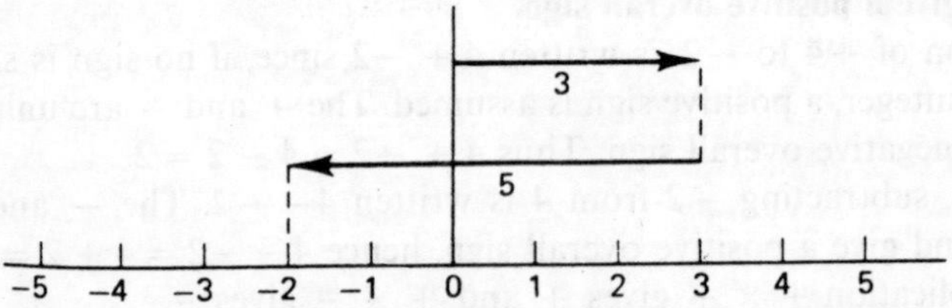

Fig. 1.3

When adding positive and negative integers, the overall sign in front of a number governs the direction of movement. For example, −6 + −10 becomes −6 − 10 or −16 and can be indicated by movement to the left. Also −4 − −10 becomes −4 + 10 or +6 and can be shown by movement to the right.

The number 234 means $(2 \times 100) + (3 \times 10) + (4 \times 1)$ and the number 561 means $(5 \times 100) + (6 \times 10) + (1 \times 1)$. Adding these numbers gives

$$(2 \times 100) + (3 \times 10) + (4 \times 1) + (5 \times 100) + (6 \times 10) + (1 \times 1)$$

and grouping like terms gives

$$(2 + 5) \times 100 + (3 + 6) \times 10 + (4 + 1) \times 1$$

or

$$700 \qquad +90 \qquad +5$$

that is, 795.

When adding or subtracting integers, the units, tens, hundreds, . . . columns must be kept underneath one another. Thus

```
        234
        561
Adding  795
```

Worked problems on arithmetic operations

Problem 1. Add 263, −146, 329 and −28

This would be written mathematically as 263 − 146 + 329 − 28

Adding the positive integers

```
        263
        329
Adding  592
```

592 is known as the **sum of the positive integers**. Adding the negative integers

```
        146
         28
Adding  174
```

174 is known as the **sum of the negative integers**. Taking the sum of the negative integers from the sum of the positive integers

```
             592
             174
Subtracting  418
```

Hence 263 − 146 + 329 − 28 = **418**

Problem 2. Subtract −26 from 284 (or find the value of 284 − −26)

When dealing with subtraction care must be taken in deciding what is to be taken from what. One way of determining this is to use the words **from** and **take**. This question then becomes **from** 284 **take** −26 and is written 284 − −26. The like signs give a positive overall sign, hence 284 − −26 = 284 + 26

```
        284
         26
Adding  310
```

Hence 284 − −26 = **310**

Problem 3. Subtract 143 from 72 (or find the value of 72 − 143)

To depict the number 72, a line 72 units long is drawn to the right. To subtract 143 from this number, a line starting at point 72 and 143 units long is drawn to the left. The answer, with reference to Fig. 1.4 is **−71 units** from 0

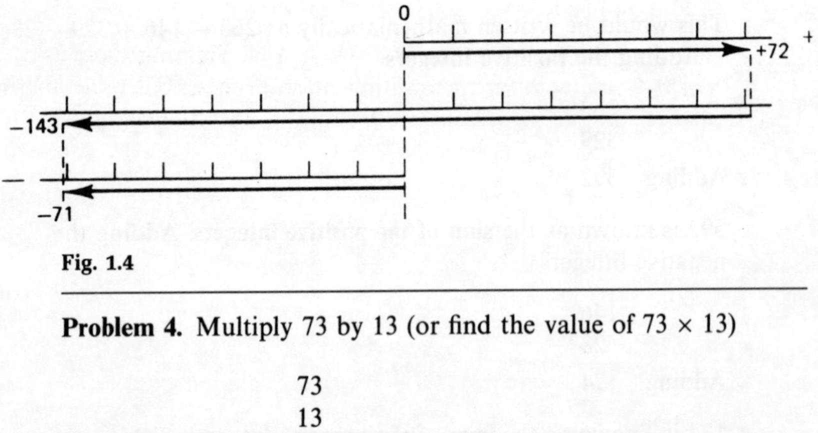

Fig. 1.4

Problem 4. Multiply 73 by 13 (or find the value of 73 × 13)

```
                     73
                     13
Multiply 73 by 3.   219   The units column figure is written under the 3.
Multiply 73 by 1.   73    The units column figure is written under the 1.
Adding              949
```

Thus 73 × 13 = **949**

Problem 5. Find the value of 215 × 63

$$\begin{array}{lrl} & 215 & \\ & 63 & \\ \hline \text{Multiply 215 by 3:} & 645 & \text{The units column figure is written under the 3.} \\ \text{Multiply 215 by 6.} & 1290 & \text{The units column figure is written under the 6 of the 63.} \\ \hline \text{Adding} & 13\,545 & \\ \hline \end{array}$$

Thus 215 × 63 = **13 545**

Problem 6. Find the value of 215 × −63

The signs are unlike and will give a negative answer. Remembering this, 215 is multiplied by 63 and gives 13 545 (from problem 5) and 215 × −63 = **−13 545** due to the unlike signs.

Problem 7. Find the value of 29 × 63 × −159

$$\begin{array}{lr} & 29 \\ & 63 \\ \hline 29 \times 3 & 87 \\ 29 \times 6 & 174 \\ \hline \text{Adding} & 1\,827 \\ \hline \end{array}$$

The problem now becomes 1 827 × −159. Treating these as positive numbers for the multiplication process, but remembering that the unlike signs will give a negative answer overall:

$$\begin{array}{r} 1\,827 \\ 159 \\ \hline 16\,443 \\ 91\,35 \\ 182\,7 \\ \hline 290\,493 \\ \hline \end{array}$$

Thus 29 × 63 × 159 = 290 493

and 29 × 63 × −159 = **−290 493**

Problem 8. Divide 9 417 by 73 (or find the value of 9 417 ÷ 73)

Yet another way of expressing this problem mathematically is to use a horizontal or oblique line, then

$$9\,417 \div 73 = \frac{9\,417}{73} \text{ or } 9\,417/73$$

(1) (4) (7) → 129

	129	
	73)9 417	(1) 73 into 94 goes once, p 1 above the 4.
(2) Multiply 73 by 1 and put under the 94.	7 3	
(4) 73 into 211 goes twice, put 2 above the 1.	2 11	(3) Subtract and bring dow the 1.
	1 46	(5) Multiply 73 by 2 and pu under the 211.
(7) 73 into 657 goes 9, put 9 above the 7.	657	(6) Subtract, bring down th 7.
	657	(8) Multiply 73 by 9 and pu under the 657.
(9) Subtract.	0	

If, when dividing one number by another, there is still a number left resulting from the last subtraction this is called the **remainder**. In this case, the 0 on the final subtraction shows that there is no remainder, that is 73 goes an exact number of times into 9 417.

Thus 9 417 ÷ 73 = **129**

Problem 9. Find the value of 6 738 ÷ 46

	146	
	46)6 738	
(2) 1 × 46.	4 6	(1) 46 into 67 goes once, put 1 up above the 7.
(3) Subtract.	2 13	(4) Bring down the 3; 46 into 213 goes 4 times Put up the 4 above the 3.
(5) 4 × 46.	1 84	
(6) Subtract.	298	(7) Bring down the 8; 46 into 298 goes 6 times Put up the 6 above the 8.
(8) 6 × 46.	276	
(9) Subtract.	22	(10) Since there are no more numbers to bring down, 22 is the remainder.

Thus 6 738 ÷ 46 = **146, remainder 22**

Problem 10. Find the value of $\frac{127\,364}{-123}$

The unlike signs show that the answer will be negative. Both numbers are treated initially, for the purpose of long division, as if they are positive.

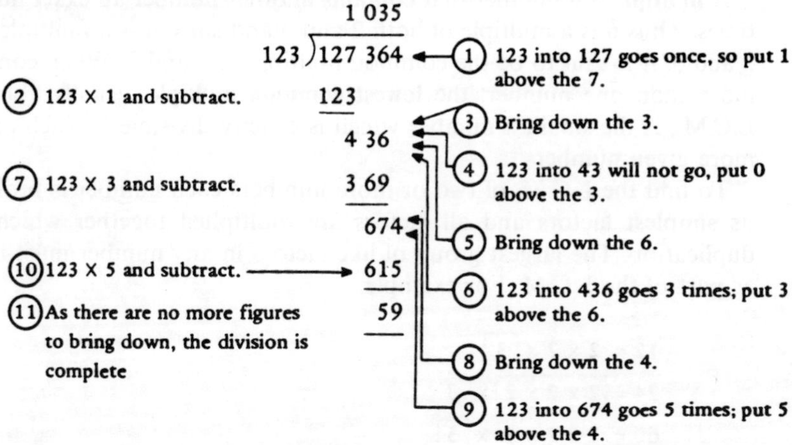

Thus 127 364 ÷ 123 = 1 035, remainder 59

And 127 364 ÷ − 123 = **− 1 035, remainder 59**

Further problems on arithmetic operations may be found in Section 1.8 (Problems 1–15), page 33.

1.2 Highest common factor and lowest common multiple

A factor is a number which divides into another number exactly. Thus both 2 and 3 are factors of 6. The numbers 2 and 3 are also factors of 12, 18 and 24 and are called common factors of all these numbers. The **highest common factor**, usually abbreviated H.C.F., is the largest number that divides into two or more numbers exactly.

To find the H.C.F. of two or more numbers, each number is broken into its simplest factors, and those which are common to all the numbers constitute the H.C.F. Numbers which have no factors apart from 1 are called **prime numbers**. Since 2, 3, 5, 7, 11, 13 and 17 . . . have no factors apart from 1, they are prime numbers. When looking for factors, numbers are expressed in terms of factors comprising these prime numbers. The simplest ones are found first; that is 2, 3, 5, 7 and so on. For example:

12 = 2 × 2 × 3		2 × 2 × 3	
48 = 2 × 2 × 2 × 2 × 3	or	2 × 2 × 3	× 2 × 2
60 = 2 × 2 × 3 × 5		2 × 2 × 3	× 5

The factors contained in each of the numbers 12, 48 and 60 are $2 \times 2 \times 3 = 12$, that is the H.C.F. of 12, 48 and 60 is 12, this being the largest number which goes exactly into 12, 48 and 60.

A **multiple** is a number that contains another number an exact number of times. Thus 6 is a multiple of both 2 and 3 and since it is a multiple of both 2 and 3, it is said to be the common multiple of 2 and 3. When considering more than one number, the **lowest common multiple**, usually abbreviated L.C.M., is the smallest number which is exactly divisible by each of two or more given numbers.

To find the L.C.M. of two or more numbers each number is broken into its simplest factors and all factors are multiplied together which avoids duplication. The largest group of like factors in any number must be taken as part of the L.C.M. For example:

$$12 = 2 \times 2 \times \boxed{3}$$
$$24 = \boxed{2 \times 2 \times 2} \times 3$$
$$60 = 2 \times 2 \times 3 \times \boxed{5}$$

The largest group of twos is in the number 24, and this covers the 2×2 factors in 12 and 60, so that they will not be required to contribute towards the L.C.M. As there is one 3 factor in each of the numbers only one of these will be required. Finally the 5 factor in 60 will also be required. The L.C.M. is $2 \times 2 \times 2 \times 3 \times 5 = 120$, the lowest number which can be exactly divided by 12, 24 and 60.

The method of finding the L.C.M. of a group of numbers one or more of which are negative is treated in a similar manner. The L.C.M. is usually expressed in terms of positive numbers. This is because, for example, the numbers 6 and 9 have factors 2×3 and 3×3 and their L.C.M. is 18. Also the numbers 6 and -9 have factors 2×3 and $-1 \times 3 \times 3$ and at first sight have an L.C.M. of $-1 \times 2 \times 3 \times 3$ or -18. However, since $\frac{18}{6} = 3$ and $\frac{18}{-9} = -2$, this shows that 18 is the L.C.M. of both 6 and -9 and the positive value of 18 is the one taken.

Worked problems on H.C.F. and L.C.M.

Problem 1. Find the H.C.F. of the numbers 42, 56 and 70

To determine the simplest factors of a number a check should first be made to see if it can be divided by 2. If it can be, this should be done and an attempt made to divide by 2 again, repeating until an odd number is left. Then a check should be made to see if it will divide by 3. This check should also be repeated as often as division is possible. This process is repeated with the prime numbers 5, 7, 11, 13, 17, 19 etc. until a prime number remains. Then

$42 = 2 \times 3 \times 7$

$56 = 2 \times 2 \times 2 \times 7$

$70 = 2 \times 5 \times 7$

The factors which are common to each of the numbers 42, 56 and 70 are $2 \times 7 = 14$. Thus **14 is the H.C.F. of these three numbers.**

Problem 2. Find the H.C.F. of 110, 286 and 330

$110 = 2 \times 5 \times 11$

$286 = 2 \times 11 \times 13$

$330 = 2 \times 3 \times 5 \times 11$

The H.C.F. is $2 \times 11 = 22$

Problem 3. Find the H.C.F. of 182, −1 547 and 4 199

$182 = 2 \times 7 \times 13$

$-1\,547 = -1 \times 7 \times 13 \times 17$

$4\,199 = 13 \times 17 \times 49$

Thus the H.C.F. of 182, −1 547 and 4 199 is **13**

Problem 4. Find the L.C.M. of the numbers 42, 56 and 140

Determining the simplest factors of each of these numbers gives

$42 = 2 \times 3 \times 7$

$56 = 2 \times 2 \times 2 \times 7$

$140 = 2 \times 2 \times 5 \times 7$

The L.C.M. is $2 \times 2 \times 2 \times 3 \times 5 \times 7 = 840$

Problem 5. Find the L.C.M. of the numbers 24, 56, 128 and 39

Determining the simplest factors of each of these numbers gives

$24 = 2 \times 2 \times 2 \times 3$

$56 = 2 \times 2 \times 2 \times 7$

$128 = 2 \times 2 \times 2 \times 2 \times 2 \times 2 \times 2$

$39 = 3 \times 13$

Selecting the largest groups of each of the prime numbers results in $2 \times 2 \times 2 \times 2 \times 2 \times 2 \times 2 \times 3 \times 7 \times 13$, or 34 944

Then the L.C.M. of 24, 56, 128 and 39 is **34 944**

Problem 6. Find the L.C.M. of −442, 663 and 1 105

$-442 = -1 \times 2 \times 13 \times 17$

$663 = 3 \times 13 \times 17$

$1\,105 = 5 \times 13 \times 17$

The L.C.M. of these three numbers is **2 × 3 × 5 × 13 × 17 or 6 630, the −1 factor being ignored.**

Further problems on H.C.F. and L.C.M. may be found in Section 1.8 (*Problems 16–25*), page 34.

1.3 Brackets and basic laws

Brackets are used in arithmetic to indicate that the operation inside the bracket must be done first. For instance, take the sum of 3 and 4 from 9 is written as

$$9 - (3 + 4) = 9 - 7 = 2$$

Since the (3 + 4) is in a bracket, this operation must be done before subtracting the result from 9.

Another use of brackets is to indicate multiplication. An integer which is next to a bracket means multiply the contents of that bracket by this integer. For instance, take twice the sum of 3 and 4 from 9 is written as

$$9 - 2(3 + 4) = 9 - 2(7) = 9 - 2 \times 7 = 9 - 14 = -5$$

Other examples are

$$3(4 + 5) = 3 \times 9 = 27$$

$$-2(6 - 3) = -2 \times 3 = -6$$

$$(8 - 3)7 = 5 \times 7 = 35$$

This last example shows that the integer can be written either in front of or behind the bracket, although it is usual to write it in front of the bracket.

When there is no integer next to the bracket, it is inferred that the integer is 1. Thus

$$(4 + 5) = 1 \times (4 + 5) = 1 \times 9 = 9$$

but

$$-(4+5) = -1 \times (4+5) = -1 \times 9 = -9$$

Adjacent brackets with no + or − sign between them indicate multiplication.

$$(2+3)(4+5) = (5)(9) = 5 \times 9 = 45$$

There are certain laws which govern basic arithmetic operations and the use of brackets. In both addition and multiplication the order of writing integers does not affect the value. For example

$2 + 3 = 5$

Also $3 + 2 = 5$

Thus $2 + 3 = 3 + 2$ [I]

For multiplication $2 \times 3 = 6$

And $3 \times 2 = 6$

Thus $2 \times 3 = 3 \times 2$ [II]

However, the same laws must *not* be applied for the operation of subtraction and division. $2 - 3$ is *not* equal to $3 - 2$ and $2 \div 3$ is *not* equal to $3 \div 2$. The position of brackets in problems containing only addition signs *or* multiplication signs does not alter the final value. For example

$2 + (3 + 4) = 2 + 7 = 9$

Also $(2 + 3) + 4 = 5 + 4 = 9$

Hence $2 + (3 + 4) = (2 + 3) + 4$ [III]

In multiplication $2 \times (3 \times 4) = 2 \times 12 = 24$

Also $(2 \times 3) \times 4 = 6 \times 4 = 24$

Then $2(3 \times 4) = (2 \times 3)4$ [IV]

These laws do *not* apply in the case of subtraction and division. $2 - (3 + 4)$ is *not* equal to $(2 - 3) + 4$. Also $(2 \div 3) + 4$ is *not* equal to $2 \div (3 + 4)$. Another way of determining the value of the expression in a bracket is given in law [V], and is sometimes helpful in evaluating brackets containing difficult numbers. Using the method of dealing with the bracket first:

$2(3 + 4) = 2 \times 7 = 14$

Alternatively $2(3 + 4) = (2 \times 3) + (2 \times 4)$

$= 6 + 8 = 14$ [V]

Frequently, in a problem, more than one operator is involved. Different answers to the problem can be obtained depending on the order in which operations are carried out. For example, to find the value of $2 \times 3 + 7$ can give an answer of $2 \times 3 = 6$ and $6 + 7 = 13$. Another way of getting an answer would be to add 3 and 7 and multiply by 2. Then $2 \times 3 + 7 = 2 \times 10 = 20$. Both of these answers cannot be correct and an **order of precedence** of operations is used. This order of operations is **B**rackets, **O**f, **D**ivision, **M**ultiplication, **A**ddition and **S**ubtraction often remembered by the initials BODMAS. Applying the order of precedence to $2 \times 3 + 7$, shows that since multiplication comes before addition, the only correct solution to this problem is $2 \times 3 + 7 = 6 + 7 = 13$, and 20 is incorrect.

Worked problems on brackets and basic laws

Problem 1. Find the value of $(6 + 7) + 8$

By law [III], the position of the brackets does not affect the answer. Thus

$(6 + 7) + 8 = \mathbf{6 + 7 + 8 = 21}$

Problem 2. Find the value of $6 - (7 + 8)$

Applying the rule that bracket terms must be dealt with first

$6 - (7 + 8) = \mathbf{6 - 15 = -9}$

Problem 3. Find the value of $6 - (7 - 8)$

$6 - (7 - 8) = 6 - (-1)$

and, since like signs give $+$,

$6 - (-1) = 6 + 1 = 7$

Alternatively $(7 - 8) = -1$

and $-1(-1) = (-1) \times (-1) = +1$

Hence $6 - (7 - 8) = 6 + 1 = 7$

Problem 4. Find the value of $(6 + 7)(8 + 9)$

Adjacent brackets mean multiply and operations within brackets must be done first. Thus

$(6 + 7)(8 + 9) = \mathbf{(13)(17) = 221}$

Problem 5. Find the value of $(16 - 7) \div 3$

This problem involves both **B**rackets and **D**ivision and remembering **BODMAS**, the brackets must be dealt with before dividing. Thus

$(16 - 7) \div 3 = \mathbf{9 \div 3 = 3}$

Problem 6. Find the value of $6 \times 4 \div 2(4 + 7)$

The operations are **M**ultiplication, **D**ivision and **B**rackets. From **BODMAS** the order is **B**rackets, **D**ivision and finally **M**ultiplication, thus

$$
\begin{aligned}
6 \times 4 \div 2(4 + 7) &= 6 \times 4 \div 2 \times 11 && \text{(B)} \\
&= 6 \times 2 \times 11 && \text{(D)} \\
&= \mathbf{132} && \text{(M)}
\end{aligned}
$$

Problem 7. Find the value of $(19 - 43) \div 6 \times 40 \div (3 + 17)$

In this problem there are two lots of brackets and two $\div$ operators. Provided the overall order dictated by **BODMAS** is adhered to, the order in which repeated operators are determined does not affect the value of the answer. Thus

$$
\begin{aligned}
(19 - 43) \div 6 \times 40 \div (3 + 17) &= -24 \div 6 \times 40 \div 20 && \text{(B)} \\
&= -4 \times 2 && \text{(D)} \\
&= \mathbf{-8} && \text{(M)}
\end{aligned}
$$

Problem 8. Find the value of

$(2 + 3) + (4 + 5) \div (10 - 7)(-3 + 2) - 16$

Applying the order of operations:

$$
\begin{aligned}
&(2 + 3) + (4 + 5) \div (10 - 7)(-3 + 2) - 16 \\
&= 5 + 9 \div 3 \times (-1) - 16 && \text{(B)} \\
&= 5 + 3 \times (-1) - 16 && \text{(D)} \\
&= 5 - 3 - 16 && \text{(M)} \\
&= \mathbf{-14} && \text{(S)}
\end{aligned}
$$

The (-1) term is usually kept in a bracket when preceded by an operator, to prevent any confusion. Thus to take -3 from 4 is written as $4 - (-3)$ and to multiply 6 by -4 is written as $6 \times (-4)$ or $6(-4)$.

Problem 9. Find the value of
$64 \div (-16) + (-7 - 12) - (-29 + 36)(-2 + 9)$

This expression becomes

$64 \div (-16) + (-19) - (7)(7)$ (B)

$= (-4) + (-19) - (7)(7)$ (D)

$= (-4) + (-19) - 49$ (M)

$= -23 - 49$ (A)

$= -72$ (S)

Further problems on brackets and basic laws may be found in Section 1.8 (Problems 26–40), page 34.

1.4 Fractions

Dividing 3 by 5 can be written as $3 \div 5$ or as $\frac{3}{5}$. In the latter form, $\frac{3}{5}$ is called a fraction and can be defined as one or more equal parts of something. For example, one-fifth of 100 metres means that the 100 metres must be divided into 5 equal parts and one of these parts is taken. Then one-fifth of 100 metres is 20 metres, obtained by dividing 100 into 5 equal parts of 20 metres and taking one of these parts. Three-fifths of 100 metres means divide 100 metres into 5 equal parts and then take 3 of these parts, that is $20 \times 3 = 60$ metres.

In every fraction the number above the line is called the **numerator** and that below the line the **denominator**. In the fraction $\frac{3}{5}$, 3 is the numerator and 5 the denominator. It is called a **proper fraction**. The number in the numerator can be larger than the number in the denominator, for example, the fraction $\frac{7}{5}$. This fraction is then called an **improper fraction** because it can be expressed as an integer plus a proper fraction. Seven-fifths comprises five-fifths plus two-fifths, and five-fifths is one. Then, mathematically:

$$\frac{7}{5} = \frac{5}{5} + \frac{2}{5} = 1\frac{2}{5}$$

To keep calculations to a minimum, it is usual to express fractions using the lowest possible numbers for both numerator and denominator. This can be achieved by dividing both numerator and denominator by the highest common factor of these two numbers when it is possible.

Now, $\frac{3}{5}$, $\frac{6}{10}$, $\frac{9}{15}$ and $\frac{12}{20}$ are all the same fraction.

For $\frac{9}{15} = \frac{3 \times 3}{3 \times 5}$ and, dividing by the H.C.F. of the numerator and denominator, $\frac{9}{15} = \frac{\not{3}^1 \times 3}{\not{3}_1 \times 5} = \frac{3}{5}$. This process is called cancelling. The fraction $\frac{32}{48}$ can be reduced to its simplest form by cancellation as follows:

$$\frac{32}{48} = \frac{\not{2}^1 \times \not{2}^1 \times \not{2}^1 \times \not{2}^1 \times 2}{\not{2}_1 \times \not{2}_1 \times \not{2}_1 \times \not{2}_1 \times 3} = \frac{2}{3}$$

Cancellation by 0 is not permitted when expressing a fraction in its simplest form, for this will make nonsense of the fraction. It could possibly be reasoned that 0 is an integer and to simplify $\frac{9}{15}$, merely multiply numerator and denominator by 0. Then $\frac{9}{15} \times \frac{0}{0} = \frac{9 \times 0}{15 \times 0} = \frac{0}{0}$, which is meaningless. Thus multiplication or division by zero will usually lead to a meaningless expression.

Addition and subtraction of fractions

To add or subtract two fractions the lowest common multiple of the denominators is found and each fraction expressed with the L.C.M. as a denominator. This is done by multiplying both numerator and denominator by the number required to give the L.C.M. as the denominator. Multiplying both numerator and denominator by the same number does not alter the value of the fraction at all. To simplify $\frac{1}{4} + \frac{1}{18}$ the procedure is as shown:

$$4 = 2 \times 2$$

$$18 = 2 \times 3 \times 3$$

giving the L.C.M. as $2 \times 2 \times 3 \times 3 = 36$ and

$$\frac{1}{4} \times \frac{9}{9} = \frac{9}{36}, \frac{1}{18} \times \frac{2}{2} = \frac{2}{36}$$

Thus

$$\frac{1}{4} + \frac{1}{18} = \frac{9}{36} + \frac{2}{36} = \frac{11}{36}$$

For subtraction

$$\frac{1}{4} - \frac{1}{18} = \frac{9}{36} - \frac{2}{36} = \frac{7}{36}$$

For mixtures of integers and fractions either improper fractions can be formed or the integers can be worked out independently of the fractions. To find the value of $3\frac{1}{4} + 5\frac{1}{2} - \frac{6}{11}$ using the improper fraction method, the terms are first turned into improper fractions.

$$3\frac{1}{4} = \frac{3 \times 4}{4} + \frac{1}{4} = \frac{12}{4} + \frac{1}{4} = \frac{13}{4}$$

$$5\frac{1}{2} = \frac{5 \times 2}{2} + \frac{1}{2} = \frac{10}{2} + \frac{1}{2} = \frac{11}{2}$$

Then

$$3\frac{1}{4} + 5\frac{1}{2} - \frac{6}{11} = \frac{13}{4} + \frac{11}{2} - \frac{6}{11}$$

The L.C.M. of the denominator is 44. Expressing each term with the L.C.M. as its denominator gives:

$$\frac{13}{4} \times \frac{11}{11} + \frac{11}{2} \times \frac{22}{22} - \frac{6}{11} \times \frac{4}{4}$$

$$= \frac{143}{44} + \frac{242}{44} - \frac{24}{44}$$

or, since each term has the same denominator, this can be written as

$$\frac{143 + 242 - 24}{44} = \frac{361}{44}$$

Now $361 \div 44$ is 8 remainder 9. Hence

$$\frac{361}{44} = 8\frac{9}{44}$$

The same problem could have been solved by treating the integers separately from the fractions.

$$3\frac{1}{4} + 5\frac{1}{2} - \frac{6}{11} = 3 + \frac{1}{4} + 5 + \frac{1}{2} - \frac{6}{11}$$

$$3 + 5 = 8$$

$$\frac{1}{4} + \frac{1}{2} - \frac{6}{11} = \frac{11 + 22 - 24}{44} = \frac{9}{44}$$

Then

$$3\frac{1}{4} + 5\frac{1}{2} - \frac{6}{11} = 8 + \frac{9}{44} = 8\frac{9}{44} \text{ as before.}$$

Multiplication of fractions

To multiply two or more fractions together the numerators are first multiplied to give a single number, and this becomes the numerator of the combined fraction. The denominators are then multiplied to give the denominator of the combined fraction. Then

$$\frac{3}{10} \times \frac{5}{7} = \frac{3 \times 5}{10 \times 7} = \frac{15}{70}$$

and, dividing both numerator and denominator by the H.C.F., 5 gives:

$$\frac{15}{70} = \frac{3}{14}$$

Questions involving the multiplication of fractions sometimes use the word 'of' instead of multiply. Then $\frac{3}{5}$ of 12 becomes

$$\frac{3}{5} \times \frac{12}{1} = \frac{3 \times 12}{5 \times 1} = \frac{36}{5} = 7\frac{1}{5}$$

Positive and negative integers can always be expressed as a fraction by dividing by 1, if this assists the calculation.

Mixed numbers must be expressed as improper fractions before multiplication.

Division of fractions

The procedure for division is to change the division sign into a multiplication sign and invert the second fraction. That is, the numerator is written for the denominator and vice versa. Then

$$\frac{3}{4} \div \frac{4}{5} = \frac{3}{4} \times \frac{5}{4} = \frac{15}{16}$$

Order of precedence

The order of performing operations is the same as for integers, namely **B**rackets, **O**f, **D**ivision, **M**ultiplication, **A**ddition and **S**ubtraction: that is, as given by BODMAS. If any of the operations appears more than once, say there were two multiplications, then the order in which these two multiplications are done is immaterial, provided the overall order dictated by BODMAS is maintained. The operation 'Of' merely means change the word 'Of' to a multiplication sign and then continue with the order of precedence.

Worked problems on fractions

Problem 1. Add $\frac{7}{12}$ and $\frac{2}{5}$

The L.C.M. of 12 and 5 is 60, thus

$$\frac{7}{12}+\frac{2}{5}=\frac{7\times 5}{12\times 5}+\frac{2\times 12}{5\times 12}$$

$$=\frac{35}{60}+\frac{24}{60}=\mathbf{\frac{59}{60}}$$

Problem 2. Evaluate $3\frac{3}{5}+2\frac{1}{10}$

Either

$$3\frac{3}{5}+2\frac{1}{10}=3+\frac{3}{5}+2+\frac{1}{10}$$

$$=5+\frac{3}{5}+\frac{1}{10}$$

Now the L.C.M. of 5 and 10 is 10

and

$$\frac{3}{5}+\frac{1}{10}=\frac{3\times 2+1}{10}=\frac{7}{10}$$

Thus

$$3\frac{3}{5}+2\frac{1}{10}=\mathbf{5+\frac{7}{10}}\text{ or }\mathbf{5\frac{7}{10}}$$

Or

$$3\frac{3}{5}+2\frac{1}{10}=\frac{3\times 5}{5}+\frac{3}{5}+\frac{2\times 10}{10}+\frac{1}{10}$$

$$=\frac{18}{5}+\frac{21}{10}$$

$$=\frac{18\times 2+21}{10}$$

$$=\frac{57}{10}=\mathbf{5\frac{7}{10}}$$

Problem 3. Find the value of $\frac{5}{8} - \frac{3}{40}$

$$\frac{5}{8} - \frac{3}{40} = \frac{5 \times 5 - 3}{40} = \frac{25 - 3}{40}$$

$$= \frac{22}{40} = \frac{2 \times 11}{2 \times 20} = \mathbf{\frac{11}{20}}$$

Problem 4. Evaluate $3\frac{4}{19} - 2\frac{7}{57}$

$$3\frac{4}{19} - 2\frac{7}{57} = (3 - 2) + \left(\frac{4}{19} - \frac{7}{57}\right)$$

$$= 1 + \frac{3 \times 4 - 7}{57} = 1 + \frac{5}{57}$$

$$= \mathbf{1\frac{5}{57}}$$

Problem 5. Find the value of $3\frac{3}{5} - 2\frac{5}{8}$

$$3\frac{3}{5} - 2\frac{5}{8} = (3 - 2) + \left(\frac{3}{5} - \frac{5}{8}\right)$$

$$= 1 + \frac{3 \times 8 - 5 \times 5}{40}$$

$$= 1 + \frac{(-1)}{40}$$

The $+ \frac{(-1)}{40}$ gives an overall minus due to unlike signs and the problem becomes $1 - \frac{1}{40}$

$$= \frac{40}{40} - \frac{1}{40} = \mathbf{\frac{39}{40}}$$

Problem 6. Simplify $2\frac{7}{16} - 3\frac{1}{5} + 1\frac{5}{8}$

$$2\frac{7}{16} - 3\frac{1}{5} + 1\frac{5}{8} = (2 - 3 + 1) + \left(\frac{7}{16} - \frac{1}{5} + \frac{5}{8}\right)$$

$$= 0 + \frac{7 \times 5 - 1 \times 16 + 5 \times 10}{80}$$

Problem 7. Evaluate $\frac{3}{8} \times \frac{4}{15}$

Multiplying the numerators together and denominators together

$$\frac{3}{8} \times \frac{4}{15} = \frac{3 \times 4}{8 \times 15} = \frac{12}{120} = \frac{1}{10}$$

In this example cancelling can be done before multiplication. Thus

$$\frac{3}{8} \times \frac{4}{15} = \frac{3}{\cancel{8}_2} \times \frac{\cancel{4}^1}{15} = \frac{\cancel{3}^1}{2 \times \cancel{15}_5} = \mathbf{\frac{1}{10}}$$

The cancelling was firstly by 4, that is, dividing both numerator and denominator by 4, and then by 3.

Problem 8. Simplify $3\frac{2}{5} \times 1\frac{3}{13} \times 2\frac{3}{8}$

The fractions must be expressed as improper fractions before multiplication can be done.

$$3\frac{2}{5} \times 1\frac{3}{13} \times 2\frac{3}{8} = \frac{17}{5} \times \frac{16}{13} \times \frac{19}{8},$$

$$\text{since } 3\frac{2}{5} = \frac{3 \times 5}{5} + \frac{2}{5} = \frac{15+2}{5} = \frac{17}{5},$$

$$1\frac{3}{13} = \frac{1 \times 13}{13} + \frac{3}{13} = \frac{16}{13},$$

$$\text{and } 2\frac{3}{8} = \frac{2 \times 8}{8} + \frac{3}{8} = \frac{16+3}{8} = \frac{19}{8}$$

$$\frac{17}{5} \times \frac{\cancel{16}^2}{13} \times \frac{19}{\cancel{8}_1} = \frac{17 \times 2 \times 19}{5 \times 13}$$

When all the numbers in either the numerator or the denominator are prime numbers, no further cancelling is possible. Thus

$$\mathbf{\frac{17 \times 2 \times 19}{5 \times 13}} = \frac{646}{65} = \mathbf{9\frac{61}{65}}$$

Problem 9. Evaluate $\frac{3}{5} \div \frac{27}{25}$

The rule is to invert the dividing fraction and change the division to a multiplication sign

$$\text{Then } \frac{3}{5} \div \frac{27}{25} = \frac{\cancel{3}^1}{\cancel{5}_1} \times \frac{\cancel{25}^5}{\cancel{27}_9} = \mathbf{\frac{5}{9}}$$

Problem 10. Find the value of $5\frac{1}{6} \div \frac{62}{16}$

Division can only be performed by using improper fractions.

$$\text{Then } 5\frac{1}{6} \div \frac{62}{16} = \frac{31}{6} \div \frac{62}{16}$$

$$= \frac{\cancel{31}^{1}}{\cancel{6}_{3}} \times \frac{\cancel{16}^{8\,4}}{\cancel{62}_{21}} = \frac{4}{3} = \mathbf{1\frac{1}{3}}$$

Problem 11. Simplify $\frac{3}{4} + \frac{11}{5} \times \frac{3}{8} - \frac{3}{10}$

Several operations are present therefore apply BODMAS to obtain the order of precedence.

$$\frac{3}{4} + \frac{11}{5} \times \frac{3}{8} - \frac{3}{10} = \frac{3}{4} + \frac{33}{40} - \frac{3}{10} \qquad \text{(M)}$$

$$= \frac{30 + 33 - 12}{40} \qquad \text{(A)}$$

$$= \frac{63 - 12}{40} = \frac{51}{40} = \mathbf{1\frac{11}{40}} \qquad \text{(S)}$$

Problem 12. Simplify $(\frac{5}{2} - \frac{1}{4}) \div \frac{3}{(\frac{4}{3} + \frac{2}{5})} \times \frac{7}{12}$

Applying BODMAS, the brackets must be worked out before doing the divisions and finally the multiplication. The brackets are sometimes omitted when writing fractions as the numerator or denominator of a term. If this is done then it must always be assumed that they exist. For example, the second term $\frac{3}{\left(\frac{4}{3} + \frac{2}{5}\right)}$ could also be written as $\frac{3}{\frac{4}{3} + \frac{2}{5}}$, but it must then be assumed that $\frac{4}{3} + \frac{2}{5}$ is a bracketed term.

$$\text{Then } \left(\frac{5}{2} - \frac{1}{4}\right) \div \frac{3}{\left(\frac{4}{3} + \frac{2}{5}\right)} \times \frac{7}{12} = \frac{9}{4} \div \frac{3}{\frac{26}{15}} \times \frac{7}{12}$$

$$= \frac{9}{4} \div \frac{45}{26} \times \frac{7}{12}$$

$$= \frac{9^1}{4} \times \frac{\not{26}^{13}}{\not{45}_5} \times \frac{7}{12_6}$$

$$= \mathbf{\frac{91}{120}}$$

Further problems on fractions may be found in Section 1.8 (Problems 41–69), page 35.

1.5 Ratio and proportion

All fractions are ratios. Ratio can be defined as the number of times which one quantity is contained in another quantity **of the same kind**. Ratios are usually used to compare the relative size, quantity or cost of two things. For example, to find the ratio of 25 days to 1 year, expressing each of these quantities in the same units, gives

$$\frac{25 \text{ days}}{365 \text{ days}} = \frac{5}{73}$$

which can be written as 5:73. The : means 'is to' and the ratio of 25 days to 1 year is as 5 is to 73. Ratios can apply to more than two quantities. When mixing the materials for a concrete floor, the ratio of sand to gravel to cement is 2:5:1, that is 2 parts of sand to 5 parts of gravel to 1 part of cement. The volume of sand in 1 cubic metre of the mix is given by

$$\frac{2 \text{ parts of sand}}{(2+5+1) \text{ parts of the whole}} = \frac{2}{8} = \frac{1}{4} \text{ of the whole.}$$

Then since the whole is 1 cubic metre, the amount of sand is $\frac{1}{4}$ cubic metre.

A proportion is formed when the ratio of two quantities is equal to the ratio of a second pair of quantities. There are three ways of expressing a proportion, for example, since $\frac{10}{4} = \frac{5}{2}$

(i) 10 is to 4 as 5 is to 2
(ii) 10:4::5:2 (where :: means 'as')
and (iii) $\frac{10}{4} = \frac{5}{2}$

Worked problems on ratio and proportion

Problem 1. Divide £192 in the ratio of 9 to 7

There are a total of $(9 + 7) = 16$ parts.

One part will be $\dfrac{192}{16} = £12$

Nine parts will be $£12 \times 9 = \mathbf{£108}$

and seven parts will be $£12 \times 7 = \mathbf{£84}$

Having divided a quantity in a certain ratio, a check should be made to make sure that the parts add up to the original amount. In this case $108 + 84 = 192$ showing that it is unlikely that an error has been made.

Problem 2. Divide 2 601 bricks in the ratio 3:5:9

There are a total of $(3 + 5 + 9) = 17$ parts

One part will be $\dfrac{2\,601}{17} = 153$ bricks

Hence

3 parts will be $3 \times 153 = 459$ bricks
5 parts will be $5 \times 153 = 765$ bricks

and

9 parts will be $9 \times 153 = 1\,377$ bricks

(**Check:** $459 + 765 + 1\,377 = 2\,601$)

Then for 2 601 bricks **3:5:9::459:765:1 377**

Problem 3. A piece of metal bar 120 cm long is cut in the ratio of 7:3. Find the length of the two pieces.

There are a total of $(7 + 3) = 10$ parts

Length of 1 part $= \dfrac{120}{10} = 12$ cm

Length of 7 parts $= 12 \times 7 =$ **84 cm**

and length of 3 parts $= 12 \times 3 =$ **36 cm**

(**Check:** $84 + 36 = 120$)

Problem 4. Three men paint a house in 5 days. Find how long it will take two men to paint the house.

Three men paint a house in 5 days. Since 1 man will take three times as long, then 1 man will paint the house in $5 \times 3 = 15$ days. Two men will do the work in half the time it takes 1 man to do it, hence

2 men will paint the house in $\frac{15}{2}$ or $7\frac{1}{2}$ days.

Problem 5. Express 75p as a ratio of £2.50

Since £2.50 = 250p, and as ratio problems can only be done when working in quantities of the same kind, the problem becomes $\frac{75}{250} = \frac{3}{10}$

Thus **75p:250p::3:10**

Problem 6. A bonus is paid to 4 men in the ratio 9:11:13:15. If the total bonus is £240, find how much each man receives.

There are a total of $(9 + 11 + 13 + 15) = 48$ parts to the bonus. Then one part will be $\frac{240}{48} = £5$, and the required amounts are $9 \times 5 = £45$; $11 \times 5 = £55$; $13 \times 5 = £65$; and $15 \times 5 = £75$. Thus for £240 9:11:13:15::**£45:£55:£65:£75**

(**Check:** $45 + 55 + 65 + 75 = 240$)

Further problems on ratio and proportion may be found in Section 1.8 (Problems 70–75), page 35.

1.6 Decimals

The decimal system of numbers is the one in everyday use based on the numbers 0 to 9. The number 23.61 is called a **decimal fraction** and means $(2 \times 10) + (3 \times 1) + (6 \times \frac{1}{10}) + (1 \times \frac{1}{100})$.

To add 17.4 and 5.8 becomes

$$\left[(1 \times 10) + (7 \times 1) + \left(4 \times \frac{1}{10}\right)\right] + \left[(5 \times 1) + \left(8 \times \frac{1}{10}\right)\right]$$

and on collecting like terms this becomes

$$(1 \times 10) + (7 + 5) \times 1 + (4 + 8) \times \frac{1}{10}$$

or $10 + 12 + \frac{12}{10}$

i.e. $10 + 12 + 1\frac{2}{10}$

or $23\frac{2}{10}$ or 23.2

The importance of the **decimal point** for the exact placing of numbers can be seen by adding these numbers as shown:

$$\begin{array}{r} 17.4 \\ 5.8 \\ \hline 23.2 \\ \hline \end{array}$$

The decimal point (a dot) is used to distinguish between multiplication by 1 (units), and multiplication by $\frac{1}{10}$ (tenths of a unit).

Addition

Adding two or more decimal fractions is a similar process to adding integers, the only difference being the positioning of the number before addition. Decimal fractions are arranged for addition in a column with each of the decimal points beneath one another. The numbers 23.4, 0.07 and 128.714 can be added as shown.

$$\begin{array}{lr} & 23.4 \\ & 0.07 \\ & 128.714 \\ \hline \text{Adding} & 152.184 \\ \hline \end{array}$$

Subtraction

To take 73.71 from 186.317 the decimal points are arranged beneath one another and subtraction carried out, as for integers. Then

$$\begin{array}{lr} & 186.317 \\ & 73.71 \\ \hline \text{Subtracting} & 112.607 \\ \hline \end{array}$$

Multiplication

Again, the same procedure is used as when multiplying integers but in this case care must be taken in placing the decimal point in the correct position in the answer. Two decimal fractions are multiplied together ignoring the decimal points and when an answer has been obtained, a count is taken of the total number of figures to the right of the decimal point in both the numbers being multiplied. These are added together and the answer will contain this number of figures to the right of the decimal point. For example, to multiply 36.81 by 2.4, ignore the decimal points and treat these numbers as integers.

$$\begin{array}{r} 3\,681 \\ 24 \\ \hline 14\,724 \\ 73\,62 \\ \hline 88\,344 \\ \hline \end{array}$$

The numbers 36.$\underline{81}$ and 2.$\underline{4}$ contain 3 figures altogether on the right-hand side of their decimal points. Hence the answer must contain 3 figures altogether to the right of its decimal point. Then

$$36.\underline{81} \times 2.\underline{4} = 88.\underline{344}$$

Division

To divide one decimal fraction by another the denominator of the decimal fraction is made into an integer by multiplying the denominator by 10 or by a multiple of 10. The numerator must be multiplied by the same number. For example, $0.06 \div 0.12$ can be written as $\dfrac{0.06}{0.12}$

Now

$$\frac{0.06}{0.12} \times \frac{100}{100} = \frac{6}{12} = \frac{1}{2}$$

Since both numerator and denominator have been muliplied by the same number, the answer obtained by dividing 0.06 by 0.12 will be the same as dividing 6 by 12.

To convert a proper fraction to a decimal fraction, the numerator is divided by the denominator. Noughts can be added after the decimal point without affecting the value of the number. Thus $\frac{1}{2}$ becomes

$$\begin{array}{r} 0.5 \\ 2\overline{)1.000} \end{array}$$

that is $\frac{1}{2} = 0.5$

To convert a decimal fraction to a proper fraction, the part of the decimal fraction to the right of the decimal point is divided by 10 or by a multiple of 10. The number 23.64 has 0.64 to the right of the decimal point. This is made up of $6 \times \frac{1}{10} + 4 \times \frac{1}{100}$ i.e. $\frac{6}{10} + \frac{4}{100}$. Now

$$\frac{6}{10} + \frac{4}{100} = \frac{60+4}{100} = \frac{64}{100}$$

and by cancellation

$$\frac{64}{100} = \frac{16}{25}$$

Then

$$23.64 = 23\frac{16}{25}$$

Significant figures and decimal places

A number which can be expressed exactly as a decimal fraction is called a **terminating decimal**. In the previous example $23\frac{16}{25} = 23.64$ exactly and is a terminating decimal. Many proper fractions when expressed as decimal fractions do not terminate and these are called **non-terminating decimals**. To convert $\frac{7}{3}$ to a decimal fraction, 7 is divided by 3.

$$3\overline{)7.000\,000\,00}\quad 2.333\,33\cdots$$

The result of this division would continue indefinitely and mathematically, the answer would be written $2.\dot{3}$, called 'two point three recurring'. The dot over the three indicates that 3 recurs or repeats indefinitely. Since many decimal fractions are non-terminating, errors are introduced in calculations when these are used, but the extent of the error can be stated.

If the number $2\frac{1}{3}$ is required to an accuracy of 4 figures only, it would be written as 2.333 and the words 'correct to 3 decimal places' would be added to the answer. These words indicate that 3 figures are written after the decimal point. To express the number $\frac{2}{7}$ correct to 4 decimal places, 2 would be divided by 7 giving $0.285\,71\cdots$ and the answer would be terminated after the digit 7. Then $\frac{2}{7} = 0.285\,7$ correct to 4 decimal places. An alternative way of expressing accuracy is to specify the number of figures to be written. If the accuracy of $\frac{7}{3}$ is required to 4 figures, then the words used are 'correct to 4 significant figures'. Thus $\frac{7}{3} = 2.333$ correct to 4 significant figures.

When writing a decimal fraction to a certain degree of accuracy, the last digit in the answer is corrected depending on the value of the digit following it. If this digit is 0, 1, 2, 3 or 4, the last digit in the answer is left as it is. However, if it is 5, 6, 7, 8 or 9, the last digit in the answer is increased by one. Then an answer of $0.285\,714\,3\cdots$ would be written 0.286 correct to 3 decimal places, since the digit 7 following the 5 is in the group 5, 6, 7, 8 or 9. Similarly the answer would be 0.285 71 correct to 5 significant figures, since the digit 4 following the 1 is in the group 0, 1, 2, 3 or 4, indicating no increase in the last digit of the answer.

When determining the number of significant figures, the first digit on the left which signifies something starts the count. Noughts written before a number starts do not signify anything. Thus 0.003 174 = 0.003 17 correct to 3 significant figures, since the noughts in the $\frac{1}{10}$th and $\frac{1}{100}$th places do not signify anything. However, noughts within a number are significant and 20.007 17 = 20.007 2 correct to 6 significant figures, or 20.0 correct to 3 significant figures.

Worked problems on decimals

Problem 1. Add 3.27, 821.7, 9.042 7, 7 263 and 0.000 46

Each number is written so that the decimal points are in a column down the page and they are then added.

$$\begin{array}{r@{.}l} 3&27 \\ 821&7 \\ 9&042\,7 \\ 7\,263&0 \\ 0&000\,46 \\ \hline \mathbf{8\,097}&\mathbf{013\,16} \end{array}$$

Adding **8 097.013 16**

Problem 2. Take 73.71 from 286.714

The numbers are written so that the decimal points are beneath one another and then subtracted. Care is taken to see what is being taken from what.

$$\begin{array}{r} 286.714 \\ 73.710 \\ \hline \mathbf{213.004} \end{array}$$

Subtracting **213.004**

Problem 3. Evaluate 683 − 21.347

Noughts can be added after the decimal point as required; this does not alter the value of the number.

$$\begin{array}{r} 683.000 \\ 21.347 \\ \hline \mathbf{661.653} \end{array}$$

Subtracting **661.653**

Problem 4. Subtract 0.005 from 0.000 513

As with integers, 0.000 513 − 0.005 = −1(0.005 − 0.000 513). Thus subtracting the smaller number from the larger and multiplying by (−1) gives

$$\begin{array}{lr} & 0.005\,000 \\ & 0.000\,513 \\ \hline \text{Subtracting} & 0.004\,487 \\ \hline \end{array}$$

Remembering to multiply by (−1) gives

0.000 513 − 0.005 = **−0.004 487**

Problem 5. Find the value of 0.006 × 26.53

Multiplying the numbers as if integers (6 × 2 653)

$$\begin{array}{lr} & 2\,653 \\ & 6 \\ \hline \text{Multiplying} & 15\,918 \\ \hline \end{array}$$

To determine the position of the decimal point in the answer, count the total number of figures to the right of the decimal point and add together. Then 0.006 × 26.53 gives (3 + 2) = 5 figures to the right of the decimal point and so the answer must have 5 figures to the right of its decimal point. Thus 0.006 × 26.53 = **0.159 18**

Problem 6. Evaluate 270 × 0.006 3

As for problem 5, multiply 270 by 63

$$\begin{array}{r} 270 \\ 63 \\ \hline 810 \\ 16\,20 \\ \hline 17\,010 \\ \hline \end{array}$$

270 × 0.006 3 gives (0 + 4) figures after the decimal point, hence
270 × 0.006 3 = **1.701 0**

Problem 7. Divide 46.3 by 7 and give the answer correct to 4 decimal places.

Since 4 decimal place accuracy is required in the answer, an answer having 5 figures after the decimal point is necessary for the last figure correction. The decimal point in the answer is written above the decimal point in the number being divided.

$$\begin{array}{r} 6.614\,28 \\ 7\overline{)46.300\,00} \end{array}$$

Since the 5th figure after the decimal point is in the group 5, 6, 7, 8 or 9, the fourth figure after the decimal point is increased by one. Thus 46.3 ÷ 7 = **6.614 3** correct to 4 decimal places.

Problem 8. Divide 27.46 by 0.07 expressing the answer correct to 5 significant figures.

The answer must be worked to 6 figures to give 5-figure correction. Multiplying each quantity by 100 gives

$$\frac{27.46}{0.07} \times \frac{100}{100} = \frac{2\,746}{7}$$

$$7\overline{)2\,746.000}\quad = 392.285$$

Since the 6th figure in the answer is in the group 5, 6, 7, 8 or 9, the 5th figure is increased by one.
Hence 27.46 ÷ 0.07 = **392.29**

Problem 9. Evaluate $\frac{0.281}{0.03}$

Multiplying numerator and denominator by 100 gives $\frac{28.1}{3}$

$$3\overline{)28.100\,0}\quad = 9.366\,6\cdots$$

and 6 will recur in the answer as long as division is continued.

Thus $\frac{0.281}{0.03} = \mathbf{9.3\dot{6}}$

Problem 10. Convert 0.017 6 to a proper fraction.

The number 0.017 6 comprises

$$0 \times \frac{1}{10} + 1 \times \frac{1}{100} + 7 \times \frac{1}{1\,000} + 6 \times \frac{1}{10\,000}$$

Converting to $\frac{1}{10\,000}$ this gives

$$0 + \frac{100}{10\,000} + \frac{70}{10\,000} + \frac{6}{10\,000}$$

or $0.017\,6 = \frac{176}{10\,000} = \mathbf{\frac{11}{625}}$ by cancellation.

As a general rule, one figure after the decimal point is expressed in $\frac{1}{10}$ths, 2 figures as $\frac{1}{100}$ths, 3 figures as $\frac{1}{1\,000}$ths and so on.

Problem 11. Convert 23.82 to an improper fraction.

$23.82 = 23 + \frac{82}{100}$ since there are 2 figures after the decimal point. Then, by cancellation,

$$\mathbf{23.82} = 23 + \frac{41}{50}$$

$$= \frac{23 \times 50 + 41}{50} = \frac{1\,150 + 41}{50}$$

$$= \mathbf{\frac{1\,191}{50}}$$

Problem 12. Express $\frac{9}{32}$ as a decimal fraction.

$\frac{9}{32}$ means 9 divided by 32. Hence

```
    0.281 25
32)9.000 00
    64
    260
    256
      40
      32
       80
       64
       160
       160
```

Thus $\frac{9}{32} = \mathbf{0.281\,25}$

Further problems on decimals may be found in Section 1.8 (Problems 76–92), page 36.

1.7 Percentages

The word percentage implies a common standard of 100, and per cent, written %, means 'for every 100' or per 100. Thus in all percentages, a fraction is formed whose denominator is 100. Thus 62% becomes $\frac{62}{100}$ or $\frac{31}{50}$ and 35% becomes $\frac{35}{100}$ or $\frac{7}{20}$.

To convert a proper fraction into a percentage, the numerator and denominator must be multiplied by a suitable number to make the denominator 100. Thus $\frac{3}{5}$ becomes $\frac{3 \times 20}{5 \times 20} = \frac{60}{100}$ or 60%.

Decimal fractions can readily be converted to percentages. The decimal fraction 0.37 is

$$\frac{3}{10} + \frac{7}{100} = \frac{3 \times 10 + 7}{100} = \frac{37}{100}$$

That is, 0.37 is 37%. Thus to express a decimal fraction as a percentage the procedure is to multiply by 100 and this means moving the decimal point 2 places to the right. Then 0.125 is 12.5% and 0.017 is 1.7%. To express a percentage as a decimal fraction can also be readily done,

for $6\% = \frac{6}{100} = 0.06$ and

$$27\% = \frac{27}{100} = 0.27$$

When converting proper fractions to a percentage, conversion to a decimal fraction, as a first step, is often the easiest method. For example, to convert $\frac{9}{17}$ to a percentage, 9 is divided by 17

$$\begin{array}{r} 0.529\,41 \\ 17\overline{)9.000\,00} \\ \underline{85} \\ 50 \\ \underline{34} \\ 160 \\ \underline{153} \\ 70 \\ \underline{68} \\ 20 \\ \underline{17} \\ 3 \end{array}$$

That is $\frac{9}{17}$ is 0.529 4 correct to 4 decimal places. Hence $\frac{9}{17} = 52.94\%$ correct to 2 decimal places.

Worked problems on percentages

Problem 1. Find 5% of 140 metres.

1% of 140 m is $\frac{1}{100}$th of 140 m or $\frac{140}{100} = 1.4$ m

Therefore 5% of 140 m is **1.4 × 5 = 7 metres.**

Problem 2. Express $\frac{3}{11}$ as a percentage, correct to 3 significant figures.

Expressing $\frac{3}{11}$ as a decimal fraction

$$11\overline{)3.0000}\quad 0.2727$$

Correcting the 3rd figure gives $\frac{3}{11} = 0.273$ correct to 3 significant figures. Multiplying by 100 to turn the decimal fraction into a percentage gives $\frac{3}{11} =$ **27.3%** correct to 3 significant figures.

Problem 3. Express 18 seconds as a percentage of 1 minute.

One minute is 60 s and 18 s is $\frac{18}{60}$ of 60 s. $\frac{18}{60} = \frac{3}{10}$ by cancellation and $\frac{3}{10} = 0.3$. Multiplying by 100 to find the percentage gives 18 seconds as **30%** of 1 minute.

Problem 4. An alloy consists of 65% by weight of zinc, the remainder being copper. Find the weight of copper in 300 grams of alloy.

One gram of alloy contains $\frac{65}{100}$ g of zinc and $\frac{100-65}{100} = \frac{35}{100}$ g of copper. Then 300 grams of alloy contains $\frac{35}{100} \times 300$ g of copper

or **105 grams** of copper.

Further problems on percentages may be found in the following section (1.8) (Problems 93–100), page 36.

1.8 Further problems

Arithmetic operations

1. Add 2 961, 743, 82 615 and 27 [86 346]
2. Take 73 from 262 [189]
3. From 8 249 take 93 714 [−85 465]
4. Subtract 3 116 from 7 271 [4 155]
5. Add 264, −78, 3 142, 683 and −2 117 [1 894]
6. Find the value of −2 198 + 1 371 + 823 − 317 − 4 142 [−4 463]
7. Multiply 746 by 11 [8 206]
8. Multiply 236 by −13 [−3 068]

9. Multiply −4 312 by 9 [−38 808]
10. Find the value of −143 × −15 [2 145]
11. Evaluate 4 × −53 × 6 [−1 272]
12. Divide 736 by 16 [46]
13. Find the value of 29 040 divided by 120 [242]
14. Evaluate $\frac{4\,450}{13}$ [$342\frac{4}{13}$]
15. Evaluate 1 626/11 [$146\frac{9}{11}$]

H.C.F. and L.C.M.

16. Find the H.C.F. of the numbers 42, 154 and 231 [7]
17. Find the H.C.F. of the numbers 30, 105 and 165 [15]
18. Find the H.C.F. of 37, 74, 148 and 296 [37]
19. Find the H.C.F. of 196, 350, 770 and −910 [14]
20. Find the H.C.F. of 616, −308, 3 696 and −13 860 [308]
21. Find the L.C.M. of the numbers 36 and 210 [1 260]
22. Find the L.C.M. of 30, 105 and 165 [2 310]
23. Find the L.C.M. of 42, 154 and 231 [462]
24. Find the L.C.M. of 196, 350, 770 and −910 [700 700]
25. Find the L.C.M. of −308, 616, 3 696 and −13 860 [55 440]

Brackets and basic laws

26. Find the value of 53 + (61 + 70) [184]
27. Find the value of 53 − (61 + 70) [−78]
28. Evaluate (53 − 61) + 70 [62]
29. Evaluate 53 − (61 − 70) [62]
30. Evaluate (53 − 61) − 70 [−78]
31. Find the value of (6 + 9)(15 − 7) [120]
32. Evaluate (6 − 9)(15 + 7) [−66]
33. Simplify (17 − 5) ÷ 4 [3]
34. Simplify (173 − 52) ÷ 11 [11]
35. Evaluate (8 + 7)4 ÷ 2(9 + 6) [450]
36. Find the value of 12 × 60 ÷ 3(26 − 21) [1 200]
37. Find the value of (3 × 4)(12 × 4) ÷ (16 + 3 − 7) [48]
38. Evaluate (7 × 6) ÷ 14 + 23 − 2 × 39 ÷ 3 [0]
39. Evaluate 17 × 3 − (−16 + 19)(86 − 22) ÷ 8 + 7 [34]
40. Find the value of
4 × 6 ÷ (3 − 7) + (12 − 3) ÷ 3 − (16 ÷ 2) ÷ (3 − 7)(7 − 5) [1]

Fractions

Simplify the following:

41. $\frac{1}{4}+\frac{1}{5}$ $[\frac{9}{20}]$
42. $\frac{3}{20}+\frac{7}{12}$ $[\frac{11}{15}]$
43. $2\frac{3}{4}+\frac{3}{10}$ $[3\frac{1}{20}]$
44. $3\frac{5}{8}+2\frac{1}{9}+\frac{7}{16}$ $[6\frac{25}{144}]$
45. $\frac{1}{7}-\frac{1}{9}$ $[\frac{2}{63}]$
46. $\frac{4}{9}-\frac{4}{15}$ $[\frac{8}{45}]$
47. $6\frac{5}{7}-4\frac{7}{8}$ $[1\frac{47}{56}]$
48. $\frac{1}{4}-\frac{1}{7}+\frac{3}{16}$ $[\frac{33}{112}]$
49. $\frac{7}{16}+\frac{3}{5}-\frac{19}{20}$ $[\frac{7}{80}]$
50. $3\frac{1}{5}-2\frac{5}{6}+6\frac{3}{15}$ $[6\frac{17}{30}]$
51. $26\frac{1}{3}+33\frac{3}{5}-61\frac{7}{15}$ $[-1\frac{8}{15}]$
52. $6\frac{3}{22}-3\frac{1}{55}-2\frac{4}{5}$ $[\frac{7}{22}]$
53. $\frac{1}{3}\times\frac{2}{5}$ $[\frac{2}{15}]$
54. $\frac{4}{5}\times\frac{3}{8}\times\frac{15}{16}$ $[\frac{9}{32}]$
55. $\frac{3}{13}\times\frac{39}{11}\times\frac{55}{19}$ $[2\frac{7}{19}]$
56. $3\frac{1}{4}\times\frac{19}{26}$ $[2\frac{3}{8}]$
57. $4\frac{2}{9}\times 1\frac{4}{5}\times 1\frac{10}{13}$ $[13\frac{29}{65}]$
58. $\frac{3}{5}\div\frac{4}{7}$ $[1\frac{1}{20}]$
59. $\dfrac{\frac{3}{4}}{\frac{5}{16}}$ $[2\frac{2}{5}]$
60. $1\frac{7}{8}\div 1\frac{15}{16}$ $[\frac{30}{31}]$
61. $\dfrac{1\frac{3}{4}}{2\frac{3}{5}}\div\frac{10}{13}$ $[\frac{7}{8}]$
62. $3\frac{1}{2}\div\frac{3}{7}\div 3\frac{1}{3}$ $[2\frac{9}{20}]$
63. $3\frac{1}{4}+\frac{3}{5}\times 1\frac{1}{6}$ $[3\frac{19}{20}]$
64. $2\frac{1}{2}-\frac{3}{4}\times 1\frac{1}{7}$ $[1\frac{9}{14}]$
65. $(3\frac{1}{4}+2\frac{3}{8})\times 4\frac{4}{7}$ $[25\frac{5}{7}]$
66. $2\frac{2}{15}(2\frac{1}{2}-3\frac{1}{8})$ $[-1\frac{1}{3}]$
67. $\dfrac{2\frac{1}{3}+\frac{3}{4}}{4\frac{2}{5}-3\frac{17}{30}}$ $[3\frac{7}{10}]$
68. $3-\dfrac{(\frac{1}{6}\times\frac{3}{4})}{(\frac{1}{2}\div\frac{3}{5})}+\frac{7}{8}$ $[3\frac{29}{40}]$
69. $\frac{1}{3}$ of $\dfrac{1}{(\frac{1}{3}\div\frac{5}{8})}\times\dfrac{(4\frac{1}{4}-3\frac{2}{5})}{(2\frac{1}{4}-1\frac{3}{16})}$ $[\frac{1}{2}]$

Ratio and proportion

70. Divide 221 in the ratio 4:9. [68:153]
71. Express 16 months as a ratio of 2 years. [2:3]

72. A piece of wood 8 metres long is cut in the ratio 9:7. Determine the length of the two pieces of wood. [$4\frac{1}{2}$ m; $3\frac{1}{2}$ m]
73. A plot of land is divided between 3 men in the ratio 13:9:2. If the plot of land is 1 440 square metres, determine how much land each man gets. [780 m^2; 540 m^2; 120 m^2]
74. Two ploughmen can plough a field in 10 days. Find how long it would take 5 ploughmen to plough the same field working at the same rate. [4 days]
75. To supply a small factory with energy costs £580 per week when it is open for 5 days a week, for 8 hours a day. Find how much it will cost to supply the factory with energy if the working day is extended to $9\frac{1}{2}$ hours. [£688.75]

Decimals

76. Add together 137.917 7, 1 271.73, 4.027 and 0.014 5 [1 413.689 2]
77. Evaluate $0.74 + 137 + 8.371 + 0.0074 + 387.7$ [533.818 4]
78. Find the value of $147 - 33.28$ [113.72]
79. Evaluate $0.0067 - 0.000324$ [0.006 376]
80. Find the value of $27.281 - 149.1124$ [−121.831 4]
81. Evaluate 427.8×0.0064 [2.737 92]
82. Evaluate 0.0182×0.674 [0.012 266 8]
83. Find the value of 75.814×27.63 [2094.740 82]
84. Find the value of 0.000746×3281.4 [2.447 924 4]
85. Evaluate $25.71 \div 0.6$ [42.85]
86. Find the value of $67.418 \div 36.71$ correct to 5 significant figures. [1.836 5]
87. Evaluate $0.255 \div 0.0015$ [170]
88. Determine the value of $237.41 \div 0.74$ correct to 3 decimal places. [320.824]
89. Express $\frac{37}{63}$ as a decimal fraction correct to 4 significant figures. [0.587 3]
90. Express $3\frac{26}{35}$ as a decimal fraction correct to 4 decimal places. [3.742 9]
91. Express 0.343 75 as a proper fraction. [$\frac{11}{32}$]
92. Express 7.562 5 as an improper fraction. [$\frac{121}{16}$]

Percentages

93. Find 17% of 3 kilograms. [0.51 kg]
94. Determine 38% of £47. [£17.86]
95. Express $\frac{5}{13}$ as a percentage correct to three significant figures. [38.5%]

96. A pay rise of $4\frac{1}{2}\%$ is paid to a man earning £150 a week. Determine how much he will then earn. [£156.75]

97. A deduction of $7\frac{1}{2}\%$ is given for paying a bill of £126.80 promptly. Find out how much will have to be paid if the bill is settled promptly. [£117.29]

98. A pile of coal weighing 2 000 kilograms is divided between three people, so that one receives 15%, another 45% and the third person the remainder. Find the weight of coal each person receives. [300 kg; 900 kg; 800 kg]

99. A piece of elastic 0.7 metres long stretches by 73% when a weight is attached to it. Determine the new length. [1.211 m]

100. A company employing 800 people increases its staff to 830. Determine the percentage increase in staff. [3.75%]

Chapter 2

Indices and standard form

2.1 Definitions

The system of numbers which uses the numbers 0 to 9 is called the **decimal or denary system.** There are ten figures or digits in this system and it is said to have a **radix** ten. If a counting system used only the numbers 0, 1, 2, 3 and 4 it would have radix five. Comparatively recently, and associated with the growth of the computer industry, the **binary system** of numbers, having radix two and using the two digits 0 and 1, has come more into general use.

When a number is multiplied by itself, it is said to be **squared**. Thus for the number 3 the square is obtained by multiplying 3 by itself. Three squared is 3×3 and is written 3^2. This process is also called **raising to a power** and three squared is also called three raised to the power two. Numbers can be multiplied by themselves more than once and three raised to the power four is written 3^4 where

$$3^4 = 3 \times 3 \times 3 \times 3 = 81$$

and seven raised to the power five is 7^5 where

$$7^5 = 7 \times 7 \times 7 \times 7 \times 7 = 16\,807$$

When raising to a power of three, the number is said to be **cubed**, but special names are not normally used for powers greater than three. The number which has to be multiplied by itself is called the **base** and the number of times the number is repeated is called the **index**. Thus 5^6 has a base of 5 and an index of 6.

The **reciprocal** of an integer or decimal fraction is found by writing the number as the denominator of a fraction and giving the numerator the value 1. The reciprocal of 5 is 1/5 and the reciprocal of 3.128 is $\frac{1}{3.128}$. The reciprocal of a proper or improper fraction is found by interchanging the positions of the numbers in the numerator and denominator; hence the reciprocal of 3/4 is 4/3, and of 81/5 is 5/81.

The **square root** of a positive number can be found by looking for a base which when multiplied by itself gives the positive number. For the number 4, for example, a base must be selected which when multiplied by itself is equal to 4. Since $2 \times 2 = 4$, the base is 2. Thus the square root of 4 is 2 and is written $\sqrt{4} = 2$.

There is also another base of -2, since $(-2) \times (-2) = 4$; hence the square root of 4 is also -2. There will always be two square roots of a number and so $\sqrt{4} = \pm 2$, called 'plus or minus two'.

Since $3 \times 3 = (-3) \times (-3) = 9$, then

$$\sqrt{9} = \pm 3, \text{ that is } +3 \text{ or } -3$$

2.2 Indices

To manipulate numbers which are raised to a power and to simplify calculations there are some fundamental rules or laws which can be applied. These are known as the **laws of indices** and those relating to arithmetic expressions are given below.

Law 1

When multiplying a number raised to a power by **the same number** raised to a power, the indices (plural of index) must be added. For example:

$$5^3 \times 5^4 = (5 \times 5 \times 5) \times (5 \times 5 \times 5 \times 5) = 5^7, \text{ and}$$

$$5^{(3+4)} = 5^7, \text{ that is, } 5^3 \times 5^4 = 5^{(3+4)} = 5^7$$

Applying the same law, $17^2 \times 17^4 = 17^{2+4} = 17^6$

When a number is written without an index, the index is 1 and so

$$5 \times 5^3 = 5^1 \times 5^3 = 5^{1+3} = 5^4, \text{ and}$$

$$13^2 \times 13^5 \times 13 = 13^{(2+5+1)} = 13^8$$

Law 2

When dividing a number raised to a power by the same number raised to a power, the indices must be subtracted. For example, by the cancellation method

$$\frac{5^6}{5^4} = \frac{5 \times 5 \times 5 \times 5 \times 5 \times 5}{5 \times 5 \times 5 \times 5} = 5^2$$

$$5^{6-4} = 5^2, \text{ that is } \frac{5^6}{5^4} = 5^{6-4} = 5^2$$

Law 3

When a number raised to a power is raised to a further power, the indices must be multiplied. For example:

$$(5^2)^3 = 5^2 \times 5^2 \times 5^2 = 5^{2+2+2} = 5^6 \text{ by law 1. And by law 3}$$

$$(5^2)^3 = 5^{2 \times 3} = 5^6$$

Law 4

Any number raised to the power nought is equal to one. This rule can be verified as follows:

$$\frac{5^3}{5^3} = 5^{3-3} = 5^0 \text{ by law 2, but by cancellation}$$

$$\frac{5^3}{5^3} = \frac{5 \times 5 \times 5}{5 \times 5 \times 5} = 1. \text{ Hence } \frac{5^3}{5^3} = 5^0 = 1$$

Similarly it can be shown that any other number raised to the power nought is equal to 1.

Law 5

A number raised to a negative power is the reciprocal of that number raised to the same positive power. Since $\frac{5^2}{5^4} = 5^{2-4} = 5^{-2}$ by law 2, and

$$\frac{5^2}{5^4} = \frac{5 \times 5}{5 \times 5 \times 5 \times 5} = \frac{1}{5 \times 5} = \frac{1}{5^2} \text{ then } 5^{-2} = \frac{1}{5^2}$$

Similarly $13^{-3} = \frac{1}{13^3}$ and $7^{-1} = \frac{1}{7}$

Law 6

When a number is raised to a fractional power the denominator of the fraction is the root of the number and the numerator is the power.

Thus $8^{\frac{2}{3}} = \sqrt[3]{8^2} = 2^2 = 4$

(Note that it does not matter whether the third root of 8 is found first, or whether 8 squared is found first – the same answer will result.)

Similarly, $9^{\frac{1}{2}} = \sqrt{9^1} = \pm 3$

Thus when $\frac{1}{2}$ is the fractional power a square root is indicated and is written as $\sqrt{\ }$ (and not $\sqrt[2]{\ }$).

Worked problems on indices

Problem 1. Simplify $3^3 \times 3^5$ and express the answer in power form.

Applying law 1, when multiplying the same quantities the indices are added, which gives

$3^3 \times 3^5 = 3^{3+5} = 3^8$

Problem 2. Simplify $6^2 \times 6^9 \times 6$ and express the answer in power form.

Any base with no index indicates an index of 1.

By law 1, $\mathbf{6^2 \times 6^9 \times 6 = 6^{2+9+1} = 6^{12}}$

Problem 3. Evaluate $2^3 \times 2^4 \times 2$

By law 1, $2^3 \times 2^4 \times 2 = 2^{3+4+1} = 2^8$

and $2^8 = 2 \times 2 \times 2 \cdots$ to 8 terms = **256**

Problem 4. Find the value of $\dfrac{3^5}{3^2}$

By law 2, $\dfrac{3^5}{3^2} = 3^{5-2} = 3^3$

and $3^3 = 3 \times 3 \times 3 = \mathbf{27}$

Problem 5. Simplify $\dfrac{13^{17}}{13^9}$ and express the answer in power form.

By law 2, $\dfrac{13^{17}}{13^9} = 13^{17-9} = \mathbf{13^8}$

Problem 6. Evaluate $3^3 \times 3^4 \div 3^6$

This can be written as $\dfrac{3^3 \times 3^4}{3^6}$

$= \dfrac{3^{3+4}}{3^6}$ by law 1

$= \dfrac{3^7}{3^6} = 3^{7-6}$ by law 2

$= 3$

Problem 7. Evaluate $(7^2 \times 7^4) \div (7 \times 7^3)$

Applying law 1 for multiplication and then writing in fraction form

$$(7^2 \times 7^4) \div (7 \times 7^3) = \frac{7^{2+4}}{7^{1+3}} = \frac{7^6}{7^4}$$

Applying law 2, $\frac{7^6}{7^4} = 7^{6-4} = 7^2 = \mathbf{49}$

Problem 8. Evaluate $(3^2)^2$

By law 3 when raising to a power the indices are multiplied, so

$(3^2)^2 = 3^{2\times 2} = 3^4 = \mathbf{81}$

Problem 9. Evaluate $\frac{4^3 \times 4^{10}}{4^8 \times 4^7}$

By laws 1 and 2

$$\frac{4^3 \times 4^{10}}{4^8 \times 4^7} = \frac{4^{13}}{4^{15}} = 4^{13-15} = 4^{-2}$$

and by law 5

$$4^{-2} = \frac{1}{4^2} = \mathbf{\frac{1}{16}}$$

Problem 10. Evaluate $\frac{17^7 \times (17^2)^4}{(17^3)^5}$

By law 3, $\frac{17^7 \times (17^2)^4}{(17^3)^5} = \frac{17^7 \times 17^8}{17^{15}}$

By law 1 $= \frac{17^{15}}{17^{15}}$

By law 2 $= 17^{15-15} = 17^0$

and by law 4, 17^0 $= \mathbf{1}$

Problem 11. Find the value of $\frac{3^2 \times 2^5}{2^3 \times 3^3}$

The laws of indices apply to the same base. As this problem can be written as

$$\frac{3^2}{3^3} \times \frac{2^5}{2^3}$$

the laws can be applied to each of these two bases independently. Then

$$\frac{3^2}{3^3} \times \frac{2^5}{2^3} = 3^{2-3} \times 2^{5-3}$$

$$= 3^{-1} \times 2^2 = \frac{2^2}{3} = \mathbf{\frac{4}{3}}$$

Problem 12. Evaluate (*a*) $4^{\frac{1}{2}}$ (*b*) $27^{\frac{2}{3}}$ (*c*) $16^{\frac{3}{4}}$

From law 6:

(*a*) $4^{\frac{1}{2}} = \sqrt{4} = \mathbf{\pm 2}$

(*b*) $27^{\frac{2}{3}} = \sqrt[3]{27^2} = 3^2 = \mathbf{9}$

(*c*) $16^{\frac{3}{4}} = \sqrt[4]{16^3} = \pm 2^3 = \mathbf{\pm 8}$

Problem 13. Evaluate (*a*) $16^{-\frac{1}{4}}$ (*b*) $81^{-0.25}$ (*c*) $\left(\frac{1}{2^2}\right)^{-1}$

From laws 5 and 6:

(*a*) $16^{-\frac{1}{4}} = \frac{1}{16^{\frac{1}{4}}} = \frac{1}{\sqrt[4]{16}} = \frac{1}{\pm 2} = \mathbf{\pm\frac{1}{2}}$

(*b*) $81^{-0.25} = 81^{-\frac{1}{4}} = \frac{1}{81^{\frac{1}{4}}} = \frac{1}{\sqrt[4]{81}} = \mathbf{\pm\frac{1}{3}}$

(*c*) $\left(\frac{1}{2^2}\right)^{-1} = \left(\frac{1}{4}\right)^{-1} = \left(\frac{4}{1}\right)^{+1} = \mathbf{4}$

Problem 14. Evaluate $\frac{(3^2)^{\frac{3}{2}} \times (8^{\frac{1}{3}})^2}{(4^3)^{\frac{1}{2}} \times 9^{-\frac{1}{2}} \times 3^3}$, taking positive square roots only.

Using laws 3, 5 and 6,

$$\frac{(3^2)^{\frac{3}{2}} \times (8^{\frac{1}{3}})^2}{(4^3)^{\frac{1}{2}} \times 9^{-\frac{1}{2}} \times 3^3} = \frac{3^{(2 \times \frac{3}{2})} \times 8^{(\frac{1}{3} \times 2)}}{4^{(3 \times \frac{1}{2})} \times \frac{1}{9^{\frac{1}{2}}} \times 3^3} = \frac{3^3 \times 8^{\frac{2}{3}}}{4^{\frac{3}{2}} \times \frac{1}{9^{\frac{1}{2}}} \times 3^3}$$

$$= \frac{3^3 \times \sqrt[3]{8^2}}{\sqrt{4^3} \times \frac{1}{\sqrt{9}} \times 3^3} = \frac{2^2}{2^3 \times \frac{1}{3}} = \frac{1}{2 \times \frac{1}{3}}$$

$$= \frac{3}{2} = \mathbf{1\frac{1}{2}}$$

Problem 15. Simplify $\dfrac{9^4 \times 8^2 \times 7^3}{7^2 \times 8^4 \times 9^5}$

This expression can be re-written as

$$\frac{9^4}{9^5} \times \frac{8^2}{8^4} \times \frac{7^3}{7^2} = 9^{4-5} \times 8^{2-4} \times 7^{3-2}$$

$$= 9^{-1} \times 8^{-2} \times 7^1$$

$$= \frac{7}{8^2 \times 9} = \mathbf{\frac{7}{576}}$$

Problem 16. Evaluate $\dfrac{2^2 \times 3^3 + 3^2 \times 2^3}{3^4 \times 2^5}$

(Remember that the fraction $\dfrac{4+7}{11}$, for example, can be split into two fractions, $\dfrac{4}{11} + \dfrac{7}{11}$). Applying this principle gives:

$$\frac{2^2 \times 3^3 + 3^2 \times 2^3}{3^4 \times 2^5} = \frac{2^2 \times 3^3}{3^4 \times 2^5} + \frac{3^2 \times 2^3}{3^4 \times 2^5}$$

$$= 2^{2-5} \times 3^{3-4} + 3^{2-4} \times 2^{3-5}$$

$$= 2^{-3} \times 3^{-1} + 3^{-2} \times 2^{-2}$$

$$= \frac{1}{2^3 \times 3} + \frac{1}{3^2 \times 2^2}$$

$$= \frac{1}{24} + \frac{1}{36} = \frac{3+2}{72}$$

$$= \mathbf{\frac{5}{72}}$$

Problem 17. Simplify $\dfrac{5^{-6} \times 13^7}{7 \times 5^{-4} \times 13^3}$ and express the answer in index form.

Negative powers can either be treated directly by the laws of indices, in which case

$$\frac{5^{-6} \times 13^7}{7 \times 5^{-4} \times 13^3} = \frac{5^{-6}}{5^{-4}} \times \frac{13^7}{13^3} \times \frac{1}{7}$$

$$= \frac{5^{-6+4} \times 13^{7-3}}{7}$$

$$= \frac{5^{-2} \times 13^4}{7} = \mathbf{\frac{13^4}{5^2 \times 7}}$$

or they can be converted into positive indices by law 5 and

$$\frac{5^{-6} \times 13^7}{7 \times 5^{-4} \times 13^3} = \frac{5^4 \times 13^7}{7 \times 5^6 \times 13^3}$$

by taking the reciprocal of the negative indices. Then as previously shown

$$\frac{5^4 \times 13^7}{7 \times 5^6 \times 13^3} = \frac{13^{7-3}}{5^{6-4} \times 7}$$

$$= \mathbf{\frac{13^4}{5^2 \times 7}}$$

Problem 18. Evaluate $\dfrac{2^5 \times 3^4}{2^2 \times 3^3 + 2^4 \times 3^2}$

To simplify the calculations, each term may be divided by the highest common factor of the three terms. By doing this, each term is reduced to its smallest integer number without altering the value of the expression. The H.C.F. is $2^2 \times 3^2$ and gives:

$$\frac{2^5 \times 3^4}{2^2 \times 3^3 + 2^4 \times 3^2} = \frac{\dfrac{2^5 \times 3^4}{2^2 \times 3^2}}{\dfrac{2^2 \times 3^3}{2^2 \times 3^2} + \dfrac{2^4 \times 3^2}{2^2 \times 3^2}} = \frac{2^{5-2} \times 3^{4-2}}{2^{2-2} \times 3^{3-2} + 2^{4-2} \times 3^{2-2}}$$

$$= \frac{2^3 \times 3^2}{2^0 \times 3^1 + 2^2 \times 3^0} = \frac{8 \times 9}{1 \times 3 + 4 \times 1} = \frac{72}{7} = \mathbf{10\tfrac{2}{7}}$$

Problem 19. Find the value of $\dfrac{3^3 \times 4^4}{2^3 \times 9^2 + 4^2 \times 27}$

This expression can be written in terms of bases 2 and 3 only, since 4 can be written as 2^2, 9 as 3^2 and 27 as 3^3. Thus

$$\frac{3^3 \times 4^4}{2^3 \times 9^2 + 4^2 \times 27} = \frac{3^3 \times (2^2)^4}{2^3 \times (3^2)^2 + (2^2)^3 \times 3^3}$$

$$= \frac{3^3 \times 2^8}{2^3 \times 3^4 + 2^4 \times 3^3}$$

The H.C.F. of all the terms is $2^3 \times 3^3$ and dividing each term by this gives

$$\frac{\dfrac{3^3 \times 2^8}{2^3 \times 3^3}}{\dfrac{2^3 \times 3^4}{2^3 \times 3^3} + \dfrac{2^4 \times 3^3}{2^3 \times 3^3}} = \frac{2^5}{3+2} = \frac{32}{5} = 6\tfrac{2}{5}$$

Problem 20. Find the value of $\dfrac{\left(\dfrac{3}{8}\right)^5 - \left(\dfrac{16}{3}\right)^{-3}}{\left(\dfrac{4}{3}\right)^{-4}}$

When proper or improper fractions are raised to a power, both numerator and denominator are raised to the power shown. Thus $\left(\dfrac{3}{8}\right)^5$ means $\dfrac{3^5}{8^5}$. To change a negative index to a positive index, the position of the numbers in the numerator and denominator are interchanged. $\left(\dfrac{16}{3}\right)^{-3}$ becomes $\dfrac{3^3}{16^3}$ and $\left(\dfrac{4}{3}\right)^{-4}$ becomes $\left(\dfrac{3}{4}\right)^4$ or $\dfrac{3^4}{4^4}$. The problem can be rewritten with positive indices as

$$\frac{\left(\dfrac{3}{8}\right)^5 - \left(\dfrac{3}{16}\right)^3}{\left(\dfrac{3}{4}\right)^4} = \frac{\dfrac{3^5}{8^5} - \dfrac{3^3}{16^3}}{\dfrac{3^4}{4^4}}$$

And, in terms of bases 2 and 3 only, as

$$\frac{\dfrac{3^5}{2^{15}} - \dfrac{3^3}{2^{12}}}{\dfrac{3^4}{2^8}}$$

The L.C.M. of 2^{15} and 2^{12} is 2^{15} and replacing the division by multiplication gives

$$\frac{3^5 - 2^3 \times 3^3}{3^4} \times \frac{2^8}{2^{15}}$$

and, by dividing each term in the first expression by 3^3, this becomes

$$\frac{3^2 - 2^3}{3} \times \frac{2^8}{2^{15}} = \frac{3^2 - 2^3}{3} \times \frac{1}{2^7}$$

$$= \frac{9 - 8}{3 \times 128} = \mathbf{\frac{1}{384}}$$

Further problems on indices may be found in Section 2.4 (*Problems 1–23*), page 52.

2.3 Standard form

To multiply a decimal fraction by 10 the decimal point is moved one place to the right, by 100 two places to the right and so on. To divide a decimal fraction by 10, the decimal point is moved one place to the left and to divide by 100, two places to the left. The value of a number is unaltered if the number is both multiplied and divided by the same number. For example, the number 3 is not altered if multiplied by 1 000 **and** divided by 1 000, for $3 \times 10^3 \div 10^3 = \frac{3 \times 10^3}{10^3} = 3$. When solving problems containing decimal or other fractions, the fractions can be expressed in decimal fraction form with one figure only in front of the decimal point by multiplying or dividing the number by 10 raised to some power. When this way of writing a number is used it is said to be written in **standard form**. Thus a number written in standard form is a number between 1 and 9.9 multiplied by 10 raised to a power.

To write 43.7 in standard form, for example, it is first divided by 10 by moving the decimal point one place to the left to give 4.37. But it must now be multiplied by 10 to retain the value of the original number. So, $43.7 = 4.37 \times 10$ when written in standard form.

Again, to write 0.043 7 in standard form, it is mutliplied by 100 or 10^2 by moving the decimal point two places to the right and then divided by 100 (or multiplied by 10^{-2}) to retain its original value. Thus $0.0437 = 4.37 \times 10^{-2}$ when written in standard form.

Writing a number in standard form enables a quick check to be made on the approximate value of a calculation to make sure an error in the position of the decimal point has not occurred. Also a similar principle is used to denote the size of certain physical quantities. The SI system of units has

Table 2.1 Powers of ten in common use to keep numbers to a reasonable size, their names and abbreviations.

When multiplying a number by	The prefix used is	The abbreviation used is
10^6	mega	M
10^3	kilo	k
10^{-1}	deci	d
10^{-2}	centi	c
10^{-3}	milli	m
10^{-6}	micro	μ
10^{-12}	pico	p

adopted the metre as its basic unit of linear measure (length or distance). To measure the distance between two towns, thousands or tens of thousands of metres would be required, whereas the length of a small insect such as an ant would be expressed in thousandths of a metre. Since length and distance can vary so much, large distances are measured in kilometres or metres $\times\ 10^3$. Table 2.1 gives some of the powers of 10 used to express numbers as a reasonable size, together with the abbreviations used for these powers of 10 and the name given to them.

To measure the power output from a large modern alternator in a power station, megawatts (MW) are used, but the power to drive a small transistor radio would be measured in milliwatts (mW). The distance between London and Birmingham would be stated in kilometres (km) but the distance between the ends of a pencil would be measured in centimetres (cm). These units are selected to keep numbers to a reasonable size. Other units used in this text such as velocity, whose SI unit is metres per second, will be written as $m\ s^{-1}$ or m/s and acceleration, having an SI unit of metres per second squared, will be written as $m\ s^{-2}$ or m/s^2.

When a number is written in standard form, the number is called the **mantissa** and the factor by which it is multiplied the **exponent**. Thus 4.3×10^5 has a mantissa of 4.3 and an exponent of 10^5. Addition and subtraction of numbers in standard form can be achieved by adding the mantissae provided the exponent is the same for each of the numbers being added. For example:

$$4 \times 10^2 + 5.6 \times 10^2 = 9.6 \times 10^2$$

This can be verified by writing the numbers as integers, for

$$4 \times 10^2 + 5.6 \times 10^2 = 400 + 560 = 960$$

Also $$9.6 \times 10^2 = 960$$

hence $$4 \times 10^2 + 5.6 \times 10^2 = 9.6 \times 10^2$$

When the exponents are not the same it is usually better to write the numbers in decimal fraction form before adding or subtracting.

The laws of indices are used when multiplying or dividing numbers given in standard form. For example:

$$(3 \times 10^3) \times (5 \times 10^2) = (3 \times 5) \times (10^{3+2}) = 15 \times 10^5 = 1.5 \times 10^6$$

Similarly,
$$\frac{8 \times 10^5}{2 \times 10^3} = \left(\frac{8}{2}\right) \times (10^{5-3}) \quad = 4 \times 10^2$$

Worked problems on standard form

Problem 1. Change the following numbers to standard form:
(*a*) 73.82 (*b*) 526 (*c*) 83 470

(*a*) To express 73.82 in standard form (that is, with one figure to the left of the decimal point), it is divided by 10. Then $73.82 \div 10 = 7.382$. To compensate for dividing by 10 and to give the number its original value it must now be multiplied by 10. Thus

$$73.82 = \mathbf{7.382 \times 10}$$

(*b*) To obtain standard form, 526 must be divided by 100 and multiplied by 100 to retain its original value. Then

$$526 = \frac{526}{100} \times 100 = \mathbf{5.26 \times 10^2}$$

(*c*) 83 470 has to be divided by 10^4 to obtain a mantissa having only one figure to the left of the decimal point and multiplied by 10^4 to retain its original value. Thus

$$83\,470 = \frac{83\,470}{10^4} \times 10^4 = \mathbf{8.347 \times 10^4}$$

Problem 2. Express the following numbers in standard form:
(*a*) 0.3716 (*b*) 0.002 (*c*) 0.000 071 3

(*a*) To express 0.371 6 in standard form, 0.371 6 is multiplied by 10 to give 3.716 but also divided by 10 to retain its original value. Thus

$$0.371\,6 = \frac{3.716}{10} = \mathbf{3.716 \times 10^{-1}}$$

since $\frac{1}{10} = 10^{-1}$ by law 5 of indices.

(*b*) To express 0.002 in standard form, multiply by 10^3 to obtain the mantissa, and then divide by 10^3 (or multiply by 10^{-3}) to retain the original value. Thus

$$0.002 = \frac{0.002 \times 10^3}{10^3} = \frac{2.0}{10^3} = \mathbf{2.0 \times 10^{-3}}$$

(*c*) 0.000 071 3 must be multiplied by 10^5 and divided by 10^5. Thus

$$0.000\,071\,3 = \frac{0.000\,071\,3 \times 10^5}{10^5}$$

$$= \frac{7.13}{10^5} = \mathbf{7.13 \times 10^{-5}}$$

Problem 3. Write the following numbers, which are in standard form, as integers or decimal fractions:
(*a*) $1.371\,2 \times 10^2$ (*b*) 6.67×10^{-3} (*c*) 9.871×10^4
(*d*) 5.57×10^0

(*a*) $1.371\,2 \times 10^2 = 1.371\,2 \times 100 = \mathbf{137.12}$

(*b*) $6.67 \times 10^{-3} = \frac{6.67}{10^3} = \frac{6.67}{1\,000} = \mathbf{0.006\,67}$

(*c*) $9.871 \times 10^4 = 9.871 \times 10000 = \mathbf{98\,710}$

(*d*) $5.57 \times 10^0 = 5.57 \times 1 =$ **5.57** (law 4 of indices states that any number raised to a power of 0 is 1).

Problem 4. Rewrite the following statements without using powers of ten:
(*a*) the speed of light is 2.998×10^8 metres per second;
(*b*) the Crab Nebula is 6×10^3 light years from earth; and
(*c*) the reciprocal of 20.116 8 is 4.971×10^{-2}

(*a*) The exponent 10^8 indicates moving the decimal place eight places to the right, thus $2.998 \times 10^8 = 29\,800000$ metres per second, and inserting this value in the statement gives:
the speed of light is 299 800 000 metres per second or 299 800 km per second

(*b*) $6 \times 10^3 = 6\,000$, the 10^3 indicating moving the decimal point three places to the right. Thus
the Crab Nebula is 6 000 light years from earth

(*c*) Since the exponent 10^{-2} means divide by 100 or move the decimal point two places to the left, the statement becomes:
the reciprocal of 20.116 8 is 0.049 71

Problem 5. Express the following in standard form, correct to 4 significant figures:
(*a*) $\frac{9}{57}$ (*b*) $7\frac{7}{32}$ (*c*) $76\frac{3}{4}$

(*a*) Since $\frac{9}{57} = 0.157\,89\ldots$, this can be written in standard form as $1.578\,9\ldots \times 10^{-1}$ and correcting to 4 significant figures gives:
$\frac{9}{57} = \mathbf{1.579 \times 10^{-1}}$ correct to 4 significant figures.

(*b*) $\frac{7}{32} = 0.218\,75$, then $7\frac{7}{32} = 7.218\,75$ or 7.219 correct to 4 significant figures. By the definition of standard form, this should be written as a mantissa multiplied by an exponent, that is 7.219×10^0, but the 10^0 which is equal to 1 is usually omitted. Thus
$7\frac{7}{32} = \mathbf{7.219}$ correct to 4 significant figures.

(*c*) $76\frac{3}{4} = 76.75$ and writing this in standard form gives:
$76\frac{3}{4} = \mathbf{7.675 \times 10}$

Problem 6. Express the following as proper or improper fractions:
(*a*) 7.5×10^{-1} (*b*) 5.625×10^{-2} (*c*) 3.375×10^2

(*a*) $7.5 \times 10^{-1} = 0.75 = \frac{75}{100}$. Dividing numerator and denominator by 25 gives:

$$7.5 \times 10^{-1} = \mathbf{\frac{3}{4}}$$

(*b*) $5.625 \times 10^{-2} = 0.056\,25$ or $\frac{5\,625}{100\,000}$. By cancellation

$$5.625 \times 10^{-2} = \mathbf{\frac{9}{160}}$$

(*c*) $3.375 \times 10^2 = 337.5$ or $337\frac{1}{2}$. Expressing this as an improper fraction gives:

$$3.375 \times 10^2 = \mathbf{\frac{675}{2}}$$

Problem 7. Evaluate:
(*a*) $6.71 \times 10^4 + 7.76 \times 10^4$; (*b*) $7.32 \times 10^{-3} - 6.77 \times 10^{-3}$; (*c*) $5.47 \times 10^{-2} - 6.16 \times 10^{-2}$; and (*d*) $3.67 \times 10^2 + 4.21 \times 10^3$; and express answers in standard form.

(*a*) Since the exponents are the same

$$6.71 \times 10^4 + 7.76 \times 10^4 = (6.71 + 7.76)10^4$$
$$= \mathbf{1.447 \times 10^5}$$

(*b*) $7.32 \times 10^{-3} - 6.77 \times 10^{-3}$

$$= (7.32 - 6.77)10^{-3}$$
$$= 0.55 \times 10^{-3}$$

and expressing the answer in standard form

$0.55 \times 10^{-3} = \mathbf{5.5 \times 10^{-4}}$

(*c*) $5.47 \times 10^{-2} - 6.16 \times 10^{-2}$

$$= (5.47 - 6.16)10^{-2}$$

$$= -0.69 \times 10^{-2}$$

and written in standard form this becomes $\mathbf{-6.9 \times 10^{-3}}$

(*d*) To evaluate $3.67 \times 10^2 + 4.21 \times 10^3$, since the exponents are different, straight addition of the mantissae cannot be done. However, 3.67×10^2 is the same value as 0.367×10^3 and having made the exponents the same, the mantissae can now be added. Thus,

$0.367 \times 10^3 + 4.21 \times 10^3 = 4.577 \times 10^3$

An alternative approach would be to write the expression in non-standard form, when

$$3.67 \times 10^2 + 4.21 \times 10^3 = 367 + 4210$$

$$= 4577$$

$$= \mathbf{4.577 \times 10^3} \text{ in standard form.}$$

Problem 8. Evaluate (*a*) $(5.75 \times 10^3) \times (6 \times 10^2)$

(*b*) $\dfrac{2.25 \times 10^5}{9 \times 10^2}$, expressing answers in standard form.

(*a*) $(5.75 \times 10^3) \times (6 \times 10^2) = (5.75 \times 6) \times (10^{3+2})$

$$= 34.5 \times 10^5 = \mathbf{3.45 \times 10^6}$$

(*b*) $\dfrac{2.25 \times 10^5}{9 \times 10^2} = \left(\dfrac{2.25}{9}\right) \times (10^{5-2}) = 0.25 \times 10^3 = \mathbf{2.5 \times 10^2}$

Further problems on standard form may be found in the following Section 2.4 (Problems 24–37), page 55.

2.4 Further problems

Indices

1. Simplify and express the answer in index form
(*a*) $5^3 \times 5^2$ (*b*) 13×13^4 (*c*) $2^2 \times 2^3 \times 2^4$ (*d*) $7^2 \times 7$
(*e*) $8^5 \times 8^2 \times 8 \times 8^4$
(*a*) [5^5] (*b*) [13^5] (*c*) [2^9] (*d*) [7^3] (*e*) [8^{12}]

2. Find the value of
(*a*) $6^2 \times 6^3$ (*b*) $2^2 \times 2^7 \times 2^3$ (*c*) $9^3 \times 9^3$
(*a*) [7776] (*b*) [4096] (*c*) [531 441]

3. Simplify and express the answer in index form
(*a*) $\frac{5^8}{5^3}$ (*b*) $\frac{9^{10}}{9^7}$ (*c*) $\frac{8^6}{8}$ (*d*) $13^6 \div 13^2$ (*e*) $17^7 \div 17^6$
(*a*) [5^5] (*b*) [9^3] (*c*) [8^5] (*d*) [13^4] (*e*) [17 or 17^1]

4. Find the value of
(*a*) $13^{15} \div 13^{13}$ (*b*) $\frac{2^{10}}{2^6}$ (*c*) $\frac{5^6}{5^3}$

(a) [169] (b) [16] (c) [125]

5. Simplify and express the answer in index form
(*a*) $(2^3)^2$ (*b*) $(8^4)^3$ (*c*) $(8^3)^4$ (*d*) $(31^6)^3$ (*e*) $(14^7)^5$
(*a*) [2^6] (*b*) [8^{12}] (*c*) [8^{12}] (*d*) [31^{18}] (*e*) [14^{35}]

6. Find the value of
(*a*) $(2^2)^3$ (*b*) $(3^2)^2$ (*c*) $(3^1)^4$
(*a*) [64] (*b*) [81] (*c*) [81]

7. Evaluate (*a*) $25^{\frac{1}{2}}$ (*b*) $8^{-\frac{1}{3}}$ (*c*) $\left(\frac{4}{9}\right)^{\frac{1}{2}}$ $\left[(a)\ \pm 5 \quad (b)\ \frac{1}{2} \quad (c)\ \pm\frac{2}{3}\right]$

8. Evaluate (*a*) $\left(\frac{1}{3^2}\right)^{-1}$ (*b*) $81^{\frac{3}{4}}$ (*c*) $32^{-0.4}$ $\left[(a)\ 9 \quad (b)\ 27 \quad (c)\ \frac{1}{4}\right]$

9. Evaluate

(*a*) $\frac{9^2 \times 9 \times 9^4}{9 \times 9^5}$ (*b*) $(13^3 \times 13^5) \div (13^7 \times 13)$ (*c*) $\frac{2^3 \times 2^{12}}{2^{17}}$

(*a*) [9] (*b*) [1] (*c*) [$\frac{1}{4}$]

10. Simplify and express in index form
(*a*) $(3^2 \times (3^2)^3) \div (3 \times 3^4)$ (*b*) $(17^2)^5 \times (17^5)^2$

(*c*) $\frac{(6^2)^7 \times (6^3)^6}{(6^8)^4}$

(*a*) [3^3] (*b*) [17^{20}] (*c*) [1^1 or 1]

11. Simplify and express in index form

(*a*) $\frac{4^3 \times 5^2}{4^2 \times 5^3}$ (*b*) $\frac{3^7 \times 8^4 \times 13^6}{8^5 \times 13^7 \times 3^9}$

(*a*) [$\frac{4}{5}$] (*b*) $\left[\frac{1}{3^2 \times 8 \times 13}\right]$

12. Evaluate (*a*) $\dfrac{3 \times 8^2 + 5 \times 8^4}{4 \times 8^3}$ (*b*) $\dfrac{4^6 \times 6^4 + 6^7 \times 4^3}{4^5 \times 6^6}$

$\left[(a)\ 10\tfrac{3}{32} \quad (b)\ \dfrac{35}{72}\right]$

13. Evaluate $\dfrac{8^{\frac{2}{3}} \times 16^{-0.25}}{2^3 \times (\frac{1}{4})^{-1}}$ $\left[\dfrac{1}{16}\right]$

14. Evaluate $\dfrac{27^{\frac{2}{3}} \times 4^{-\frac{1}{2}}}{2^{-5} \times (\frac{1}{3})^{-2}}$, taking positive square roots only [16]

15. Find the value of

(*a*) $\dfrac{2^{-4} \times 3 \times 4^3}{3^{-3} \times 4^2 \times 2^{-6}}$ (*b*) $\dfrac{3 \times 7^8 \times 9^{-6} \times 7^{-6}}{5 \times 9^{-5} \times 7}$

(*c*) $\dfrac{6^{-3} \times 8^3}{7 \times 6^{-4} \times 8^4}$ $\left[(a)\ 1\,296 \quad (b)\ \dfrac{7}{15} \quad (c)\ \dfrac{3}{28}\right]$

16. Find the value of the following expressions in decimal fraction form correct to 4 significant figures.

(*a*) $\dfrac{4^2 \times 5^3 + 5^2 \times 4^3}{4^4 \times 5^4}$ (*b*) $\dfrac{3 \times 6^2 + 6 \times 8^3}{3^2 \times 8^2}$

(*c*) $\dfrac{4 \times 7^{-4} + 7^{-3}}{6 \times 7^{-2}}$ [(*a*) 0.022 50 (*b*) 5.521 (*c*) 0.037 41]

17. Find the value of

(*a*) $\dfrac{4^6 \times 5^7}{5^6 \times 4^7 + 4^5 \times 5^8}$ (*b*) $\dfrac{3 \times 4^3}{4 \times 5^2 + 3^2 \times 4^2}$

(*c*) $\dfrac{2^3 \times 3^2 \times 4}{2^2 \times 3^3 + 3 \times 4^2}$ $\left[(a)\ \dfrac{20}{41} \quad (b)\ \dfrac{48}{61} \quad (c)\ \dfrac{24}{13} \text{ or } 1\dfrac{11}{13}\right]$

18. Evaluate (*a*) $\dfrac{5^3 \times 7^4}{25^2 \times 3 \times 49^2}$

(*b*) $\dfrac{6^3 \times 4^2 \times 13}{2^4 \times 3^3 \times 13^2 + 2^3 \times 3^2 \times 13}$ $\left[(a)\ \dfrac{1}{15} \quad (b)\ \dfrac{48}{79}\right]$

Simplify the following and express in index form with positive indices.

19. $\dfrac{3^{-4} \times 2^3 \times 5^{-4}}{2^{-2} \times 3^{-3} \times 5}$ $\left[\dfrac{1}{3}\left(\dfrac{2}{5}\right)^5\right]$

20. $\dfrac{2^4 \times 3^{-3} \times 4^2 + 2^{-2} \times 3^2 \times 4^{-3}}{9^2 \times 8^3}$ $\left[\dfrac{2^{16} + 3^5}{2^{17} \times 3^7}\right]$

21. $\dfrac{(2^3)^4 \times (3^2)^2}{16^2 \times 9^3}$ $\left[\dfrac{16}{9} \text{ or } 1\dfrac{7}{9}\right]$

22. $\dfrac{3^{-2} \times 5^4 \times 6^{-7}}{\left(\dfrac{2}{5}\right)^3 \times \left(\dfrac{5}{3}\right)^{-2}}$ $\left[\dfrac{5^9}{2^{10} \times 3^{11}}\right]$

23. $\dfrac{\left(\dfrac{7}{8}\right)^{-2} - \left(\dfrac{1}{4}\right)^4}{\left(\dfrac{1}{16}\right)^2 \times \left(\dfrac{1}{32}\right)^{-3}}$ $\left[\dfrac{2^{14} - 7^2}{2^{15} \times 7^2}\right]$

Standard form

In problems 24 to 28 express the numbers in standard form:

24. (*a*) 47.44 (*b*) 83.6 (*c*) 91.274 (*d*) 387.7 (*e*) 671.772
(*a*) [4.744×10] (*b*) [8.36×10] (*c*) [9.1274×10]
(*d*) [3.877×10^2] (*e*) [6.71772×10^2]

25. (*a*) 563 (*b*) 7 210 (*c*) 630 000 000 (*d*) 76 271.85
(*a*) [5.63×10^2] (*b*) [7.21×10^3] (*c*) [6.3×10^8]
(*d*) [7.627185×10^4]

26. (*a*) 0.375 (*b*) 0.14 (*c*) 0.6 (*d*) 0.002 6 (*e*) 0.003 02
(*a*) [3.75×10^{-1}] (*b*) [1.4×10^{-1}] (*c*) [6×10^{-1}]
(*d*) [2.6×10^{-3}] (*e*) [3.02×10^{-3}]

27. (*a*) 0.000 001 7 (*b*) 0.000 101 5 (*c*) 0.100 02 (*d*) 0.070 73
(*a*) [1.7×10^{-6}] (*b*) [1.015×10^{-4}] (*c*) [1.0002×10^{-1}]
(*d*) [7.073×10^{-2}]

28. (*a*) $63\frac{7}{8}$ (*b*) $\frac{3}{20}$ (*c*) $468\frac{4}{5}$ (*d*) $\frac{1}{500}$
(*a*) [6.3875×10] (*b*) [1.5×10^{-1}] (*c*) [4.688×10^2]
(*d*) [2×10^{-3}]

In problems 29–31 change the numbers from standard form to integers or decimal fractions:

29. (*a*) 3.72×10^2 (*b*) 6.2174×10^{-2} (*c*) 1.1004×10^3
(*d*) 3.27×10^4 (*e*) 8.27×10^{-1}
(*a*) [372] (*b*) [0.062 174] (*c*) [1 100.4] (*d*) [32 700]
(*e*) [0.827]

30. (*a*) 5.21×10^0 (*b*) 3×10^{-6} (*c*) 1.4771×10^{-3}
(*d*) 5.87×10
(*a*) [5.21] (*b*) [0.000 003] (*c*) [0.001 477 1] (*d*) [58.7]

31. (*a*) 7.176×10^6 (*b*) 9.98×10^{-4} (*c*) 4×10^{-5}
(*d*) 4.0001×10^5
(*a*) [717 600 0] (*b*) [0.000 998] (*c*) [0.00004] (*d*) [400 010]

32. Change the following numbers from standard form to proper or improper fractions:
(*a*) 9.375×10^{-2} (*c*) 1.8735×10^2 (*c*) 5.625×10^{-1}
(*d*) 3.2475×10^2
(*a*) [$\frac{3}{32}$] (*b*) [$\frac{3\,747}{20}$] (*c*) [$\frac{9}{16}$] (*d*) [$\frac{1\,299}{4}$]

33. Evaluate and express the answer in standard form:
(*a*) $3.774 \times 10^{-2} + 7.28 \times 10^{-2}$ (*b*) $6.3 \times 10^3 + 5.381 \times 10^3$
(*c*) $1.476 \times 10^{-6} - 1.471 \times 10^{-6}$
(*d*) $3.576 \times 10^4 - 4.211 \times 10^4$
(*a*) [1.1054×10^{-1}] (*b*) [1.1681×10^4]
(*c*) [5×10^{-9}] (*d*) [-6.35×10^3]

34. Find the value of the following, giving the answer in standard form:
(*a*) $1.874 \times 10^{-2} + 2.227 \times 10^{-3}$
(*b*) $5.27 \times 10^{-10} + 8.37142 \times 10^{-10}$
(*c*) $3.8771 \times 10^{-4} - 7.2873 \times 10^{-4}$
(*d*) $9.71 \times 10^2 - 9.998 \times 10^3$
(*a*) [2.0967×10^{-2}] (*b*) [1.364142×10^{-9}]
(*c*) [-3.4102×10^{-4}] (*d*) [-9.027×10^3]

35. Rewrite the following statements without using powers of 10:
(*a*) the freezing temperature of copper is 1.3576×10^3 Kelvin [1 357.6 K]
(*b*) one kilowatt hour has the same energy as 3.6×10^6 joules [3 600000 joules]
(*c*) the reciprocal of 1.609×10^3 is 6.214×10^{-4} [1609; 0.000 621 4]
(*d*) the volume of one fluid ounce is 2.841×10^{-5} cubic metres [0.000028 41 m³]
(*e*) the square root of 4×10^{-4} is $\pm 2 \times 10^{-2}$ [0.000 4, ±0.02]

In problems 36 and 37, evaluate giving the answers in standard form:

36. (*a*) $(4.75 \times 10^2)(8 \times 10^3)$ (*b*) $3 \times (4.4 \times 10^3)$
[(*a*) 3.8×10^6 (*b*) 1.32×10^4]

37. (*a*) $\dfrac{8 \times 10^{-3}}{5 \times 10^{-5}}$ (*b*) $\dfrac{(4.5 \times 10^3)(3 \times 10^{-2})}{2.7 \times 10^4}$
[(*a*) 1.6×10^2 (*b*) 5×10^{-3}]

Chapter 3

Calculations

3.1 Errors, accuracy and approximate values

In order to find out how long it takes a person to walk one kilometre at constant speed, several methods may be adopted. A reasonable walking speed is 5 or 6 kilometres per hour; using this data, a result of 10 to 12 minutes can be calculated. So in integer values, it will take, say, 11 minutes to walk one kilometre. For a more accurate result, a distance of one kilometre can be marked, using the recording meter on a car as a measuring device. A person can be timed by a wrist watch walking this distance and the result obtained may be accurate correct to one decimal place. By using a stop watch and carefully measuring the distance using surveying equipment a result which is correct to two decimal places may be achieved. Ultimately, using a device calibrated at the National Physics Laboratory and an atomic clock which only varies from the correct time by about 5 seconds in 700 years, a result which is correct to several decimal places can be achieved.

In all these cases the result is correct to a stated accuracy only and is not the exact time taken by a person to walk 1 kilometre. In all problems in which the measurement of distance, time, weight, or other quantities occurs, an exact answer cannot be given, only an answer which is correct to a stated degree of accuracy. To take account of this an **error due to measurement** is said to exist.

When numbers are expressed as decimal fractions, an error frequently exists. The proper fraction $\frac{9}{17}$ is equal to $0.529\,411\,7\cdots$ but will be expressed as, say, 0.529 correct to 3 significant figures or 0.529 4 correct to 4 decimal places. The error introduced in this case is called a **rounding off error**.

Errors can occur in calculations. When evaluating an expression errors can be made when performing arithmetic operations such as multiplication, division, addition and subtraction. For example, if when multiplying 27.63 by 62.7 the answer obtained is 1 731.401 then an error would have been made since $27.63 \times 62.7 = 1\,732.401$. This type of error is known as a **blunder**.

Another kind of error may occur due to the incorrect positioning of the decimal point after a calculation has been completed, and this is called an **order of magnitude error**. If, when determining the value of $\dfrac{274.3}{0.067\,1}$ the result is found to be 408.79 instead of 4 087.9, correct to 5 significant figures, then an order of magnitude error will have been made.

To take account of measurement errors it is usual to limit answers so that the result given **is not more than one significant figure greater than the least accurate number given in the data**. For example, it is required to find the resistance of a resistor in an electrical circuit when a voltage of 1.34 volts results in a current flow of 7.3 amperes. The relationship is $R = \frac{V}{I}$. Thus

$$R = \frac{1.34}{7.3} = 0.183\,561\,6 \cdots \text{ ohms.}$$

The least accurate number in the data is 7.3 A, since its value is only given correct to two significant figures. Thus the answer should not be expressed to more than a three significant figure accuracy. Hence

$$R = \frac{1.34}{7.3} = 0.184 \text{ ohms}$$

would be the normal way of writing this answer.

Blunders and order of magnitude errors can be reduced by determining approximate values of calculations. There is no hard and fast rule for doing this, but generally the problem is reduced to a mental arithmetic type problem by expressing numbers correct to one or two significant figures only and reducing the calculation to standard form if necessary. Then the approximate value of

$$\frac{289.63 \times 0.047\,1}{73.824} \text{ could be } \frac{3 \times 10^2 \times 5 \times 10^{-2}}{7 \times 10}$$

Using the laws of indices this becomes $\frac{15}{7} \times 10^{-1} \approx 0.2$. That is, an answer of about 0.2 will be expected. On performing the calculation, if an answer of 0.184 8 is obtained, it will indicate that the calculation has probably been done correctly. An answer of 0.482 7 will suggest that a blunder has occurred or an answer of 0.018 48 that an order of magnitude error has occurred. If either of these last two answers are obtained, the calculation should be checked and repeated if necessary.

Answers which do not seem feasible must be checked and repeated as necessary. If as a result of a calculation the area of a triangle is found to be thousands of square kilometres, or the height of a man tens of metres, then it is likely that an error has occurred. For example, if 5 kg of sugar cost 125p and it is required to find out how much 2 kg cost, then an answer of $125 \times \frac{5}{2}$ or 312.5p is obviously incorrect since it must cost less to buy 2 kg than 5 kg.

When determining an answer to a calculation the points listed below should be considered to ascertain whether the result obtained is likely to be correct and expressed to a reasonable degree of accuracy:

(*a*) the answer should not be more accurate than the data on which it is based;
(*b*) errors can exist in all problems based on measurements;
(*c*) errors do exist in all problems where rounding off has occurred;
(*d*) blunders and order of magnitude errors should be eliminated as far as possible by determining the approximate value of a calculation; and
(*e*) a check should be made to see whether the answer obtained seems feasible.

Worked problems on errors

Problem 1. The area of a rectangle $A = bh$. The base, b, of the rectangle when measured was found to be 4.63 centimetres and the height, h, 9.4 centimetres. Determine the area of the rectangle.

Area $= bh = 9.4 \times 4.63 = 43.522\ \text{cm}^2$.

The approximate value is $9 \times 5 = 45\ \text{cm}^2$, so there are no obvious blunder or order of magnitude errors. However, it is not usual in a measurement type question to state the answer to an accuracy greater than one significant figure more than the least accurate number in the data; this is 9.4 cm, so the result should not have more than 3 significant figures.
Thus
Area = **43.5 cm²**
would be the normal way of expressing the result.

Problem 2. State which type of error or errors have been made in the following problems:

(*a*) $\dfrac{37.4}{16.8 \times 0.071} = 1.335$ correct to 3 decimal places;

(*b*) $16.814 \times 0.003\,8 = 0.036\,89$ correct to 4 significant figures;
(*c*) In the formula $C = 2\pi r$, r was measured as 1.72 cm, giving a value for C of 10.808 5 correct to 4 decimal places when π was taken to be 3.142; and
(*d*) $127.8 \times 1.632 = 208.6$

(*a*) $\dfrac{37.4}{16.8 \times 0.071}$ is approximately equal to $\dfrac{4 \times 10}{2 \times 10 \times 7 \times 10^{-2}}$ or

$\frac{4}{14} \times 10^2$, that is about 30. **Thus two errors have occurred, both a blunder and an order of magnitude error.**

(*b*) 16.814×0.0038 is approximately equal to $1.7 \times 10 \times 4 \times 10^{-3}$, that is about 7×10^{-2} or 0.07. **Thus a blunder has been made in the calculation.**

(*c*) $C = 2\pi r = 2 \times 3.142 \times 1.72$. The answer should be expressed to 4 significant figures only (one more than the least accurate data figure, 1.72). **Hence the error is that of introducing a greater accuracy than exists in the data.**

(*d*) $127.8 \times 1.632 = 208.5696$, **hence a rounding off error has occurred.** This answer should have stated $127.8 \times 1.632 = 208.6$ correct to 4 significant figures.

Further problems on errors may be found in Section 3.5 (Problems 1–10), page 73.

3.2 Using a calculator

The most modern aid to calculations is the pocked-sized electronic calculator. With one of these, calculations can be performed quickly and accurately correct to about nine significant figures. The scientific type of calculator has made the use of tables and logarithms largely redundant. Since detailed instructions on the use of electronic calculators are given in leaflets supplied with the calculator, and since there are many types which all differ in their operation, no attempt will be made to describe their use here.

To help you to become competent at using your calculator check that you agree with the answers to the following problems.

Worked problems on using a calculator

Problem 1. Evaluate the following, correct to 4 significant figures:
(*a*) $5.7839 + 0.02977$ (*b*) $16.8321 - 11.9487$ (*c*) 23.96×0.013892

(*a*) $5.7839 + 0.02977 = 5.81367 = \mathbf{5.814}$, correct to 4 significant figures

(*b*) $16.8321 - 11.9487 = 4.8834 = \mathbf{4.883}$, correct to 4 significant figures

(*c*) $23.96 \times 0.013892 = 0.3328523 = \mathbf{0.3329}$, correct to 4 significant figures

---

Problem 2. Evaluate the following, correct to 4 decimal places:
(*a*) $36.47 \times 92.73 \times 0.01359$ (*b*) $\dfrac{4.739}{21.78}$ (*c*) $\dfrac{1}{3}(71.94 \times 0.0169)$

(*a*) $36.47 \times 92.73 \times 0.01359 = 45.9595195\cdots = \mathbf{45.9595}$, correct to 4 decimal places (*b*) $\dfrac{4.739}{21.78} = 0.217\,584\,9\cdots = \mathbf{0.217\,6}$, correct to 4 decimal places

(*c*) $\dfrac{1}{3}(71.94 \times 0.016\,9) = 0.405\,262 = \mathbf{0.405\,3}$, correct to 4 decimal places

---

Problem 3. Evaluate the following, correct to 3 decimal places:
(*a*) $\dfrac{1}{47.68}$ (*b*) $\dfrac{1}{0.026\,4}$ (*c*) $\dfrac{1}{6.49} + \dfrac{1}{1.93}$

(*a*) Using the $\dfrac{1}{x}$ (i.e. reciprocal) function of a calculator

$\dfrac{1}{47.68} = 0.020\,973\,15\cdots = \mathbf{0.021}$, correct to 3 decimal places

(*b*) $\dfrac{1}{0.026\,4} = 37.878\,787\,8\cdots = \mathbf{37.879}$, correct to 3 decimal places

(*c*) $\dfrac{1}{6.49} + \dfrac{1}{1.93} = 0.672\,217\,9\cdots = \mathbf{0.672}$, correct to 3 decimal places

---

Problem 4. Evaluate the following, expressing the answers in standard form, correct to 4 significant figures:
(*a*) $(0.004\,73)^2$ (*b*) $536.9 - (6.31 + 2.84)^2$ (*c*) $47.91^2 - 28.47^2$

(*a*) $(0.004\,73)^2 = 2.237\,29 \times 10^{-5} = \mathbf{2.237 \times 10^{-5}}$, correct to 4 significant figures

(*b*) $536.9 - (6.31 + 2.84)^2 = 453.177\,5 = 4.531\,775 \times 10^2$
$= \mathbf{4.532 \times 10^2}$, correct to 4 significant figures

(*c*) $47.91^2 - 28.47^2 = 1\,484.827\,2 = \mathbf{1.485 \times 10^3}$, correct to 4 significant figures

---

Problem 5. Evaluate the following, correct to 3 decimal places:

(*a*) $\dfrac{(2.46)^2}{0.053\,1}$ (*b*) $\left(\dfrac{4.28}{1.79}\right)^2 + \left(\dfrac{5.27}{2.48}\right)^2$ (*c*) $\dfrac{17}{3.9^2 - 2.7^2}$

(*a*) $\dfrac{(2.46)^2}{0.053\,1} = 113.966\,101\cdots = \mathbf{113.966}$, correct to 3 decimal places (*b*) $\left(\frac{4.28}{1.79}\right)^2 + \left(\frac{5.27}{2.48}\right)^2 = 10.232\,799\,8\cdots = \mathbf{10.233}$, correct to 3 decimal places

(*c*) $\frac{17}{3.9^2 - 2.7^2} = 2.146\,464\,64\cdots = \mathbf{2.146}$, correct to 3 decimal places

---

Problem 6. Evaluate the following, correct to 4 significant figures: (*a*) $\sqrt{7.391}$ (*b*) $\sqrt{73.91}$ (*c*) $\sqrt{793.1}$

(*a*) $\sqrt{7.391} = \pm 2.718\,639\,36\cdots = \mathbf{\pm 2.719}$, correct to 4 significant figures

(*b*) $\sqrt{73.91} = \pm 8.597\,092\,53\cdots \mathbf{\pm 8.597}$, correct to 4 significant figures

(*c*) $\sqrt{739.1} = \pm 27.186\,393\,6\cdots = \mathbf{\pm 27.19}$, correct to 4 significant figures

---

Problem 7. Evaluate the following, correct to 3 decimal places, taking positive values of square roots only:
(*a*) $\sqrt{0.006\,389}$ (*b*) $\sqrt{56.82} - \sqrt{39.81}$ (*c*) $\sqrt{(1.729\,4 \times 10^4)}$

(*a*) $\sqrt{0.006\,389} = 0.079\,931\,2\cdots = \mathbf{0.080}$, correct to 3 decimal places

(*b*) $\sqrt{56.82} - \sqrt{39.81} = 1.228\,387\,59\cdots = \mathbf{1.228}$, correct to 3 decimal places

(*c*) $\sqrt{(1.729\,4 \times 10^4)} = \sqrt{172\,94} = 131.506\,653\cdots = \mathbf{131.507}$, correct to 3 decimal places

---

Problem 8. Evaluate the following, correct to 4 significant figures:
(*a*) $(3.79)^3$ (*b*) $(0.912\,7)^4$ (*c*) $\sqrt{(74.29^2 - 58.62^2)^3}$

(*a*) $(3.79)^3 = 54.439\,93\cdots = \mathbf{54.44}$, correct to 4 significant figures

(*b*) $(0.912\,7)^4 = 0.693\,924\,46\cdots = \mathbf{0.693\,9}$, correct to 4 significant figures

(*c*) $\sqrt{(74.29^2 - 58.62^2)^3} = 95\,047.343\,1\cdots = \mathbf{95\,050}$, correct to 4 significant figures

---

Problem 9. Evaluate the following, correct to 3 significant figures:

(*a*) $\sqrt{\left[\frac{(5.92)^2}{(21.81)(\sqrt{11})}\right]}$ (*b*) $\sqrt[3]{(54.943)}$

(*c*) $\sqrt{(7.849^2 + 6.217^3 + 3.178^4)}$

(*a*) $\sqrt{\left[\dfrac{(5.92)^2}{(21.81)(\sqrt{11})}\right]} = 0.696\,058\,43\cdots = \mathbf{0.696}$, correct to 3 significant figures

(*b*) $\sqrt[3]{(54.943)} = 3.801\,638\,25\cdots = \mathbf{3.80}$, correct to 3 significant figures

(*c*) $\sqrt{(7.849^2 + 6.217^3 + 3.178^4)} = 20.097\,368\,9\cdots = \mathbf{20.1}$, correct to 3 significant figures

---

Problem 10. Evaluate the following, expressing the answers in standard form, correct to 4 decimal places.

(*a*) $(4.127 \times 10^{-3})^2$ (*b*) $\left(\dfrac{1.986 \times 10^2 \times 8.41 \times 10^{-3}}{3.176}\right)^4$

(*c*) $\sqrt{(1.843 \times 10^{-4})}$

(*a*) $(4.127 \times 10^{-3})^2 = 1.703\,212\cdots \times 10^{-5} = \mathbf{1.7032 \times 10^{-5}}$, correct to 4 decimal places

(*b*) $\left(\dfrac{1.986 \times 10^2 \times 8.41 \times 10^{-3}}{3.176}\right)^4 = 0.076\,485\,47\cdots$

$= \mathbf{7.6485 \times 10^{-2}}$, correct to 4 decimal places

(*c*) $\sqrt{(1.843 \times 10^{-4})} = 0.013\,575\,71\cdots = \mathbf{1.3576 \times 10^{-2}}$, correct to 4 decimal places

---

Further problems on using a calculator may be found in Section 3.5 (Problems 11–48), page 73.

3.3 Use of conversion tables and charts

It is often necessary to make calculations from various conversion tables and charts. Examples include imperial to metric unit conversions, currency exchange rates, bus or train timetables and production schedules. Some of these are demonstrated in the following worked problems.

Worked problems on conversion tables and charts

Problem 1. Table 3.1 shows some approximate metric to imperial conversions. Use the table to determine:
(*a*) the number of millimetres in 18 inches,
(*b*) a speed of 30 miles per hour in kilometres per hour,
(*c*) the number of miles in 500 km,
(*d*) the number of kilograms in 25 pounds weight,

Table 3.1

Length:	2.54 cm	= 1 inch
	1.61 km	= 1 mile
Weight:	1 kg	= 2.2 lb (1 lb = 16 oz)
Capacity:	1 litre	= 1.76 pints (8 pints = 1 gallon)

(*e*) the number of pounds and ounces in 36 kilograms (correct to the nearest ounce),
(*f*) the number of litres in 12 gallons, and
(*g*) the number of gallons in 50 litres.

(*a*) 18 inches = 18 × 2.54 cm = **45.72 cm**
45.72 cm = 45.72 × 10 mm = **457.2 mm**
(*b*) 30 m.p.h. = 30 × 1.61 km h^{-1} = **48.3 km $\mathbf{h^{-1}}$**

(*c*) $500\text{ km} = \frac{500}{1.61}\text{miles} = \mathbf{310.6\ miles}$

(*d*) $25\text{ lb} = \frac{25}{2.2}\text{ kg} = \mathbf{11.36\ kg}$

(*e*) 36 kg = 36 × 2.2 lb = 79.2 lb
0.2 lb = 0.2 × 16 oz = 3.2 oz = 3 oz, correct to the nearest ounce. Thus, 36 kg = **79 lb 3 oz**, correct to the nearest ounce.
(*f*) 12 gallons = 12 × 8 pints = 96 pints

$96\text{ pints} = \frac{96}{1.76}\text{ litres} = \mathbf{54.55\ litres}$

(*g*) 50 litres = 50 × 1.76 pints = 88 pints

$88\text{ pints} = \frac{88}{8}\text{ gallons} = \mathbf{11\ gallons.}$

Problem 2.

Table 3.2

France	£1 = 8.0 francs (f)
Germany	£1 = 2.38 Deutschmarks (Dm)
Italy	£1 = 2300 lira (l)
U.S.A.	£1 = 1.50 dollars ($)
Spain	£1 = 170 pesetas (pes)

Currency exchange rates for five countries are shown in Table 3.2. Calculate (*a*) how many French francs £14.50 will buy, (*b*) the number of German Deutschmarks which can be bought for

£62, (*c*) the pounds sterling which can be exchanged for 55 200 lira, (*d*) the number of American dollars which can be purchased for £83.50, and (*e*) the pounds sterling which can be exchanged for 1 615 pesetas.

(*a*) £1 = 8.0 francs, hence £14.50 = 14.50 × 8.0 francs = **116 f**
(*b*) £1 = 2.38 Deutschmarks, hence
£62 = 62 × 2.38 Dm = **147.56 Dm**

(*c*) £1 = 2 300 lira, hence 55 200 lira $= \pounds\frac{55\,200}{2\,300} = \mathbf{\pounds 24}$

(*d*) £1 = 1.50 dollars, hence £83.50 = 83.50 × 1.50 dollars = **$125.25**

(*e*) £1 = 170 pesetas, hence 1 615 pesetas $= \pounds\frac{1\,615}{170} = \mathbf{\pounds 9.50}$

Problem 3. A production schedule used in the manufacture of a machine tool is as shown below. Each process requires the work of one man.

(*a*) How many man-hours are required to produce the machine tool?
(*b*) How long does it take to make the tool?
(*c*) The men employed in processes A, B and C are paid £5 per hour whereas those employed in processes D, E and F are paid £6 per hour. Materials for the production of one tool cost £50. If each tool is sold for £262.50 determine the percentage profit of each tool based on the sale price.

Process	Start hour	End hour
A	1	5
B	4	12
C	2	7
D	7	9
E	8	11
F	12	14

(*a*) Since process A starts in the first hour and ends in the fifth hour then 5 hours are required altogether for the process. This is obtained by (end hour − start hour + 1). Similarly, process C takes (7 − 2 + 1), i.e. 6 hours to complete.

Hence:

Length of time for process A = 5 hours
B = 9 hours
C = 6 hours
D = 3 hours
E = 4 hours
F = 3 hours

Total time for production: = 30 hours

Since each process requires the work of one man, then **30 man-hours** are required for the production of the tool.

(*b*) The machine tool is completed after process F (i.e. the highest 'end-hour' value), i.e. after **14 hours**.

(*c*) For each machine tool produced:

Men on process A earn 5 × 5 = £25.00
B earn 9 × 5 = £45.00
C earn 6 × 5 = £30.00
D earn 3 × 6 = £18.00
E earn 4 × 6 = £24.00
F earn 3 × 6 = £18.00

Total labour costs = £160

Cost of labour + materials = £160 + £50 = £210.
If each machine tool is sold for £262.50 then £(262.50 − 210), i.e. £52.50 profit is made on each. Percentage

$$\text{profit} = \frac{52.50}{262.5} \times 100 = \mathbf{20\%}.$$

Problem 4. Determine the following information for the BR train timetable shown in Tables 3.3 and 3.4:

(*a*) At what time should a man catch a train at London Euston to enable him to be in Edinburgh by 2.30 p.m.?

(*b*) A lady leaves Rugby at 7.27 a.m. and travels to Glasgow Central.

(i) How long does the journey take?

and (ii) if the journey is 300 miles, what is the average speed of the train?

(*c*) A man wishes to travel from London Euston to Carlisle and he cannot leave before 12 noon. Which train should he catch for the quickest journey?

London — West Midlands — Northwest England — Scotland

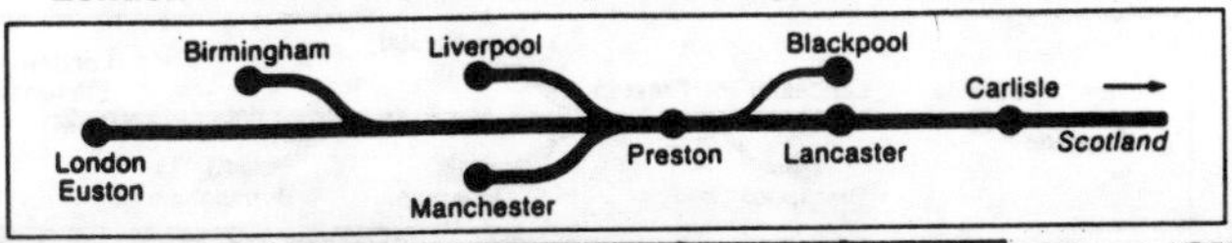

Mondays to Fridays	ICR ⊗	ICR ⊗	ICR ⊠	IC ⊠Ⓢ	ICR ⊗	ICR ■⊗	IC ⊠🚗	IC ✕	ICR	ICR ✕	ICR ●	IC	ICR ✕	ICR ●	ICR ■⊗
London Euston	—	—	0625	0725	—	—	0825	0925	—	1025	—	1125	1225	—	—
Watford Junction	—	—	0641u	0741u	—	—	0842u	0942u	—	—	—	1111	1242u	—	—
Milton Keynes Central	—	—	0703	—	—	—	0907	1005	—	—	—	1159	—	—	—
Rugby	—	0548	0727	—	—	—	—	—	—	—	—	—	1326	1206	—
Coventry	—	0601	0715p	—	—	0758	—	—	—	—	1031	—	—	1231	1255
Birmingham Intern'l ⊕	—	0619	—	—	—	0811	—	—	1009	—	1049	—	—	1249	1308
Birmingham New Street arrive	—	—	—	—	—	0826	—	—	1021	—	1110	—	—	1311	1325
Birmingham New Street depart	—	0655	—	—	—	0835	—	—	1025	—	1125	—	—	1325	1340
Wolverhampton	—	0711	—	—	—	0852	—	—	1043	—	1143	—	—	1343	1357
Stafford	—	0726	0815	—	—	0908	—	—	1058	—	1158	—	—	1328d	—
Crewe	—	0750	0839	0914	—	—	1031	1119	1125	—	—	1314	—	1425	1435
Warrington Bank Quay	—	0809	0857	0932	—	—	1051	1138	1144	—	—	1332	—	1444	1454
Liverpool Lime Street (d)	0553q	0712	0812	—	0907	—	1012	—	1037	—	1137	1237	1337	—	—
Wigan North Western	0633q	0821	0909	0944	0941	—	1105	1150	1156	—	1247	1344	1418d	1456	1505
Manchester Piccadilly (d)	0614	0730	0800	—	0900	1024	—	—	1100	1130	1200	—	1400	—	—
Preston	0709	0837	0927	1004	1014	1118	1123	1207	1217	1243	1306	1404	1500	1515	1523
Blackpool North	—	0923	1018	1053	—	1157	1205	—	1253	—	1354	1458	—	1554	1618
Lancaster	0728	0855	0945	—	1032	1136	1143	—	1233	1301	1323	1450	1518	1532	1543
Oxenholme: Lakes	0748	—	1001	—	1048	1152	1159	—	1249	—	—	—	1535	1548	1557
Penrith	0813	—	1027	—	1113	—	1227	—	1314	—	—	—	1601	1614	—
Carlisle	0848	0949	1047	—	1133	1234	1305v	—	1334	1354	1418	—	1623	1648x	1640
Dumfries	—	1036	1150	—	—	1328	—	—	—	1442	1527	—	—	—	1824
Stranraer Harbour	—	1352k	—	—	—	—	—	—	—	1755	—	—	—	—	—
Motherwell	1008s	1052s	—	—	—	—	—	—	1438s	—	—	—	1728s	1752s	—
Glasgow Central	1026	1113	1205	—	—	—	—	—	1455	1512	—	—	1748	1815	—
Edinburgh	1032b	—	—	—	1305	1355	1428	—	—	—	1544	—	—	1925	1800
Stirling	1135c	1253n	1353n	—	1405	1505	1535	—	1555n	—	1705	—	1853n	—	1905
Perth	1222n	1322n	1422n	—	1453	—	1651	—	1624n	1719n	1758	—	1927n	2022n	2053
Inverness	1509c	—	—	—	1718g	—	1904	—	—	—	2005	—	—	—	2307
Dundee	1224c	1348n	1448n	—	—	1525	1609	—	1650n	1743n	1713	—	—	—	1940
Aberdeen	1339c	1500n	1605n	—	—	1641	1725	—	1809n	1855n	1830	—	—	—	2054
Fort William	—	1622n	—	—	—	—	—	—	—	—	—	—	2208n	—	—

Mondays to Fridays	IC	ICR ■	ICR	ICR	ICR ⊗	ICR ⊠Ⓢ	ICR	IC PⓈ	IC	IC ✕	IC ✕	IC	R
London Euston	1325	—	1425	1525	—	1625	—	1725	1805	1830	1925	2150	2345
Watford Junction	1341u	—	—	1511	—	1641u	—	—	1821u	1846u	1941u	2206u	—
Milton Keynes Central	—	—	—	1559u	—	—	—	—	1746	1832	2004u	2228	—
Rugby	—	—	—	—	—	—	—	—	1905	1928	—	—	—
Coventry	—	1414	—	—	1605	—	—	—	—	—	—	—	—
Birmingham Intern'l ⊕	—	1425	—	—	1627	—	1722	—	—	—	—	—	—
Birmingham New Street arrive	—	1440	—	—	1642	—	1736	—	—	—	—	—	—
Birmingham New Street depart	—	1450	—	—	1655	—	1755	—	1858	—	—	—	—
Wolverhampton	—	1507	—	—	1711	—	1812	—	1915d	—	—	—	—
Stafford	—	—	—	1624d	1726	—	1827	—	1944	—	—	2339	—
Crewe	1519	1544	—	1714	1750	—	1852	1915	—	2026	2133	0005	0214
Warrington Bank Quay	1537	1603	—	1732	1809	1825	1910	1933	—	2045	2152	0028	—
Liverpool Lime Street (d)	—	1512	—	1637	1722	—	1817	—	—	—	2052	—	—
Wigan North Western	1549	1615	—	1744	1821	—	1922	1945	—	2057	2204	0042	—
Manchester Piccadilly (d)	—	1530	—	1648r	1730	1750r	1830	—	—	—	2100	—	—
Preston	1609	1634	1643	1802	1837	1850	1939	2005s	2044	2117	2222	0105	0304
Blackpool North	1654	—	1724	1838	1922	1928	2023	2039	2123	2223	2323	—	—
Lancaster	—	1651	1701	1820	1855	1908	1957	2028	2103	—	2240	—	—
Oxenholme: Lakes	—	1707	1717	1836	1911	—	2013	—	2119	—	2258	—	—
Penrith	—	1733	—	1902	—	—	2039	—	2146	—	2324	—	—
Carlisle	—	1753	1759	1922	1953	2002	2101	—	2209	—	2347	—	0454
Dumfries	—	—	1925	—	—	2102	2156	—	2327	—	—	—	0604
Stranraer Harbour	—	—	—	—	—	0034k	—	—	—	—	—	—	0943k
Motherwell	—	1900s	1905s	—	—	2104s	2159s	—	—	—	—	—	0608s
Glasgow Central	—	1920	1924	2040	—	2123	2220	—	—	—	—	—	0627
Edinburgh	—	2030	—	—	2118	—	—	—	—	—	—	—	0750n
Stirling	—	—	2053n	2153n	2305	2312n	0017n	—	—	—	—	—	0729n
Perth	—	—	2127n	2222n	2347	—	0056n	—	—	—	—	—	0811n
Inverness	—	—	—	—	—	—	—	—	—	—	—	—	1030n
Dundee	—	—	2153n	2252n	2355	—	—	—	—	—	—	—	0851n
Aberdeen	—	—	—	0013n	—	—	—	—	—	—	—	—	1009n
Fort William	—	—	—	—	—	—	—	—	—	—	—	—	1155n

Scotland — Northwest England — West Midlands — London

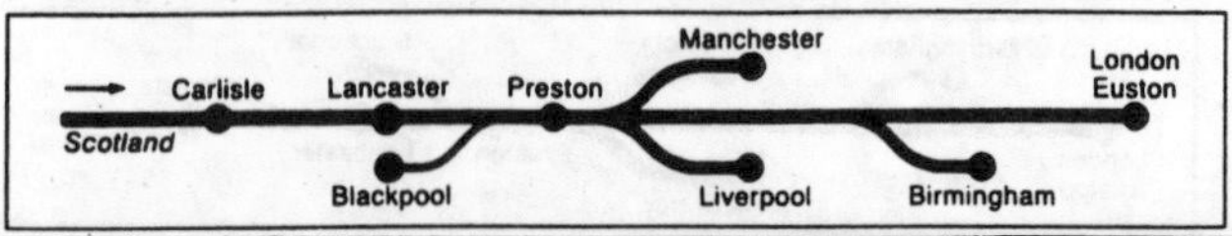

Mondays to Fridays	IC ☒Ⓢ	IC PⓈ	IC ☒Ⓢ	IC	IC	ICR	IC ✕	ICR ■	ICR ✕	ICR ⊛	ICR ●	IC ✕	ICR ⊛	ICR	ICR ■⊛
Fort William	—	—	—	—	—	—	—	—	—	—	—	—	—	—	—
Aberdeen	—	—	—	—	—	—	—	—	0512n	—	0625n	—	0710n	—	0755
Dundee	—	—	—	—	—	—	—	—	0637n	0722	0742n	—	0836n	—	0906
Inverness	—	—	—	—	—	—	—	—	—	—	—	—	0645r	—	0750
Perth	—	—	—	—	—	0515n	—	0610n	0700n	0708	0807n	—	0901n	—	0950
Stirling	—	—	—	—	—	0550n	—	0650n	0737n	0810	0840n	—	0936n	—	1027
Edinburgh	—	—	—	—	—	—	—	0710t	—	0924	—	—	—	—	1136
Glasgow Central	—	—	—	—	—	0705	—	0822	0912	—	0950	—	1045	1118	—
Motherwell	—	—	—	—	—	0720u	—	0828u	—	—	1005u	—	—	—	—
Stranraer Harbour	—	—	—	—	—	—	—	—	—	—	0703k	—	—	—	—
Dumfries	—	—	—	0617	—	0736	—	0841	—	—	1006	—	1050	1134	—
Carlisle	—	—	0600	0713	—	0823	—	0950	1030	1047	1114	—	1201	1232	1251
Penrith	—	—	0617	0730	—	0840	—	1006	1047	1103	1130	—	—	—	—
Oxenholme: Lakes	—	—	0643	0756	—	0907	—	1032	1113	1129	1156	—	1239	—	1330
Lancaster	—	0640	0659	0812	—	0923	1013	1048	1114	1144	1211	—	1254	1323	1345
Blackpool North	—	0608	0650	0800	0830	0900	1006	1030	1108	1125	1148	1207	1230	1308	1330
Preston	0615	0708	0730	0845	0915	0945	1045	1110	1145	1205	1234	1245	1317	1345	1407
Manchester Piccadilly (a)	—	—	0853	0953	—	1053	—	—	1253	1323	1353	—	1423	1453	1523
Wigan North Western	0629	0723	0745	0914a	0930	1014a	1059	1125	1214a	1219	1248	1259	—	1414a	1420
Liverpool Lime Street (a)	—	—	0835	0954	—	1055	—	—	1254	1315	1354	—	—	1454	1515
Warrington Bank Quay	0641	0735	0757	—	0942	—	1111	1137	—	1230	1300	1311	1339	—	1432
Crewe	0708	—	0820	0926	1011	1026	1134	1201	—	1254	1324	1334	1402	—	1457
Stafford	0751a	—	0849a	0947	—	1109a	1155	—	—	1328a	—	1409a	—	—	—
Wolverhampton	0810	—	0907	—	—	1136	—	1237	—	1330	1401	—	1438	—	1532
Birmingham New Street *arrive*	0834	—	0926	—	—	1155	—	1256	—	1351	1422	—	1500	—	1551
Birmingham New Street *depart*	—	—	—	—	—	—	—	1306	—	1359	1430	—	—	—	1606
Birmingham Intern'l ⊕	0852	—	0945	—	—	—	—	1316	—	1415	1458	—	1528	—	1616
Coventry	0904	—	—	—	—	—	—	1327	—	—	1508	1511p	1538	—	1628
Rugby	0807	—	0914	—	1107	—	1233	—	—	—	—	1433	—	—	—
Milton Keynes Central	—	—	—	1047	—	—	1306e	—	—	—	—	1457	—	—	—
Watford Junction	—	0929s	1002s	1137	—	1202s	—	—	—	—	—	1525s	—	—	—
London Euston	0913	0945	1022	1130	1211	1222	1339	—	1410	—	—	1545	—	1611	—

Mondays to Fridays	ICR ●	IC ✕	ICR ■⊛	ICR	IC	ICR ⊛	ICR ✕	ICR ⊛	ICR ☒🚗	ICR ✕	ICR ⊛	ICR	R
Fort William	—	—	0805n	—	—	—	—	—	—	1157n	—	—	1742n
Aberdeen	0912	—	0926n	—	—	1117	1135n	1235n	1216	1335n	1420	1436n	2030n
Dundee	1023	—	1045n	—	—	1227	1245n	1345n	1327	1445n	1530	1545n	2145n
Inverness	—	—	—	—	—	—	1015n	—	—	1218n	—	—	2000n
Perth	—	—	1108n	—	—	1127	1308n	1408n	—	1508n	1500	1608n	2217n
Stirling	1032	—	1139n	—	—	1302	1339n	1439n	1432	1539n	1602	1720n	2249n
Edinburgh	1150	—	—	—	—	1422	—	—	1555	—	1726	—	—
Glasgow Central	—	—	1245	1317	—	—	1505	1550	—	1700	—	1840	2355
Motherwell	—	—	—	—	—	—	1520u	—	—	1715u	—	1855u	0011u
Stranraer Harbour	—	—	—	1100	—	—	—	—	—	—	—	—	2115k
Dumfries	—	—	1311	1346	—	1431	—	—	1646	1726	—	1913	0001
Carlisle	1320	—	1401	1432	—	1545	1624	1706	1733z	1818	1851	2009	0151
Penrith	—	—	1417	—	—	1601	1641	1722	—	—	1907	2026	—
Oxenholme: Lakes	—	—	1444	—	—	1626	1707	1747	—	1856	1932	2052	—
Lancaster	1410	—	1459	1523	1602	1641	1723	1802	—	1912	1947	2108	—
Blackpool North	—	1405	1448	1508	1608	—	1700	1730	1800	1857	1930	2053	—
Preston	1433	1445	1528	1545	1645	1702	1745	1824	1845	1935	2009	2130	0314
Manchester Piccadilly (a)	1553	—	1618	1653	—	1823	1853	—	1953	2053	—	2230	—
Wigan North Western	1447	1459	—	1559	1659	1726	—	—	1859	—	2023	2207q	—
Liverpool Lime Street (a)	1554	—	—	1654	—	1812	—	—	—	2116	2130	2254q	—
Warrington Bank Quay	—	1511	—	1611	1711	—	—	—	1912	—	2034	—	—
Crewe	1519	1534	—	1634	1734	—	1827	1906	1936	2016	2059	—	0404
Stafford	1541	—	1722	—	1754	—	—	—	1958	—	2118	—	—
Wolverhampton	1600	—	1741	—	1834	—	—	1940	—	—	2134	—	—
Birmingham New Street *arrive*	1618	—	1801	—	1854	—	—	2000	—	—	2200	—	—
Birmingham New Street *depart*	1630	—	1806	—	—	—	—	—	—	—	—	—	—
Birmingham Intern'l ⊕	1652	—	1816	—	—	—	—	2028	—	—	2237	—	—
Coventry	1704	—	1829	—	1912p	—	—	2038	2109p	—	2254	—	—
Rugby	—	1629	—	1729	1843e	—	—	2012e	—	—	2306	—	—
Milton Keynes Central	—	1654	—	—	1908e	—	1945	—	2102	2128	—	—	—
Watford Junction	—	1735e	—	—	1919s	—	2013s	—	2131s	2156s	—	—	0603s
London Euston	—	1739	—	1835	1939	—	2033	—	2151	2216	—	—	0638

(*d*) On a day-trip a lady travels from Milton Keynes Central to Penrith.
  (i) What is the earliest train she can take?
  (ii) What is the latest train she can catch to return?
  (iii) How long will she have in Penrith?
  (iv) If the journey from Milton Keynes to Penrith is 250 miles and the cost of the return ticket is £89, how much is this per mile travelled?

(*a*) 2.30 p.m. is 1430 on the 24 hour clock. After locating Edinburgh near the bottom of the first half of Table 3.3, move across horizontally to the nearest time before 1430, which is 1428. Move upwards from 1428 to London Euston and it shows 0825. Hence the man needs to catch the **8.25 a.m.** train from Euston to arrive at Edinburgh before 2.30 p.m.

(*b*) (i) Locating the 0727 train from Rugby in Table 3.3 and moving down to Glasgow Central shows that the train arrives at 1205. The journey therefore takes **4 hours 38 minutes.**

(ii) Average speed of train $= \dfrac{\text{distance travelled}}{\text{time taken}}$

$$= \frac{300 \text{ miles}}{4\frac{38}{60} \text{ hours}}$$

$$= \mathbf{64.75 \text{ m.p.h.}}$$

(*c*) The quickest journey from London Euston to Carlisle after noon is the journey with the least stops which, from Table 3.3 is the 1425 from Euston arriving in Carlisle at 1759, i.e., a journey of **3 hours 34 minutes.**

(*d*) (i) The earliest train from Milton Keynes Central to Penrith is, from Table 3.3, the **0703 train** arriving in Penrith at 1027.
  (ii) The latest train from Penrith to Milton Keynes is, from Table 3.4, the **1641 train**, arriving in Milton Keynes at 1945.
  (iii) The lady is in Penrith from 1027 until 1641, i.e., **6 hours 14 minutes.**
  (iv) If the distance from Milton Keynes to Penrith is 250 miles then the total (return) journey is 500 miles.

Cost of journey per mile $= \dfrac{£89}{500 \text{ miles}}$

$$= \frac{8900\text{p}}{500 \text{ miles}}$$

$$= \mathbf{17.8\ p/mile}$$

Further problems on conversion tables and charts may be found in Section 3.5 (Problems 49 to 53), page 77.

3.4 Logarithms

With the use of calculators firmly established, logarithmic tables are now rarely used for calculations. However, the theory of logarithms is important for there are several scientific and engineering laws that involve the rules of logarithms.

If a number y can be written in the form a^x, then the index x is called the 'logarithm of y to the base a',
i.e., **if $y = a^x$ then $x = \log_a y$**

Thus, since $1000 = 10^3$, then $3 = \log_{10} 1000$
This may be checked using the 'log' function on your calculator.

Common logarithms

Logarithms having a base of 10 are called **common logarithms** and $\log_{10}$ is often abbreviated to 'lg'.

The following values may be checked by using a calculator:
lg 15.4 = 1.187 5 . . ., lg 378.2 = 2.577 . . . and lg 0.024 1 = −1.617 9 . . .

Napierian logarithms

Logarithms having a base of e (where 'e' is a mathematical constant approximately equal to 2.718 3) are called hyperbolic, Napierian or natural logarithms, and $\log_e$ is often abbreviated to 'ln'.

The following values may be checked by using a calculator:
ln 4.73 = 1.553 9 . . ., ln 278.4 = 5.629 0 . . . and ln 0.764 2 = −0.268 9 . . .

Rules of logarithms

There are three rules of logarithms, which apply to any base.

Rule 1. To multiply two numbers:

$$\log (A \times B) = \log A + \log B$$

The following may be checked by using a calculator:

$$\lg 10 = 1$$

Also $\lg 5 + \lg 2 = 0.698\,97 \cdots + 0.301\,029 \cdots = 1$
Hence $\lg(5 \times 2) = \lg 10 = \lg 5 + \lg 2$

Rule 2. To divide two numbers:

$$\boxed{\log \frac{A}{B} = \log A - \log B}$$

The following may be checked using a calculator:

$$\ln \frac{5}{2} = \ln 2.5 = 0.916\,29 \cdots$$

Also $\ln 5 - \ln 2 = 1.609\,43 \cdots - 0.693\,14 \cdots = 0.916\,29 \cdots$

Hence $\ln \frac{5}{2} = \ln 5 - \ln 2$

Rule 3. To raise a number to a power

$$\boxed{\log a^n = n \log A}$$

The following may be checked using a calculator:

$$\lg 5^2 = \lg 25 = 1.397\,94 \cdots$$

Also $2 \lg 5 = 2(0.698\,97 \cdots) = 1.397\,94 \cdots$
Hence $\lg 5^2 = 2 \lg 5$

These rules, or laws, of logarithms are further explained in *Technician Mathematics 2*.

Worked problems on logarithms

Problem 1. Evaluate (*a*) $\log_3 9$ (*b*) $\log_{10} 10$ (*c*) $\log_{16} 8$

(*a*) Let $x = \log_3 9$, then $3^x = 9$ from the definition of a logarithm,

i.e., $3^x = 3^2$

from which, $x = 2$. Hence $\mathbf{\log_3 9 = 2}$

(*b*) Let $x = \log_{10} 10$, then $10^x = 10$ from the definition of a logarithm,

i.e., $10^x = 10^1$

from which, $x = 1$
Hence $\mathbf{\log_{10} 10 = 1}$ (which may be checked using a calculator)

(*c*) Let $x = \log_{16} 8$, then $16^x = 8$ from the definition of a logarithm,

i.e., $(2^4)^x = 2^3$, i.e., $2^{4x} = 2^3$,

from which, $4x = 3$ and $x = \frac{3}{4}$. Hence $\mathbf{\log_{16} 8 = \frac{3}{4}}$

Problem 2. Evaluate the following using a calculator, giving answers correct to 4 significant figures.
(*a*) lg 7.962 (*b*) lg 241.6 (*c*) lg 0.003 687

(*a*) lg 7.962 = 0.901 022 17 ⋯ = **0.901 0**, correct to 4 significant figures
(*b*) lg 241.6 = 2.383 096 9 ⋯ = **2.383**, correct to 4 significant figures
(*c*) lg 0.003 687 = −2.433 326 86 ⋯ = **−2.433**, correct to 4 significant figures

---

Problem 3. Evaluate the following using a calculator, giving answers correct to 3 decimal places.
(*a*) ln 17.49 (*b*) ln 0.027 6 (*c*) ln e^2

(*a*) ln 17.49 = 2.861 629 28 ⋯ = **2.862**, correct to 3 decimal places
(*b*) ln 0.027 6 = −3.589 939 50 ⋯ = **−3.590**, correct to 3 decimal places
(*c*) $\ln e^2 = \ln(7.389\,056\,09) \cdots = \mathbf{2}$
(In general, it may be shown that $\lg 10^x = x$ and $\ln e^x = x$)

Problem 4. Solve the equations:
(*a*) $\lg x = 4$ (*b*) $\log_2 x = 5$ (*c*) $\log_5 x = -2$

(*a*) If $\lg x = 4$ then $\log_{10} x = 4$ and $x = 10^4$, i.e., $\mathbf{x = 10\,000}$
(*b*) If $\log_2 x = 5$ then $x = 2^5 = \mathbf{32}$
(*c*) If $\log_5 x = -2$ then $x = 5^{-2} = \frac{1}{5^2} = \mathbf{\frac{1}{25}}$

Problem 5. Write (a) log 30 (*b*) log 450 in terms of log 2, log 3 and log 5 to any base.

(*a*) $\log 30 = \log(2 \times 15) = \log(2 \times 3 \times 5)$

$= \mathbf{\log 2 + \log 3 + \log 5}$, by the first rule of logarithms.

(*b*) $\log 450 = \log(2 \times 225) = \log(2 \times 3 \times 75)$

$= \log(2 \times 3 \times 3 \times 25) = \log(2 \times 3 \times 3 \times 5 \times 5)$

$= \log(2 \times 3^2 \times 5^2)$

$= \log 2 + \log 3^2 + \log 5^2$ by the first rule of logarithms, i.e., log 450 = **log 2 + 2 log 3 + 2 log 5**, by the third rule of logarithms.

Further problems on logarithms may be found in Section 3.5 following (Problems 54 to 65), page 79.

3.5 Further problems

Errors

1. The area of a triangle is given by $A = \frac{1}{2}bh$. If $b = 7.37$ cm and $h = 13.6$ cm, find the area of the triangle [50.12 cm^2]
2. Determine the velocity of a body v when $v = u + at$. The initial velocity $u = 14$ m s^{-1}, the acceleration $a = 9.81$ m s^{-2} when the time t is 24 s. [249 m s^{-1}]
3. The pressure p and volume V of a gas are constant, such that $pV = c$. Find c when $p = 105\,600$ Pa and $V = 0.43$ m^3. [45 400 Pa m^3]

In problems 4–10 state the type of error or errors which have been made if, indeed, there is an error present.

4. $73 \times 68.247 = 4\,982.03$ [Accuracy error, add 'correct to 2 decimal places']
5. $18 \times 0.08 \times 6 = 86.4$ [Order of magnitude error]
6. $\dfrac{47.3}{9 \times 0.071} = 5\,301.2$ correct to 1 decimal place [Blunder]
7. $\dfrac{16.7 \times 0.02}{41} = 8.15 \times 10^{-3}$ correct to 3 significant figures [No error]
8. The force, $P = mf$, where m is the mass and f the acceleration. The value of P was found to be 1 775.61 N when m was 181 kg and f was 9.81 m s^{-2}. [Measured values, hence $P = 1\,776$ N]
9. $6.73 \div 0.006 = 4\,211.\dot{6}$ [Blunder]
10. $\dfrac{3.1 \times 0.008}{37.6 \times 0.347} = 9.1007 \times 10^{-4}$ [Blunder]

Using a calculator

In problems 11–14, evaluate correct to 4 significant figures.

11. (*a*) 4.721 96 + 0.056 18 (*b*) 23.9172 − 17.869 5

(*c*) $\dfrac{17.429 + 11.873\,1}{6.411\,2 - 4.733\,8}$

(*a*) [4.778] (*b*) [6.048] (*c*) [17.47]

12. (*a*) 38.71×46.3 (*b*) $0.076\,28 \times 13.7$
(*c*) $93.82 \times 16.74 \times 0.037\,1$ (*d*) $8.417 \times 0.003\,84 \times 6\,742$
(*a*) [1 792] (*b*) [1.045] (*c*) [58.27] (*d*) [217.9]

13. (*a*) $\dfrac{28.41}{76.82}$ (*b*) $\dfrac{3.714}{0.039\,7}$ (*c*) $\dfrac{743.8}{62.49}$ (*d*) $\dfrac{0.027\,4}{837.2}$

(*a*) [0.369 8] (*b*) [93.55] (*c*) [11.90] (*d*) [3.273×10^{-5}]

14. (*a*) $\dfrac{1}{67.4}\ \dfrac{1}{7\,628}$ (*b*) $\dfrac{1}{0.024\,77}$ (*c*) $\dfrac{1}{0.024\,77}$ (*d*) $\dfrac{1}{7.241}$

(*a*) [0.014 84] (*b*) [1.311×10^{-4}] (*c*) [40.37] (*d*) [0.138 1]

In problems 15 to 17, evaluate, expressing the answers in standard form, correct to 3 decimal places.

15. (*a*) $(0.037\,14)^2$ (*b*) $784.9 - (7.923 - 4.785)^2$

(*c*) $61.48^2 - 49.31^2$ (*d*) $\left(\dfrac{11.72}{5.98}\right)^2 - \left(\dfrac{9.86}{7.14}\right)^2$

(*a*) [1.379×10^{-3}] (*b*) [7.751×10^2]
(*c*) [1.348×10^3] (*d*) [1.934]

16. (*a*) 23^2 (*b*) 67.4^3 (*c*) 824^2 (*d*) $0.024\,7^4$ (*e*) $0.005\,14^2$
(*a*) [5.290×10^2] (*b*) [3.063×10^5] (*c*) [6.789×10^5]
(*d*) [3.722×10^{-7}] (*e*) [2.642×10^{-5}]

17. (*a*) $\sqrt{7.284}$ (*b*) $\sqrt{23.84}$ (*c*) $\sqrt{4\,977}$ (d) $\sqrt{0.624}$
(*e*) $\sqrt{0.039\,28}$ (*f*) $\sqrt{0.002\,91}$
(*a*) [± 2.699] (*b*) [± 4.883] (c) [$\pm 7.055 \times 10^1$]
(*d*) [$\pm 7.899 \times 10^{-1}$] (*e*) [$\pm 1.982 \times 10^{-1}$]
(*f*) [$\pm 5.394 \times 10^{-2}$]

18. Evaluate, correct to 3 significant figures:

(*a*) $\sqrt{(41.92^2 - 37.49^2)^3}$ (*b*) $\sqrt{\left[\dfrac{(7.82)^2}{\sqrt{7} \times \sqrt{3}}\right]}$

(*c*) $\sqrt[3]{(4.2^3 - 3.2^2 - 2.2^4)}$
(*a*) [6 600] (*b*) [3.65] (*c*) [3.43]

19. Evaluate, expressing the answers in standard form, correct to 4 decimal places.

(*a*) $(3.973 \times 10^{-4})^2$ (*b*) $\left(\dfrac{4.92 \times 10^{-3} \times 2.541 \times 10^2}{5.682 \times 10^{-1}}\right)^3$

(*c*) $\sqrt{(2.463 \times 10^{-3})^3}$
(*a*) [$1.578\,5 \times 10^{-7}$ (*b*) [$1.065\,1 \times 10^1$] (*c*) [$1.222\,4 \times 10^{-4}$]

zinc by mass. In what proportions must bronze and brass be mixed to give an alloy containing 91% of copper? [1:2]

41. 18 560 bricks are required for a specific building job. If this includes 8% for cutting and wastage how many bricks are actually used for the job? [17 075]
42. Assuming a loss of bulk on mixing of 25%, how much sand and cement, mixed in a ratio of 3:1, will be required to give 147 m^3 of mortar? What will be the cost if sand costs £3.50 m^{-3} and cement costs £15.20 m^{-3}?
[Sand = 137.8 m^3; cement = 45.94 m^3; £482, £698]
43. One tonne of gypsum plaster mixed with Type 1 sand in the ratio 2:3 will cover an area of plaster board of 300 m^2 to a thickness of 6 mm. How much plaster and sand will be required to cover 188 m^2?
[Plaster = 0.626 7 tonne; sand = 0.940 0 tonne]
44. The ratio of the rise (vertical height) of a staircase to its going (horizontal distance) is 1 to 1.3. If the going of each step is 225 mm, what is the rise of a staircase of eight steps?
[1.385 m]
45. It was found on analysis that a sample of water contained 47 parts per million by mass (ppm) of carbonates and bicarbonates, 8 ppm of mineral acids, 49 ppm of calcium and 29 ppm of magnesium. What is the percentage mass of these impurities in water expressed as a total percentage?
[0.013 3%]
46. A brick has a mass of 3.200 kg when thoroughly wet and 2.800 kg when dry. What is the moisture content (percentage mass of the water to that of the dry brick) of the brick?
[14.3%]
47. A Petrograd standard of timber has a volume of 4.67 m^3. How many standards are contained in a volume of 60.71 m^3?
[13]
48. On strong heating 0.234 1 g of hydrated barium chloride loses 0.034 5 g of water of crystallisation. What is the percentage mass of water of crystallisation in the original salt?
[14.74%]

Conversion tables and charts

49. Use Table 3.1 (page 64) to determine:
(*a*) the number of millimetres in 30 inches,
(*b*) a speed of 50 m.p.h. in km h^{-1},
(*c*) the number of miles in 177.1 km,
(*d*) the number of kilograms in 44 lb,

(*e*) the number of litres in 20 gallons, and
(*f*) the number of gallons in 38 litres.

(*a*) [762 mm] (*b*) [80.5 km h^{-1}] (*c*) [110 miles]
(*d*) [20 lb] (*e*) [90.91 litres] (*f*) [8.36 gallons]

50. Use Table 3.2 (page 64) to determine:
(*a*) the pounds sterling 248 francs will buy,
(*b*) the number of Italian lira that can be exchanged for £15.50,
(*c*) the pounds sterling which can be bought for 512 Dm
(*d*) the number of Spanish pesetas which can be exchanged for £52.75, and
(*e*) the pounds sterling which can be exchanged for $213.

(*a*) [£31.00] (*b*) [35 650 lira] (*c*) [£215.13]
(*d*) [8 967.5 ptas.] (*e*) [£142]

51. Three similar holidays are advertised. The prices are: Italy 483 000 lira; Spain 40 800 pesetas; France 1 704 francs. Use Table 3.2 (page 64) to determine the least expensive holiday. [Italy £210, France £213, Spain £240]

Table 3.5

HAVANT · Leigh Park · Eastern Road · SOUTHSEA (Clarence Pier)

Weekdays								
HAVANT, Market Parade	....	....	....	0718	0755	0825	0855	0925
Park Parade, Purbrook Way	0603	0633	0703	0733	0810	0840	0910	0940
Hulbert Road, Purbrook Way	0609	0639	0709	0739	0816	0846	0916	0946
Farlington, Rectory Avenue	0617	0647	0717	0747	0824	0854	0924	0954
Eastern Road, Airport Service Road	0624	0654	0724	0754	0831	0901	0931	1001
Stanley Avenue	0629	0659	0729	0759	0836	0906	0936	1006
Milton, St. Mary's Road	0636	0706	0736	0806	0843	0913	0943	1013
Station Street, Commercial Road	0646	0716	0746	0816	0853	0923	0953	1023
SOUTHSEA, Clarence Pier	....	....	....	0823	0900	0930	1000	1030
Guildhall	*	*	*	....	....	....	....	....
PORTSMOUTH, The Hard	0652	0722	0752	....	....	....	....	....

SOUTHSEA (Clarence Pier) · Eastern Road · Leigh Park · HAVANT

Weekdays								
PORTSMOUTH, The Hard	....	....	....	0705	0735	0805	....	....
Guildhall	....	....	....	....	....	....	....	....
SOUTHSEA, Clarence Pier	....	....	....	*	*	*	0833	0910
Station Street, Commercial Road	....	....	....	0711	0741	0811	0841	0918
Milton, St. Mary's Road	0551	0621	0651	0721	0751	0821	0851	0928
Stanley Avenue	0557	0627	0657	0727	0757	0827	0857	0934
Eastern Road, Airport Service Road	0602	0632	0702	0732	0802	0832	0902	0939
Farlington, Rectory Avenue	0609	0639	0709	0739	0809	0839	0909	0946
Hulbert Road, Purbrook Way	0617	0647	0717	0747	0817	0847	0917	0954
Park Parade, Stockheath Road	0623	0653	0723	0753	0823	0853	0923	1000
HAVANT, Market Parade	....	0708	0738	0808	0838	0908	0938	1015

52. Deduce the following information from the bus timetable shown in Table 3.5:

(*a*) At what time should a man catch a bus at Park Parade to enable him to be at Southsea by 8.45 a.m.?

(*b*) At what time should a man catch a bus at Clarence Pier to arrive at Park Parade by 4.15 p.m.?

(*c*) A lady leaves Havant at 8.55 a.m. and travels to Commercial Road. She needs 30 minutes for shopping before catching the next available bus back to Havant. What is the earliest time she can be back at Havant?

(*d*) A girl leaves Havant by the 6.55 p.m. bus and travels to Southsea. How long does the journey take? If she has to be back in Havant by 9.30 p.m., what time is the last bus that she should catch from Southsea?

(*a*) [7.33 a.m.] (*b*) [3.10 p.m.] (*c*) [11.45 a.m.]
(*d*) [65 minutes; 8.10 p.m.]

Table 3.5 (*continued*)

Service 24

0955	1025	1055		25	55		1825	1855	1925	2025	2125	2205
1010	1040	1110		40	10		1840	1910	1940	2040	2140	2220
1016	1046	1116	Then	46	16		1846	1916	1946	2046	2146	2226
1024	1054	1124	at	54	24		1854	1924	1954	2054	2154	2234
1031	1101	1131	these	01	31		1901	1931	2001	2101	2201	2241
1036	1106	1136	minutes	06	36	until	1906	1936	2006	2106	2206	2246
1043	1113	1143	past	13	43		1913	1943	2013	2113	2213	2253
1053	1123	1153	each	23	53		1923	1953	2023	2123	2223	2303
1100	1130	1200	hour	30	00		1930	2000	2030	2132†	....	....
....	....	....		....	....		....	....	....	2139	*	*
....	....	....		....	....		....	....	....	2145	2229‡	2309‡

Service 24

											●	
....		....	....		....	....	....	....	....	2200	2248	....
....		....	....		....	....	....	....	....	2206	....	....
0940	Then	10	40		1810	1840	1910	2010	2050	2215†	*	....
0948	at	18	48		1818	1848	1918	2018	2058	2223	2253	....
0958	these	28	58		1828	1858	1928	2028	2108	2233	2303	....
1004	minutes	34	06	until	1834	1904	1934	2034	2114	2239	2309	....
1009	past	39	09		1839	1909	1939	2039	2119	2244	....	....
1016	each	46	16		1846	1916	1046	2046	2126	2251	....	....
1025	hour	54	24		1854	1924	1954	2054	2134	2259	....	....
1030		00	30		1900	1930	2000	2100	2140	2305‡	....	....
1045		15	45		1915	....	2015	2115	2155	....	....	....

53. A production schedule used in the manufacture of an engineering component is shown below. Stage 3 requires two men, the other stages requiring one man.

Stage	Start day	End day
1	1	3
2	3	8
3	2	9
4	5	8
5	5	11

Determine (*a*) the minimum time needed for the component to be completed,
(*b*) the number of man-days needed for the production of the component,
(*c*) the total labour cost if each man is paid £4.50 per hour and they each work a 7 hour day, and
(*d*) the percentage profit, based on the sale price, if materials and overheads cost £246 and the component is sold for £1 725.

(*a*) [11 days] (*b*) [36 man-days] (*c*) [£1 134] (*d*) [20%]

Logarithms

In problems 54 and 55, evaluate correct to 4 significant figures.

54. (*a*) lg 2.931 (*b*) lg 641.9 (*c*) lg 0.004 72
(*a*) [0.467 0] (*b*) [2.807] (*c*) [−2.326]

55. (*a*) ln 1.683 (*b*) ln 2168.3 (*c*) ln 0.0371
(*a*) [0.520 6] (*b*) [7.682] (*c*) [−3.294]

In problems 56 to 60, evaluate the given expressions.

56. (*a*) $\log_{10} 100$ (*b*) $\log_2 16$ (*a*) [2] (*b*) [4]

57. (*a*) $\log_7 343$ (*b*) lg 1000 (*a*) [3] (*b*) [3]

58. (*a*) $\log_5 125$ (*b*) $\log_2 \frac{1}{8}$ (*a*) [3] (*b*) [−3]

59. (*a*) $\log_4 8$ (*b*) $\log_{27} 3$ (*a*) [$1\frac{1}{2}$] (*b*) [$\frac{1}{3}$]

60. (*a*) $\log_{10} 10^5$ (*b*) $\ln e^7$ (*a*) [5] (*b*) [7]

In problems 61 to 63, solve the equations.

61. (*a*) $\log_{10} x = 4$ (*b*) $\lg x = 5$ (*a*) [10 000] (*b*) [100000]

62. (*a*) $\log_3 x = 2$ (*b*) $\lg x = -2$ (*a*) [9] (*b*) [0.01]

63. (*a*) $\log_8 x = \frac{-4}{3}$ (*b*) $\ln x = 4$ (*a*) [$\frac{1}{16}$] (*b*) [e^4]

In problems 64 and 65, write the given expressions in terms of log 2, log 3 and log 5 to any base.

64. (*a*) log 60 (*b*) log 300

(*a*) [2 log 2 + log 3 + log 5] (*b*) [2 log 2 + log 3 + 2 log 5]

65. (*a*) $\log\left(\frac{8 \times \sqrt{5}}{9}\right)$ (*b*) $\log\left(\frac{125 \times \sqrt[4]{16}}{\sqrt[4]{81^3}}\right)$

(*a*) [$3 \log 2 + \frac{1}{2} \log 5 - 2 \log 3$]

(*b*) [$\log 2 + 3 \log 5 - 3 \log 3$]

Chapter 4

An introduction to algebra

4.1 Introduction

Algebra is that part of mathematics in which the relations and properties of numbers are investigated by means of general symbols. It is one of the most important concepts used in the understanding of mathematics and, in the form of general equations, is used in many areas of engineering and science.

Taking the simple case of a rectangle, whose area is found by multiplying the length by the breadth, this can be expressed algebraically as $A = l \times b$ where A represents the area, l the length and b the breadth. Hence $A = l \times b$ is a convenient general expression which is true for all positive values of l and b, when determining the area of a rectangle.

4.2 Basic notation

Problems involving numbers give a particular answer, for example $2 + 3 = 5$. This is called an **arithmetic equation**.

In algebra, results are generalised and letters are used which can stand for any number. The laws of indices introduced for numbers can be applied to show that

$$2^2 \times 2^3 = 2^{2+3}$$

Similarly

$$3^2 \times 3^3 = 3^{2+3}$$

and

$$10^2 \times 10^3 = 10^{2+3}$$

Since this result applies to all numbers, the general result is

$$a^m \times a^n = a^{m+n}$$

where a, m and n can each represent any number. This is an **algebraic equation**. A comparison between arithmetic and algebra is given in Table 4.1.

When a letter is used more than once in an algebraic expression, it must represent the same value throughout the expression.

Table 4.1

Arithmetic	Algebra
Numbers 1, 2, 15, $-6, \ldots,$ have fixed values	Letters a, b, p, $-y, \ldots,$ can stand for any values
$(2 \times 3) + (4 \times 7)$ is an arithmetic expression	$2b + 4c$ is an algebraic expression. $2b$ means $2 \times b$, the $2b$ form being adopted since the letter x is often used in algebra
(2×3) is the first term of the expression and (4×7) the second term	$2b$ is the first term of the expression and $4c$ the second term

4.3 Basic operations

In algebra, addition (+), subtraction (−), multiplication (×) and division (÷) have the same meaning as in arithmetic.

In arithmetic, $3 \times 5 = 5 + 5 + 5$
and $5^3 = 5 \times 5 \times 5$

Corresponding algebraic terms are

$$3 \times a = a + a + a$$

and

$$a^3 = a \times a \times a$$

The four basic operations in algebra are

Addition $2a + a = (a + a) + a = 3a$

Subtraction $2a - a = (a + a) - a = a$

Multiplication $2a \times a = 2 \times a \times a = 2a^2$

Division $2a \div a = \dfrac{2a}{a} = 2$ by cancellation

In arithmetic, kilograms cannot be added to metres. In the same way, in algebra, different letters represent different quantities and therefore cannot be added. If 'a' represents one number and 'b' another number, when $2a$ and b are added the result is written as

$$2a + b$$

When b is taken from $2a$, the result is written as

$$2a - b$$

Multiplying $2a$ by b is written as

$$2a \times b = 2ab,$$

and dividing $2a$ by b is written

$$2a \div b = \frac{2a}{b}$$

If b represents a number, then except when $b = 1$, b^2 will represent a different number. Hence terms containing b cannot be added or subtracted from those containing b^2. Thus the addition of b and b^2 is written as $b + b^2$. Subtraction of b from b^2 is written as $b^2 - b$. For multiplication and division, the terms can be simplified. Multiplication of b by b^2 becomes

$$b \times b^2 = b \times b \times b = b^3$$

Division of b by b^2 gives

$$b \div b^2 = \frac{b}{b^2} = \frac{b}{b \times b} = \frac{1}{b}$$

by cancellation.

Worked problems on basic operations

Problem 1. Find the value of $2ab - 3bc + abc$, when $a = 2$, $b = 5$, $c = 1$

Replacing a, b and c with numbers gives:

$$2 \times 2 \times 5 - 3 \times 5 \times 1 + 2 \times 5 \times 1 = 20 - 15 + 10$$
$$= \mathbf{15}$$

Problem 2. Find the value of $3a^2bc^3$ given $a = \frac{1}{2}$, $b = 2$ and $c = 1\frac{1}{2}$

Replacing a, b and c with numbers gives:

$$3 \times (\tfrac{1}{2})^2 \times 2 \times (\tfrac{3}{2})^3 = 3 \times \tfrac{1}{2} \times \tfrac{1}{2} \times 2 \times \tfrac{3}{2} \times \tfrac{3}{2} \times \tfrac{3}{2}$$
$$= \tfrac{81}{16}$$
$$= \mathbf{5\tfrac{1}{16}}$$

Problem 3. Find the sum of $3a$, $2a$, $-a$ and $-6a$

The sum of the positive terms is $3a + 2a = 5a$
The sum of the negative terms is $a + 6a = 7a$
Taking the sum of the negative terms from the sum of the positive terms gives $5a - 7a$ or $-2a$. Thus

$$3a + 2a + (-a) + (-6a) = \mathbf{-2a}$$

The rule for this type of problem is to add all the positive terms, add all the negative terms and then to find the difference between these two sums.

Problem 4. Find the sum of $3a$, $2b$, c, $-2a$, $-3b$ and $4c$

Each letter must be dealt with individually.

For the a terms $\quad +3a - 2a = a$

For the b terms $\quad +2b - 3b = -b$

For the c terms $\quad +c + 4c = 5c$

Thus

$$\mathbf{3a + 2b + c + (-2a) + (-3b) + 4c = a - b + 5c}$$

The signs $+$, $-$, $\times$ and $\div$ are called operators, meaning that they signify which operation has to be carried out. Adjacent operators in this type of problem follow the rule:
like signs give $+$ overall ($++$ or $--$)
unlike signs give $-$ overall ($+-$ or $-+$)
Thus the 4th term from problem 4 above, $+(-2a)$, becomes $-2a$ and the 5th term $+(-3b)$ becomes $-3b$. Therefore the problem can be written

$$\mathbf{3a + 2b + c - 2a - 3b + 4c = a - b + 5c}$$

Problem 5. Find the sum of $3a - b$, $2a + c$, $2b - 4d$ and $3b - a + 2d - 2c$

This problem can be laid out as shown, forming columns for the a's, b's, c's and d's. It should be remembered that $3a - b$ means $+3a - b$.

$+3a$	$-b$			
$+2a$		$+c$		
	$+2b$		$-4d$	
$-a$	$+3b$	$-2c$	$+2d$	
$\mathbf{4a}$	$\mathbf{+4b}$	$\mathbf{-c}$	$\mathbf{-2d}$	Adding

---

Problem 6. Subtract $2a + 4b - 3c$ from $a - 2b + 8c$

Check what is being subtracted from what, i.e.

From $\quad a \quad -2b \quad +8c$

Take $\quad 2a \quad +4b \quad -3c$

Dealing with each term, the a terms become $a - 2a = -a$

the b terms give $\quad -2b - (+4b) = -2b - 4b = -6b$

and the c terms give $\quad 8c - (-3c) = 8c + 3c = 11c$. Thus

$$\begin{array}{llll} a & -2b & +8c & \\ 2a & +4b & -3c & \\ \hline -a & -6b & +11c & \text{Subtracting} \end{array}$$

An alternative approach is to think of all the signs of the 'take' expression as being changed and then to add the two expressions:

$$\begin{array}{llll} a & -2b & +8c & \\ -2a & -4b & +3c & \\ \hline \mathbf{-a} & \mathbf{-6b} & \mathbf{+11c} & \text{Adding} \end{array}$$

Problem 7. Multiply $3ab$ by $2ab^2$

$3ab$ means $3 \times a \times b$

$2ab^2$ means $2 \times a \times b \times b$. Thus

$3ab \times 2ab^2 = 3 \times a \times b \times 2 \times a \times b \times b$

Grouping like terms

$$\begin{aligned} 3ab \times 2ab^2 &= 3 \times 2 \times a \times a \times b \times b \times b \\ &= 6 \times a^2 \times b^3 \\ &= \mathbf{6a^2b^3} \end{aligned}$$

Problem 8. Simplify $2pq \times 3p^2q^2 \times (-\frac{1}{2}pq^2)$

The product of two terms with like signs is positive.
The product of two terms with unlike signs is negative.
Thus $(+a) \times (-b) = -ab$ but $(+a) \times (+b) = ab$. Similarly $(-a) \times (-b) = ab$ and $(-a) \times (+b) = -ab$. Applying this general principle to the problem:

$$2pq \times 3p^2q^2 = 6p^3q^3 \text{ (like signs) and}$$

$$6p^3q^3 \times (-\tfrac{1}{2}pq^2) = -3p^4q^5 \text{ (unlike signs). Hence}$$

$$2pq \times 3p^2q^2 \times (-\tfrac{1}{2}pq^2) = \mathbf{-3p^4q^5}$$

Problem 9. Multiply $3x + 4y - 5z$ by $x + y$

Each term in the first expression is multiplied by x, then each term in the first expression is multiplied by y and the two results are added. The usual layout is to keep similar terms in columns so that they are underneath one another as shown:

$$\begin{array}{lllll} & 3x + 4y - 5z & & & \\ & x + \ \ y & & & \\ \hline \text{Multiplying by } x \rightarrow & 3x^2 + 4xy - 5xz & & & \\ \text{Multiplying by } y \rightarrow & \quad 3xy & & + 4y^2 - 5yz & \\ \hline \text{Adding} & \mathbf{3x^2 + 7xy - 5xz + 4y^2 - 5yz} & & & \end{array}$$

Problem 10. Multiply $2p - 3q^2 + 4pq$ by $2p - 3q$

$$\begin{array}{ll} & 2p - 3q^2 + 4pq \\ & 2p - 3q \\ \hline \text{Multiplying by } 2p \rightarrow & 4p^2 - 6pq^2 + 8p^2q \\ \text{Multiplying by } -3q \rightarrow & \quad -12pq^2 \quad - 6pq + 9q^3 \\ \hline \text{Adding} & \mathbf{4p^2 - 18pq^2 + 8p^2q - 6pq + 9q^3} \end{array}$$

Problem 11. Simplify $2a \div 6ab$

This can be written as $\dfrac{2a}{6ab}$ and reduced by cancelling as in arithmetic. Thus

$$\frac{2a}{6ab} = \frac{{}^{1}2 \times \not{a}^{1}}{{}_{3}\not{6} \times \not{a}_{1} \times b} = \frac{\mathbf{1}}{\mathbf{3b}} \text{ by cancellation}$$

Problem 12. Simplify $\dfrac{3a^2bc^3}{9ab^2c^2}$

$$\frac{3a^2bc^3}{9ab^2c^2} = \frac{\mathbf{ac}}{\mathbf{3b}} \text{ by cancellation}$$

Problem 13. Express as single fractions: (*a*) $\dfrac{a}{c+d} + \dfrac{b}{c+d}$

(*b*) $\dfrac{x}{2x-1} - \dfrac{1-x}{-1+2x}$ (*c*) $\dfrac{4}{x+1} + \dfrac{x}{1+x} - 3$

(*a*) $\dfrac{a}{c+d} + \dfrac{b}{c+d} = \dfrac{\mathbf{a+b}}{\mathbf{c+d}}$

$$(b)\ \frac{x}{2x-1}-\frac{1-x}{-1+2x}=\frac{x}{2x-1}-\frac{1-x}{2x-1}=\frac{x-(1-x)}{(2x-1)}=\frac{2x-1}{2x-1}=\mathbf{1}$$

$$(c)\ \frac{4}{x+1}+\frac{x}{1+x}-3=\frac{4}{x+1}+\frac{x}{x+1}-\frac{3}{1}=\frac{4+x-3(x+1)}{x+1}$$

$$=\frac{4+x-3x-3}{x+1}=\frac{\mathbf{1-2x}}{\mathbf{x-1}}$$

Problem 14. Divide $x^3 + 3x^2y + 3xy^2 + y^3$ by $x + y$

In arithmetic, long division can be used to divide, say, 192 by 16.

$$\begin{array}{r}
12\\
16\overline{)192}\\
\underline{16}\\
32\\
\underline{32}\\
\underline{\cdot\,\cdot}
\end{array}$$

Another way of doing this problem is

$$\begin{array}{rccccc}
 & 6 & + & 5 & + & 0\\
16) & \overline{100} & \overline{+} & \overline{90} & \overline{+} & \overline{2}\\
 & 96 & & 80 & & 2\\
 & \frac{4}{16} & & \frac{10}{16} & & \frac{2}{16}
\end{array}$$

That is $\frac{192}{16} = 6\frac{4}{16} + 5\frac{10}{16} + \frac{2}{16} = 12$
An approach something along the same lines is made in algebraic division

$$\begin{array}{rl}
 & \overset{①}{x^2} + \overset{③}{2xy} + \overset{⑤}{y^2}\\
x+y\,) & \overline{x^3 + 3x^2y + 3xy^2 + y^3}\\
 & \underline{x^3 + x^2y}\\
 & \quad 2x^2y + 3xy^2\\
 & \quad \underline{2x^2y + 2xy^2}\\
 & \qquad\quad xy^2 + y^3\\
 & \qquad\quad \underline{xy^2 + y^3}\\
 & \qquad\quad \underline{\cdot\quad\ \ \cdot}
\end{array}$$

② $x^2(x+y)$ and subtracting

④ $2xy(x+y)$ and subtracting

⑥ $y^2(x+y)$ and subtracting

① x into x^3 go… x^2, put x^2 u…

③ Bring down $3xy^2$. x int… $2x^2y$ goes 2… put $2xy$ up…

⑤ Bring down… x into xy^2… y^2, put y^2 u…

Thus $(x^3 + 3x^2y + 3xy^2 + y^3) \div (x + y) = \mathbf{x^2 + 2xy + y^2}$
A check can be made on this answer by multiplying $x^2 + 2xy + y^2$ by $x + y$ and seeing if it is equal to $x^3 + 3x^2y + 3xy^2 + y^3$

Problem 15. Simplify $\dfrac{a^3 - b^3}{a - b}$

$$
\begin{array}{lllll}
 & & a^2 + ab + b^2 & & \\
 & a - b\,\big) & a^3 + 0 + 0 - b^3 & & a \text{ into } a^3 \text{ goes } a^2 \\
a^2(a-b) & & a^3 - a^2b & & \\
\text{Subtract} & & \quad a^2b + 0 & & a \text{ into } a^2b \text{ goes } ab \\
ab(a-b) & & \quad a^2b - ab^2 & & \\
 & \text{Subtract} & \qquad ab^2 - b^3 & & a \text{ into } ab^2 \text{ goes } b^2 \\
b^2(a-b) & & \qquad ab^2 - b^3 & & \\
 & \text{Subtract} & \qquad \cdot \quad \cdot & &
\end{array}
$$

Thus $\dfrac{a^3 - b^3}{a - b} = \mathbf{a^2 + ab + b^2}$

The noughts shown in the $a^3 - b^3$ expression are not normally shown, but have been included to clarify the subtraction process and to keep similar terms in their respective columns.

Problem 16. Divide $6m^3 - 4m^2n + 2n^3$ by $2m - n$

$$
\begin{array}{ll}
 & 3m^2 - \tfrac{1}{2}mn - \tfrac{1}{4}n^2 \\
2m - n\,\big) & 6m^3 - 4m^2n \qquad\qquad + 2n^3 \\
 & 6m^3 - 3m^2n \\
 & \quad - m^2n \\
 & \quad - m^2n + \tfrac{1}{2}mn^2 \\
 & \qquad - \tfrac{1}{2}mn^2 \\
 & \qquad - \tfrac{1}{2}mn^2 + \tfrac{1}{4}n^3 \\
 & \qquad\qquad - \tfrac{1}{4}n^3 + 2n^3
\end{array}
$$

Thus $\dfrac{6m^3 - 4m^2n + 2n^3}{2m - n} = \mathbf{3m^2 - \dfrac{mn}{2} - \dfrac{n^2}{4}}$ **and the remainder is** $\mathbf{(-\tfrac{1}{4}n^3 + 2n^3)}$ **or** $\mathbf{+\tfrac{7}{4}n^3}$

Further problems on basic operations may be found in Section 4.7 (Problems 1–18), page 104.

4.4 Laws of indices

The laws of indices, introduced for arithmetic in Chapter 2, can now be generalised to apply to algebra. These rules are shown below in algebraic form, where *a*, *m* and *n* are any number.

$$a^m = a \times a \times a \times \cdots \text{ to } m \text{ terms}$$

$$a^n = a \times a \times a \times \cdots \text{ to } n \text{ terms}$$

Laws:

$$\boxed{a^m a^n = a^{m+n}} \qquad \text{[I]}$$

For example $a^3 a^5 = a^{3+5} = a^8$

$$\boxed{\frac{a^m}{a^n} = a^{m-n}} \qquad \text{[II]}$$

For example $\dfrac{a^6}{a^2} = a^{6-2} = a^4$

$$\boxed{(a^m)^n = a^{mn}} \qquad \text{[III]}$$

For example $(a^3)^5 = a^{3 \times 5} = a^{15}$

$$\boxed{a^{\frac{m}{n}} = \sqrt[n]{a^m}} \qquad \text{[IV]}$$

For example $a^{7/2} = \sqrt{a^7}$

$$\boxed{a^{-n} = \frac{1}{a^n}} \qquad \text{[V]}$$

For example $a^{-3} = \dfrac{1}{a^3}$

and expressing $a^4 \times a^{-7}$ with a positive index gives

$$a^{4-7} = a^{-3} \qquad \text{(by law [I])}$$

$$= \frac{1}{a^3} \qquad \text{(by law [V])}$$

$$\boxed{a^0 = 1} \qquad \text{[VI]}$$

For example $\frac{a^2a^3}{a^5} = \frac{a^{2+3}}{a^5}$ (by law [I])

$$= \frac{a^5}{a^5}$$

$$= a^{5-5} \quad \text{(by law [II])}$$

$$= a^0$$

$$= 1 \quad \text{(by law [VI])}$$

Worked problems on laws of indices

Problem 1. Simplify $a^2b^3c \times ab^2c^3$ and evaluate when $a = 1$, $b = \frac{1}{2}$ and $c = 2$

Grouping like terms this becomes $a^2 \times a \times b^3 \times b^2 \times c \times c^3$ and since $a = a^1$ and $c = c^1$, using law [I] this becomes

$$a^{(2+1)} \times b^{(3+2)} \times c^{(1+3)} = a^3 \times b^5 \times c^4$$

$$= \boldsymbol{a^3b^5c^4}$$

When $a = 1$, $b = \frac{1}{2}$ and $c = 2$, $a^3b^5c^4 = (1)^3\left(\frac{1}{2}\right)^5(2)^4$

$$= \frac{2^4}{2^5} = \mathbf{\frac{1}{2}}$$

Problem 2. Simplify $p^{\frac{1}{2}}q^2r^{-3} \times p^{1/3}q^{\frac{1}{2}}r^{-\frac{1}{2}}$ and evaluate when $p = 64$, $q = \frac{1}{4}$ and $r = 1$

This becomes by law (I), $p^{(1/2+1/3)}q^{(2+1/2)}r^{(-3-1/2)} = p^{5/6}q^{5/2}r^{-7/2}$

When $p = 64$, $q = 4$ and $r = 1$, $p^{5/6}q^{5/2}r^{-7/2} = (64)^{5/6}\left(\frac{1}{4}\right)^{5/2}(1)^{-7/2}$

$$= \sqrt[6]{(64)^5}\sqrt{\left(\frac{1}{4}\right)^5}(1) = (2)^5\left(\frac{1}{2}\right)^5 = \mathbf{1}$$

Problem 3. Simplify $\dfrac{a^2b^3c^4}{ab^2c}$ and evaluate when $a = 4$, $b = 3$ and $c = \dfrac{1}{2}$

Using the second law of indices

$$\frac{a^2}{a} = a^{2-1}, \frac{b^3}{b^2} = b^{3-2} \text{ and } \frac{c^4}{c} = c^{4-1}$$

Thus $\dfrac{a^2b^3c^4}{ab^2c} = a^{2-1}b^{3-2}c^{4-1} = \boldsymbol{abc^3}$

When $a = 4$, $b = 3$ and $c = \dfrac{1}{2}$, $abc^3 = (4)(3)\left(\dfrac{1}{2}\right)^3 = \dfrac{(4)(3)}{8} = \mathbf{1\frac{1}{2}}$

Problem 4. Simplify $\dfrac{x^{\frac{1}{2}}y^2z^{\frac{1}{3}}}{x^{\frac{1}{4}}y^{\frac{1}{2}}z^{\frac{1}{6}}}$

Using law [II], this becomes

$x^{\frac{1}{2}-\frac{1}{4}}y^{2-\frac{1}{2}}z^{\frac{1}{3}-\frac{1}{6}}$ or $\boldsymbol{x^{\frac{1}{4}}y^{\frac{3}{2}}z^{\frac{1}{6}}}$

Problem 5. Simplify $\dfrac{x^2y^3 + x^{\frac{1}{2}}y}{x^{1/3}y^{1/3}}$

Algebraic expressions of the form $\dfrac{a+b}{c}$ can be split into $\dfrac{a}{c} + \dfrac{b}{c}$

Thus

$$\frac{x^2y^3 + x^{\frac{1}{2}}y}{x^{1/3}y^{1/3}} = \frac{x^2y^3}{x^{1/3}y^{1/3}} + \frac{x^{\frac{1}{2}}y}{x^{1/3}y^{1/3}}$$

$$= x^{2-1/3}y^{3-1/3} + x^{1/2-1/3}y^{1-1/3}$$

$$= \boldsymbol{x^{5/3}y^{8/3} + x^{1/6}y^{2/3}}$$

Problem 6. Simplify $\dfrac{a^2b}{ab^2 - ab}$

The highest common factor of each of the three terms comprising the numerator and denominator is found. This is ab. Dividing each term by ab gives:

$$\frac{a^2b}{ab^2 - ab} = \frac{\dfrac{a^2b}{ab}}{\dfrac{ab^2}{ab} - \dfrac{ab}{ab}}$$

$$= \frac{\boldsymbol{a}}{\boldsymbol{b - 1}}$$

Problem 7. Simplify $\dfrac{p^2q}{pq^2 - p^{\frac{1}{2}}q}$ and evaluate when $p = 9$, and $q = \dfrac{1}{2}$, taking positive values of square roots only.

A common factor approach must be made where addition or subtraction of terms occurs in the denominator. The highest common factor of the three terms is $p^{\frac{1}{2}}q$. Thus

$$\frac{p^2q}{pq^2 - p^{\frac{1}{2}}q} = \frac{\dfrac{p^2q}{p^{\frac{1}{2}}q}}{\dfrac{pq^2}{p^{\frac{1}{2}}q} - \dfrac{p^{\frac{1}{2}}q}{p^{\frac{1}{2}}q}} = \frac{\boldsymbol{p^{3/2}}}{\boldsymbol{p^{\frac{1}{2}}q - 1}}$$

When $p = 9$ and $q = \dfrac{1}{2}$, $\dfrac{p^{3/2}}{p^{\frac{1}{2}}q - 1} = \dfrac{(9)^{3/2}}{(9)^{\frac{1}{2}}\left(\dfrac{1}{2}\right) - 1} = \dfrac{\sqrt{(9)^3}}{[\sqrt{(9)}]\left(\dfrac{1}{2}\right) - 1}$

$$= \frac{3^3}{(3)\left(\dfrac{1}{2}\right) - 1} = \frac{27}{\dfrac{1}{2}} = \mathbf{54}$$

Problem 8. Simplify $(a^3)^{\frac{1}{2}}(b^2)^3$

Using law [III], the expression becomes $a^{3 \times \frac{1}{2}}b^{2 \times 3}$. Thus $(a^3)^{\frac{1}{2}}(b^2)^3 = \boldsymbol{a^{3/2}b^6}$

Problem 9. Simplify $\dfrac{(m^2n)^3}{(m^{\frac{1}{2}}n^{1/3})^4}$

The bracket indicates that each letter in the bracket must be raised to the power shown outside. Using law [III] the numerator becomes $m^{2 \times 3}n^{1 \times 3}$ or m^6n^3 and the denominator becomes

$m^{\frac{1}{2}+4}n^{\frac{1}{3}\times 4}$ or $m^2n^{4/3}$. Hence

$$\frac{(m^2n)^3}{(m^{\frac{1}{2}}n^{1/3})^4} = \frac{m^6n^3}{m^2n^{4/3}}$$

$$= m^{6-2}n^{3-4/3} \text{ by law [II]}$$

$$= m^4n^{5/3}$$

Problem 10. Simplify $a^2\sqrt{b}\sqrt{c^3} \times \sqrt{a}\sqrt[3]{b^2}c^4$ and evaluate when $a = \frac{1}{4}$, $b = 64$ and $c = 1$

By law [IV] this can be written as

$$a^2b^{\frac{1}{2}}c^{3/2} \times a^{\frac{1}{2}}b^{2/3}c^4 = a^{5/2}b^{7/6}c^{11/2}$$

It is usual in mathematics to give the answer in the same form as the question.
Hence $a^{5/2}b^{7/6}c^{11/2} = \sqrt{a^5}\sqrt[6]{b^7}\sqrt{c^{11}}$

When $a = \frac{1}{4}$, $b = 64$ and $c = 1$,

$$\sqrt{a^5}\sqrt[6]{b^7}\sqrt{c^{11}} = \sqrt{\left(\frac{1}{4}\right)^5}\sqrt[6]{(64)^7}\sqrt{(1)^{11}}$$

$$= \left(\frac{1}{2}\right)^5(2)^7(1) = \frac{2^7}{2^5} = 2^2 = \mathbf{4}$$

Problem 11. Simplify $(p^2q)(p^{-3}q^{-2})$ expressing the answer with positive indices only.

Using law [I] $(p^2q)(p^{-3}q^{-2}) = p^{-1}q^{-1}$

Applying law [V] $p^{-1}q^{-1} = \frac{\mathbf{1}}{\mathbf{pq}}$

Problem 11. Simplify $\dfrac{p^2q^4r^{\frac{1}{2}}}{(p^{3/2}q^2r^{3/2})^2}$

Using law [III], this becomes $\dfrac{p^2q^4r^{\frac{1}{2}}}{p^{3/2\times 2}q^{2\times 2}r^{3/2\times 2}} = \dfrac{p^2q^4r^{\frac{1}{2}}}{p^3q^4r^3}$

and by law [II] $p^{2-3}q^{4-4}r^{\frac{1}{2}-3} = p^{-1}q^0r^{-5/2}$
By law [VI] $q^0 = 1$

Hence

$$p^{-1}q^0r^{-5/2} = p^{-1}r^{-5/2}$$

$$= \frac{1}{pr^{5/2}}$$

Problem 13. Simplify $\dfrac{(a^2b^{\frac{1}{2}})(\sqrt{a}\sqrt[3]{b^2})}{(a^5b^4)^{\frac{1}{2}}}$

By laws [III] and [IV] this expression becomes $\dfrac{(a^2b^{\frac{1}{2}})(a^{\frac{1}{2}}b^{2/3})}{a^{5/2}b^{4/2}}$

$= a^{2+\frac{1}{2}-5/2}b^{\frac{1}{2}+2/3-2}$ laws [I] and [II])

$= a^0b^{-5/6}$

$= \dfrac{1}{b^{5/6}}$ (laws [V] and [VI])

Further problems on the laws of indices may be found in Section 4.7 (Problems 19–32), page 105.

4.5 Brackets and factorisation

The use of brackets

The addition of the sum of b and c to a may be written as

$a + (b + c)$

The subtraction of the sum of b and c from a is written as

$a - (b + c)$

The multiplication of twice the sum of a and b by the difference between c and d is written as

$2(a + b)(c - d)$

The division of the square of the sum of a and b by twice the sum of c and d is written as

$$\frac{(a + b)^2}{2(c + d)}$$

Removing brackets

An addition sign in front of a bracket does not change the signs within the bracket when they are removed, thus

$(a + b) + (c + d) = a + b + c + d$

and

$$(a + b) + (c - d) = a + b + c - d$$

A subtraction sign in front of a bracket results in all terms within the bracket being multiplied by -1, thus having their signs changed when the brackets are removed. Thus

$$(a + b) - (c + d) = a + b - c - d$$

and

$$(a + b) - (c - d) = a + b - c + d$$

A term in front of or behind a bracket indicates that each term within the bracket is multiplied by that term. Hence

$$a(b + c) = ab + ac$$

and

$$d(a + b) - 2(c - d) = ad + bd - 2c + 2d$$

and

$$(3a + b)d = 3ad + bd$$

One bracketed expression immediately followed by another bracketed expression indicates that the contents of one bracket should be multiplied by the contents of the other bracket when they are being removed. Thus, using algebraic multiplication techniques,

$$(a + b)(c + d) = ac + ad + bc + bd$$

A bracketed expression raised to some index indicates that the contents of the bracket must be multiplied by itself the number of times indicated by the power, thus

$$\begin{aligned}(p + q)^2 = (p + q)(p + q) &= p^2 + pq + pq + q^2 \\ &= p^2 + 2pq + q^2\end{aligned}$$

using algebraic multiplication techniques. When an expression contains more than one type of bracket, it is usual to remove the innermost bracket first and work outwards clearing one bracket at a time. For an expression containing two brackets

$$\begin{aligned}6\{3y - 4(2z + 4)\} &= 6\{3y - 8z - 16\} \\ &= 18y - 48z - 96\end{aligned}$$

and for an expression containing three brackets:

$$\begin{aligned}4[2 + a\{c - 3(a + 2)\}] &= 4[2 + a\{c - 3a - 6\}] \\ &= 4[2 + ac - 3a^2 - 6a] \\ &= 8 + 4ac - 12a^2 - 24a\end{aligned}$$

Factorisation

When two or more terms in an algebraic expression contain a common factor, then this factor can be shown outside a bracket. Thus

$$\begin{aligned} 4ax + 2ay &= 2a \cdot 2x + 2a \cdot y \\ &= 2a(2x + y) \end{aligned}$$

(The product point $\cdot$ is used here instead of $\times$ to indicate multiplication.) This process is called **factorisation**, $2a$ and $(2x + y)$ being the factors of the expression $4ax + 2ay$. In the expression

$$ax + 3a + 2bx + 6b$$

a is common to the first two terms and $2b$ to the last two terms. Thus the expression can be written as

$$a(x + 3) + 2b(x + 3)$$

Also $(x + 3)$ is a common factor of both the first and second terms and so

$$a(x + 3) + 2b(x + 3) = (x + 3)(a + 2b)$$

Then $(x + 3)$ and $(a + 2b)$ are the factors of the expression

$$ax + 3a + 2bx + 6b$$

Worked problems on brackets and factorisation

Problem 1. Remove the brackets and simplify the expression $(2a + b) + 3(b + c) - 2(c - d)$

Both b and c in the second bracket have to be multiplied by 3, and c and $-d$ in the third bracket by -2, when the brackets are removed. Thus
$(2a + b) + 3(b + c) - 2(c - d) = 2a + b + 3b + 3c - 2c + 2d$
and collecting together the like terms gives $2a + 4b + c + 2d$, that is $(2a + b) + 3(b + c) - 2(c - d) = \mathbf{2a + 4b + c + 2d}$

Problem 2. Simplify $x^2 - (3x - xy) - x(4y + x)$

When the brackets are removed, both $3x$ and $-xy$ in the first bracket must be multiplied by -1 and both $4y$ and x in the second bracket by $-x$. Thus
$x^2 - (3x - xy) - x(4y + x) = x^2 - 3x + xy - 4xy - x^2$
Grouping like terms gives $(x^2 - x^2) - 3x + (xy - 4xy)$
or
$-3x - 3xy$

Since $-3x$ is a common factor, this can be factorised as $-3x(1 + y)$. Thus

$x^2 - (3x - xy) - x(4y + x) = \mathbf{-3x(1 + y)}$

When forming a bracket in this way, a check should be made by multiplying out to see that the signs have been correctly determined.

Problem 3. Remove the brackets from the expression $(2p - q)(p + q^2)$

Each term in the second bracket has to be multiplied by each term in the first bracket. Thus

$$(2p - q)(p + q^2) = 2p(p + q^2) - q(p + q^2)$$
$$= \mathbf{2p^2 + 2pq^2 - pq - q^3}$$

Alternatively, algebraic multiplication can be used.

Problem 4. Remove the brackets from the expression $-3[a^2 - 2(b + c) + b^2]$

Removing the inner bracket by multiplying b and c by -2 gives

$-3[a^2 - 2b - 2c + b^2]$

Removing the outer bracket by multiplying all terms within this bracket by -3 gives

$-3a^2 + 6b + 6c - 3b^2$

Then

$-3[a^2 - 2(b + c) + b^2] = \mathbf{-3a^2 + 6b + 6c - 3b^2}$

Problem 5. Remove the brackets and simplify the expression $4a - [2\{(2a - 4b) - 2(a + 5b)\} + a]$

Starting by removing the innermost brackets gives

$4a - [2\{2a - 4b - 2a - 10b\} + a]$

The $\{\ \}$ brackets are now the inner ones and are removed next, giving

$4a - [4a - 8b - 4a - 20b + a]$

Multiplying all terms in the square brackets by -1 gives

$4a - 4a + 8b + 4a + 20b - a$

Grouping like terms together results in

$4a - [2\{(2a - 4b) - 2(a + 5b)\} + a]$ **being equal to** $\mathbf{3a + 28b}$

Problem 6. Simplify $m(3m - 2n) - 2m(3m + 4n)$

Multiplying each term in the first bracket by m and each term in the second bracket by $-2m$ gives

$3m^2 - 2mn - 6m^2 - 8mn = -3m^2 - 10mn$

Since $-m$ is a common factor of both terms, the expression can be written as $-m(3m + 10n)$, thus

$m(3m - 2n) - 2m(3m + 4n) = \mathbf{-m(3m + 10n)}$

Problem 7. Factorise (*a*) $pq - 2pr$ (*b*) $3a^2 + 15ab^2$ (*c*) $5x^2y - 10xy^2 + 25xy$

For each part of this problem, the highest common factor of the terms will become one of the factors. Thus

(*a*) $pq - 2pr = \mathbf{p(q - 2r)}$

(*b*) $3a^2 + 15ab^2 = \mathbf{3a(a + 5b^2)}$

(*c*) $5x^2y - 10xy^2 + 25xy = \mathbf{5xy(x - 2y + 5)}$

Problem 8. Factorise $ax + bx - by - ay$

The first two terms have a common factor of x and the last two terms a common factor of y. Thus

$$ax + bx - by - ay = x(a + b) + y(-a - b)$$
$$= x(a + b) - y(a + b)$$

since the second bracket also has a common factor of -1. Both terms now have a common factor of $(a + b)$, hence

$ax + bx - by - ay = \mathbf{(a + b)(x - y)}$

Problem 9. Factorise $3ax - 2ay + 3bx - 2by$

Selecting a as a common factor of the first two terms and b as a common factor of the last two terms gives

$a(3x - 2y) + b(3x - 2y)$

Since $(3x - 2y)$ is a common factor of both these terms, the expression becomes $(3x - 2y)(a + b)$. Thus

$3ax - 2ay + 3bx - 2by = \mathbf{(3x - 2y)(a + b)}$

Alternatively, selecting $3x$ as the common factor of the first and third terms and $-2y$ as the common factor of the second and last terms gives

$3x(a + b) - 2y(a + b) = (3x - 2y)(a + b)$

the same result as for selecting a and b as the factors.

Problem 10. Factorise $x^3 - 2x^2 - x + 2$

Selecting x^2 as a common factor of the first two terms gives

$x^3 - 2x^2 - x + 2 = x^2(x - 2) - x + 2$

and by making -1 a factor of the third and fourth terms the expression becomes $x^2(x - 2) - 1(x - 2)$

which is usually written $x^2(x - 2) - (x - 2)$

But $(x - 2)$ is now a common factor, hence

$x^3 - 2x^2 - x + 2 = \mathbf{(x^2 + 1)(x - 2)}$

Further problems on brackets and factorisation may be found in Section 4.7 (Problems 33–45), page 106.

4.6 Fundamental laws and precedence

The basic laws introduced in arithmetic are generalised in algebra. These are given below, where a, b and c represent any three numbers

$$\boxed{a + (b + c) = (a + b) + c} \qquad [I]$$

For example

$2m + (3m + 4m) = 2m + 7m \quad = 9m$

and $(2m + 3m) + 4m = 5m + 4m \quad = 9m$

thus $2m + (3m + 4m) = (2m + 3m) + 4m = 9m$

$$\boxed{a(bc) = (ab)c} \qquad [II]$$

For example

$2m(3m \cdot 4m) = 2m(12m^2) \quad = 24m^3$

and $(2m \cdot 3m)4m = (6m^2)4m \quad = 24m^3$

thus $2m(3m \cdot 4m) = (2m \cdot 3m)4m \quad = 24m^3$

Laws [I] and [II] are called **associative laws.**

$$\boxed{(a + b) = (b + a)} \qquad [III]$$

For example

$(2m + 3m) = 5m$

and $(3m + 2m) = 5m$

thus $(2m + 3m) = (3m + 2m) = 5m$

$$\boxed{ab = ba} \qquad [IV]$$

For example

$$2m \cdot 3m = 6m^2$$

and $\quad 3m \cdot 2m = 6m^2$

thus $\quad 2m \cdot 3m = 3m \cdot 2m = 6m^2$

Laws [III] and [IV] are called **commutative laws.**

$$\boxed{a(b + c) = ab + ac} \qquad \text{[V]}$$

For example

$$2m(3m + 4m) = \quad 2m(7m) \quad = 14m^2$$

and $\quad 2m \cdot 3m + 2m \cdot 4m = \quad 6m^2 + 8m^2 \quad = 14m^2$

thus $\quad 2m(3m + 4m) = 2m \cdot 3m + 2m \cdot 4m = 14m^2$

Law [V] is called the **distributive law.**

The rules of precedence which apply to arithmetic governing the order in which operations are performed apply equally to algebraic expressions. The order is **B**rackets, **O**f, **D**ivision, **M**ultiplication, **A**ddition and **S**ubtraction, that is, governed by BODMAS. (As with arithmetic, the word 'of' appearing in an expression should be changed to a multiplication sign after brackets have been removed.)

For example, to simplify the expression

$$(5a + 3a) - 4a \times 6a \div 7a$$

the bracketed terms must first be evaluated, then the division performed, followed by the multiplication and finally the subtraction. Thus

$$(5a + 3a) - 4a \times 6a \div 7a = 8a - 4a \times 6a \div 7a \qquad \text{(B)}$$

$$= 8a - 4a \times \frac{6a}{7a} \qquad \text{(D)}$$

$$= 8a - 4a \times \frac{6}{7}$$

$$= 8a - \frac{24a}{7} \qquad \text{(M)}$$

$$= \frac{56a - 24a}{7} = \frac{32a}{7} \qquad \text{(S)}$$

Worked problems on fundamental laws and precedence

Problem 1. Simplify $3y + 7y \times 2y - y$

The order of precedence indicates that the multiplication must be done before addition and subtraction, hence

$$3y + 7y \times 2y - y = 3y + 14y^2 - y$$
$$= 14y^2 + 2y$$
$$= \mathbf{2y(7y + 1)}$$

Problem 2. Simplify $(3y + 7y) \times 2y - y$

The order is brackets, multiplication and subtraction, hence

$$(3y + 7y)2y - y = 10y \cdot 2y - y$$
$$= 20y^2 - y$$
$$= \mathbf{y(20y - 1)}$$

Problem 3. Simplify $3y + 7y \times (2y - y)$

The order is brackets, multiplication and addition, thus

$$3y + 7y(2y - y) = 3y + 7y \times y$$
$$= 3y + 7y^2$$
$$= \mathbf{y(3 + 7y)}$$

Problem 4. Simplify $6k \div 3k + 4k - 2k$

The order is division, addition and subtraction, thus

$$6k \div 3k + 4k - 2k = \frac{6k}{3k} + 4k - 2k$$
$$= 2 + 4k - 2k$$
$$= \mathbf{2 + 2k = 2(1 + k)}$$

Problem 5. Simplify $6k \div (3k + 4k) - 2k$

The order is brackets, division and subtraction, hence

$$6k \div (3k + 4k) - 2k = 6k \div 7k - 2k$$
$$= \mathbf{\frac{6}{7} - 2k}$$

Problem 6. Simplify $6k \div (3k + 4k - 2k)$

The order is brackets and then division, thus

$$6k \div (3k + 4k - 2k) = 6k \div 5k$$
$$= \mathbf{\frac{6}{5}}$$

Problem 7. Simplify $4m + 2m \times 3m + m \div 5m - 6m$

Applying BODMAS gives the order of precedence as division, multiplication, addition and finally subtraction. Thus

$$4m + 2m \times 3m + m \div 5m - 6m = 4m + 2m \times 3m + \frac{m}{5m} - 6m$$

$$= 4m + 6m^2 + \frac{1}{5} - 6m$$

$$= \mathbf{6m^2 - 2m + \frac{1}{5}}$$

Problem 8. Simplify $(4m + 2m)3m + m \div 5m - 6m$

The order is brackets, division, multiplication, addition and subtraction, thus

$$(4m + 2m)3m + m \div 5m - 6m = 6m \times 3m + m \div 5m - 6m$$

$$= 6m \times 3m \times \frac{m}{5m} - 6m$$

$$= 18m^2 + \frac{1}{5} - 6m$$

$$= \mathbf{6m(3m - 1) + \frac{1}{5}}$$

Problem 9. Simplify $4m + 2m \times 3m + m \div (5m - 6m)$

The order is brackets, division, multiplication and addition. Thus

$$4m + 2m \times 3m + m \div (5m - 6m) = 4m + 2m \times 3m + m \div -m$$

$$= 4m + 2m \times 3m + \frac{m}{-m}$$

Now $\frac{m}{-m} = \frac{1}{-1}$ and multiplying numerator and denominator by -1 gives $\frac{1 \times -1}{-1 \times -1} = -1$, hence

$$4m + 2m \times 3m + \frac{m}{-m} = 4m + 6m^2 - 1$$

$$= \mathbf{2m(3m + 2) - 1}$$

Problem 10. Simplify $(4m + 2m)(3m + m) \div (5m - 6m)$

The order is brackets, division and multiplication, thus

$$(4m + 2m)(3m + m) \div (5m - 6m) = 6m \times 4m \div -m$$

$$= 6m \times \frac{4m}{-m}$$

$$= \frac{24m^2}{-m}$$

$$= \frac{24m}{-1} \times \frac{-1}{-1} = \mathbf{-24m}$$

Problem 11. Simplify $(2y - 5) \div 4y + 6 \times 7 - 3y$

The bracket around the $2y - 5$ shows that both $2y$ and -5 have to be divided by $4y$, and to remove the bracket the expression is written in fraction form. Thus

$$(2y - 5) \div 4y + 6 \times 7 - 3y = \frac{2y-5}{4y} + 6 \times 7 - 3y$$

$$= \frac{2y-5}{4y} + 42 - 3y$$

$$= \frac{2y - 5 + 168y - 12y^2}{4y}$$

$$= \mathbf{\frac{170y - 5 - 12y^2}{4y}}$$

Problem 12. Simplify $\frac{1}{4}$ of $3y + 8y(5y - 2y)$

Applying BODMAS, the expression becomes $\frac{1}{4}$ of $3y + 8y(3y)$ and changing 'of' to $\times$, this becomes $\frac{1}{4} \times 3y + 8y \times 3y$ or $\frac{3}{4}y + 24y^2$ that is $\mathbf{y(\frac{3}{4} + 24y)}$

Further problems on the order of precedence may be found in the following Section (4.7) (Problems 46–57), page 107.

4.7 Further problems

Basic operations

1. Find the sum of $-4c$, $3c$, $-12c$ and $6.5c$ [$-6.5c$]
2. Simplify $\frac{3}{2}p + 8p - \frac{3}{5}p - \frac{p}{3}$ [$8\frac{17}{30}p$]
3. Find the sum of $4p$, $7q$, $-16p$, $3r$, $\frac{1}{2}p$ and $-3q$ [$-11\frac{1}{2}p + 4q + 3r$]

4. Add together $3x + 4y + z$, $-2x + y - 5z$ and $5x - 3y - 10z$ $[6x + 2y - 14z]$
5. From $3x - 2y + z$ subtract $2x - 3y - 4z$ $[x + y + 5z]$
6. Subtract $\frac{1}{2}p + \frac{1}{3}q - \frac{1}{4}r$ from $\frac{1}{3}p - \frac{1}{4}q - \frac{1}{5}r$ $[-\frac{1}{6}p - \frac{7}{12}q + \frac{1}{20}r]$
7. From $0.3a - 0.4b + 6z$ subtract $0.5a + 0.6b - 0.8z$ $[-0.2a - b + 6.8z]$
8. Multiply $4xyz$ by $3x^2y^3z$ $[12x^3y^4z^2]$
9. Simplify $3a^2b \times 4ac^2 \times -2abc$ $[-24a^4b^2c^3]$
10. Multiply $3p - 6pq + 2q^2$ by $p - q$ $[3p^2 - 6p^2q + 8pq^2 - 3pq - 2q^3]$
11. Multiply $3x - 4y$ by $x^2 + 3xy + 2y^3$ $[3x^3 + 5x^2y + 6xy^3 - 12xy^2 - 8y^4]$
12. Simplify $\dfrac{abc}{3a^2b^3c^4}$ $\left[\dfrac{1}{3ab^2c^3}\right]$
13. Simplify $\dfrac{3xy^2z^3}{24xy^3z}$ $\left[\dfrac{z^2}{8y}\right]$
14. Divide $a^2 - 2ab + b^2$ by $a - b$ $[a - b]$
15. Express $\dfrac{3a}{a-2} - \dfrac{a}{a-2}$ as a single fraction $\left[\dfrac{2a}{a-2}\right]$
16. Express $\dfrac{2x}{1+x} + \dfrac{5}{x+1} - 3$ as a single fraction $\left[\dfrac{2-x}{1+x}\right]$
17. Divide $a^3 + b^3$ by $a + b$ $[a^2 - ab + b^2]$
18. Divide $x^6 - y^6$ by $x^2 - xy + y^2$ $[x^4 + x^3y - xy^3 - y^4]$

Laws of indices

19. Simplify $p^2q^3r \times pq^2r^5$ and evaluate when $p = 3$, $q = 2$ and $r = \dfrac{1}{2}$ $[p^3q^5r^6;\ 13\frac{1}{2}]$
20. Simplify $\dfrac{x^{\frac{1}{2}}y^2z^{3/2}}{y^{3/2}z}$ and evaluate when $x = 2$, $y = 2$ and $z = 9$ $[x^{\frac{1}{2}}y^{\frac{1}{2}}z^{\frac{1}{2}};\ \pm 6]$
21. Simplify $a^{\frac{1}{4}}b^2c^{-2} \times b^{-3/2}c^3$ and evaluate when $a = 16$, $b = \dfrac{1}{9}$ and $c = 6$ $[a^{\frac{1}{4}}b^{\frac{1}{2}}c;\ \pm 4]$

In problems 22 to 32, simplify the expressions given.

22. $\dfrac{p^{1/5}q^{1/4}r^{1/3}}{p^{1/6}q^{-\frac{1}{2}}r^{-2}}$ $[p^{1/30}q^{3/4}r^{7/3}]$

23. $\dfrac{k^2l + k^{\frac{1}{2}}l^{3/2}}{k^3l^4}$ $\left[\dfrac{1}{kl^3} + \dfrac{1}{(kl)^{5/2}}\right]$

24. $\dfrac{x^3y^{\frac{1}{2}}}{xy - x^2y^3}$ $\left[\dfrac{x^2}{y^{\frac{1}{2}} - xy^{5/2}}\right]$

25. $\dfrac{(abc)^3}{(a^2b^{\frac{1}{2}}c^3)^{\frac{1}{2}}}$ $[a^2b^{11/4}c^{3/2}]$

26. $(a^2)^3(b^{\frac{1}{2}})^{1/3}$ $[a^6b^{1/6}]$

27. $x^2(\sqrt{y})\sqrt[3]{z^2} + (\sqrt[3]{x})y^2z^3$ $[x^{7/3}y^{5/2}z^{11/3}]$

28. $a^2b^3c \times a^{-3}b^{-5}c^2$, expressing the answer with positive indices only $\left[\dfrac{c^3}{ab^2}\right]$

29. $\dfrac{p^{1/3}q^2r^{1/5}}{(pq^{\frac{1}{2}}r^3)^{1/3}}$ $[q^{11/6}r^{-4/5}]$

30. $\dfrac{k^3(\sqrt{l})m^{3/2} \times l\sqrt[3]{m^2}}{k^{1/3}\sqrt{l^3}}$ $[k^{8/3}m^{13/6}]$

31. $(x^{-2/3}y^{\frac{1}{2}}z^3)^{\frac{1}{2}} \div (x^{2/3}y^2z^{1/3})^{1/3}$ expressing the answer with positive indices only $\left[\dfrac{z^{25/18}}{x^{5/9}y^{5/12}}\right]$

32. $\dfrac{a^2b^{\frac{1}{2}}c^{-1/4} \times (ab)^{1/4}}{\sqrt[3]{a^2}\sqrt{b^3c^{\frac{1}{2}}}}$ $[a^{19/12}b^{-3/4}c^{-3/4}]$

Brackets and factorisation

In problems 33 to 40, remove the brackets and simplify where possible.

33. (*a*) $(p + 2q) + (3p - q)$ (*b*) $(a - 6b) + (3b - 2c)$
(*c*) $(3x + 4y) - (x - 2y)$
(*a*) $[4p + q]$ (*b*) $[a - 3b - 2c]$ (*c*) $[2x + 6y]$

34. (*a*) $(a + b - c) + (2a - 4b - 3c)$
(*b*) $(3p - 4q + r) - (2p + 3q - 6r)$
(*c*) $(4x - z) - (y + 2x - 4z)$
(*a*) $[3a - 3b - 4c]$ (*b*) $[p - 7q + 7r]$ (*c*) $[2x - y + 3z]$

35. (*a*) $3(p - q) - 2(q - p)$ (*b*) $6(a - 4b) - 4(2a + b)$
(*c*) $3(p + 2q - r) - 4(r - 2q + p)$
(*a*) $[5p - 5q]$ (*b*) $[-2a - 28b]$ (*c*) $[-p + 14q - 7r]$

36. (*a*) $(k + 2l)(k + l)$ (*b*) $(p + q)(2p - 3q)$
(*c*) $2(x - 3y)(x - y + 2z)$
(*a*) $[k^2 + 3kl + 2l^2]$ (*b*) $[2p^2 - pq - 3q^2]$
(*c*) $[2x^2 - 8xy + 4xz + 6y^2 - 12yz]$

37. (*a*) $(a+b)^2$ (*b*) $(a-2b)^2$ (*c*) $3(2a-b)^2$ (*d*) $(a-b)^3$
(*a*) $[a^2+2ab+b^2]$ (*b*) $[a^2-4ab+4b^2]$
(*c*) $[12a^2-12ab+3b^2]$ (*d*) $[a^3-3a^2b+3ab^2-b^3]$

38. (*a*) $k^2(l+m)-l(m^2-k)$ (*b*) $3p(q+r)+4r(p-q)$
(*c*) $2a(p+q)-3p(2a-2q)$
(*a*) $[k^2l+k^2m-lm^2+lk]$ (*b*) $[3pq+7pr-4qr]$
(*c*) $[-4ap+2aq+6pq]$

39. (*a*) $3a+\{b-(2a+b)\}$ (*b*) $4p-\{2q+3(p-q)+6p\}$
(*c*) $3x+4\{2x-(4x+3)\}$
(*a*) $[a]$ (*b*) $[-5p+q]$ (*c*) $[-5x-12]$

40. (*a*) $3-4\{(a+b)-(a-b)^2\}$
(*b*) $-4\{(a-2b+c)-(3c-a)\}$
(*c*) $\{(2p+4q)-(4q-r)\}-3pq+2qr$
(*a*) $[4a^2+4b^2-8ab-4a-4b+3]$ (*b*) $[-8a+8b+8c]$
(*c*) $[2p+r-3pq+2qr]$

For problems 41 to 45, factorise

41. (*a*) $ab+ac$ (*b*) p^2+pq (*c*) $6x-18$
(*a*) $[a(b+c)]$ (*b*) $[p(p+q)]$ (*c*) $[6(x-3)]$

42. (*a*) $2m^2+10mn$ (*b*) $15ab^2-30a^2b$ (*c*) $21x^2y^2-28xy$
(*a*) $[2m(m+5n)]$ (*b*) $[15ab(b-2a)]$ (*c*) $[7xy(3xy-4)]$

43. (*a*) $16a^3b^3-48a^2b^2$ (*b*) ab^2+bc-b^3
(*c*) $3p^2q-6pq^2+15pq$
(*a*) $[16a^2b^2(ab-3)]$ (*b*) $[b\{c+b(a-b)\}]$
(*c*) $[3pq(p-2q+5)]$

44. (*a*) $ax+bx+a+b$ (*b*) $ay+by+az+bz$
(*c*) $ep-fp+eq-fq$
(*a*) $[(a+b)(x+1)]$ (*b*) $[(a+b)(y+z)]$
(*c*) $[(e-f)(p+q)]$

45. (*a*) $ax-ay-by+bx$ (*b*) $ep+fp-fq-eq$
(*c*) $ak+al-bk-bl$
(*a*) $[(a+b)(x-y)]$ (*b*) $[(e+f)(p-q)]$
(*c*) $[(a-b)(k+l)]$

Order of precedence

In problems 46 to 57, simplify

46. $3a \div 4a + 7a$ $[7a+\frac{3}{4}]$

47. $3a \div (4a+7a)$ $[\frac{3}{11}]$

48. $4m-3m\times 5m+2m$ $[3m(2-5m)]$

49. $(4m-3m)5m+2m$ $[m(5m+2)]$

50. $4m-3m(5m+2m)$ $[m(4-21m)]$

51. $3x+4\div 7x+2\times 5-2x$ $\left[x+\dfrac{4}{7x}+10\right]$

52. $(3x + 4) \div 7x + 2 \times 5 - 2x$ $\left[\dfrac{3x+4}{7x} + 10 - 2x\right]$

53. $3x + 4 \div 7x + 2(5 - 2x)$ $\left[\dfrac{4}{7x} - x + 10\right]$

54. $2 \div x + 3 \div (x + 1)$ $\left[\dfrac{5x+2}{x(x+1)}\right]$

55. $a^2 - 3ab \times 2a \div 6b + ab$ $[ab]$

56. $(x + 2)(x - 3) \div (2x - 6)$ $[\frac{1}{2}(x + 2)]$

57. $6k \div (4kl + 8l \times 2k) - 2lk \div l - k$ $\left[3\left(\dfrac{1}{10l} - k\right)\right]$

Chapter 5

Simple equations

5.1 Solution of simple equations

An **equation** is simply a statement that two quantities are equal. For example,

$$1 \text{ m} = 100 \text{ cm}$$

$$0°\text{C} = 273.15 \text{ K}$$

Often, however, an equation contains an unknown quantity which is represented by a symbol, say x. Simple linear equations are those in which an unknown quantity is raised only to the power 1, i.e. $x^1 = x$ (an equation of the **first degree** is another term used to describe such equations). Terms such as x^2, x^3 etc. do not appear in simple equations. (The word 'simple' in this context does not necessarily mean 'easy' – it is merely the term given to an equation with an unknown of power unity.)

To **'solve an equation'** means 'to find the value of the unknown'. When the unknown is found, its value is said **'to satisfy'** the given equation, or, it is the **'solution of the equation'**. With simple linear equations there is only one value of the unknown which satisfies the equation. (It will be seen later that simultaneous equations and quadratic equations can have more than one value for the unknown). For example, if $x + 2 = 5$, x represents the unknown value. There is only one value of x which is correct for this equation, i.e. x must be equal to 3. Thus, $\boldsymbol{x = 3}$ **is the 'solution of the equation'**. Another way of expressing this fact is to say that $\boldsymbol{x = 3}$ **satisfies the equation** $\boldsymbol{x + 2 = 5}$. There is no other possible value of x which satisfies the given equation.

An **identity** is a relationship which is true for all values of the unknown, whereas an equation is only true for particular values of the unknown. For example, $2x - 5 = 1$ is an equation since it is only true when $x = 3$, whereas $2x \equiv 6x - 4x$ is an identity since it is true for **all** values of x. (Note that '$\equiv$' means 'is identical to').

What is the difference between an algebraic expression and an equation? The answer to this question is shown by the following:

$\dfrac{3x-6}{5x-10}$ is an algebraic expression, but $\dfrac{3x-6}{5x-10} = 3$ is an equation,

i.e. **it contains an 'equals' sign.**

The most important thing when dealing with simple equations is to remember that the equality of a given equation must be maintained. For example, if the algebraic expression to the left of the 'equals' sign is multiplied by 2, then the algebraic expression to the right of the 'equals' sign must also be multiplied by 2. Similarly if 3 is subtracted from the left-hand side (L.H.S.) of the 'equals' sign then 3 must also be subtracted from the right-hand side (R.H.S). **The equality of the equation must be maintained.**

The solution of simple linear equations has many practical applications and some of these are shown later, but firstly, to ensure a firm understanding of the various types of simple linear equations, work through the following graded worked examples in order.

The general objective with each of the following problems is to get the unknown, x, by itself on the left-hand side of the equation.

Worked problems on simple equations

Problem 1. Solve the equation $3x = 15$

Dividing each side of the equation by 3 gives

$$\frac{3x}{3} = \frac{15}{3}$$

(Note, the same operation has been applied to both L.H.S. and R.H.S. of the equation so the equality has been maintained.)

Therefore $x = 5$

Thus $x = 5$ is the solution to the equation.

The equation should always be checked. This is accomplished by substituting the solution into the **original equation**.

Check:

Since L.H.S. $= 3x$

L.H.S. $= (3)(5) = 15$

R.H.S. $= 15$

Hence the solution $x = 5$ is correct. Solutions to simple equations can always be checked in this manner.

Problem 2. Solve the equation $\frac{2x}{3} = 4$

The L.H.S. is a fraction and this can be removed by multiplying both sides of the equation by 3.

Hence $3\left(\frac{2x}{3}\right) = 3(4)$

The 3 in the numerator and the denominator of the L.H.S. will 'cancel'.

Therefore $2x = 12$
Dividing both sides of the equation by 2 gives $x = \mathbf{6}$
Check:

L.H.S. $= \frac{2}{3}(6) = 4$

R.H.S. $= 4$

Hence the solution $x = 6$ is correct.

Problem 3. Solve $x - 4 = 9$

Adding 4 to both sides of the equation gives

$x - 4 + 4 = 9 + 4$

Hence $\quad x = 9 + 4$

Therefore $x = \mathbf{13}$
Check:

L.H.S. $= 13 - 4 = 9$

R.H.S. $= 9$

Hence the solution $x = 13$ is correct.
It is often easier to look upon the above procedure as follows:
To solve $x - 4 = 9$
The '-4' on the L.H.S. of the equation may be moved to the R.H.S. of the equation **but the sign must be changed**. That is, when the '-4' on the L.H.S. is moved across the 'equals' sign to the R.H.S. it becomes a '$+4$'.

Therefore $x - 4 = 9$

becomes $x \quad = 9 + 4$

i.e. $\quad x \quad = \mathbf{13}$

Problem 4. Solve $x + 2 = 7$

As in Problem 3, the equivalent term, '$+2$', on the L.H.S. may be moved across the 'equals' sign to the R.H.S., but the positive sign must be changed to a negative sign.

Hence $\quad x + 2 = 7$

becomes $x \quad = 7 - 2$

i.e. $\quad x \quad = \mathbf{5}$

This, of course, should be checked.

Sometimes terms containing the unknown quantity appear on both sides of an equation.

Problem 5. Solve $7x + 1 = 2x + 6$

In such cases as this, the terms containing x are grouped on one side of the equation and the remaining terms grouped on the other side of the equation. When this is done, changing from one side of an equation to the other must be accompanied by a change of sign.

Hence $7x + 1 = 2x + 6$

Therefore $7x - 2x = 6 - 1$

i.e. $5x = 5$

Therefore $\frac{5x}{5} = \frac{5}{5}$

or $\mathbf{x = 1}$

Check:

L.H.S. $= 7(1) + 1 = 8$

R.H.S. $= 2(1) + 6 = 8$

Hence the solution $x = 1$ is correct.

Problem 6. Solve $3 - 4x = 2x - 15$

In order to keep the term in x positive the unknowns can be moved to the R.H.S.

Hence $3 + 15 = 2x + 4x$

or $18 = 6x$

Therefore $\frac{18}{6} = \frac{6x}{6}$

i.e. $\mathbf{x = 3}$

Check:

L.H.S. $= 3 - 4(3) = -9$

R.H.S. $= 2(3) - 15 = -9$

Hence the solution $x = 3$ is correct.

If the unknown quantities had been grouped on the L.H.S. instead of the R.H.S. the resulting equation would have been solved as follows:

$$3 - 4x = 2x - 15$$

Therefore $-4x - 2x = -15 - 3$

i.e. $-6x = -18$

Therefore $\frac{-6x}{-6} = \frac{-18}{-6}$

$$\mathbf{x = 3}$$

Obviously the same result has been achieved. However, it is suggested that if at all possible, students should try to work with positive values of the unknown, until adept at this kind of work.

Further problems on simple equations may be found in Section 5.3 (Problems 1–21), page 124.

Worked problems on simple equations involving brackets

Problem 7. Solve $4(x-4)=2$

This equation contains a bracket and before anything else this bracket must be removed. Hence

$$4x-16=2$$

Therefore

$$4x=2+16$$
$$4x=18$$
$$\frac{4x}{4}=\frac{18}{4}$$
$$x=\mathbf{4.5}$$

Check:

L.H.S. $=4(4.5-4)=4.(0.5)=2$

R.H.S. $=2$

Hence the solution $x=4.5$ is correct.

Problem 8. Solve $3(2x-6)+2(x+3)=4(x-5)$

Hence $\quad 6x-18+2x+6=4x-20$

Therefore

$$6x+2x-4x=-20+18-6$$
$$4x=-8$$
$$\frac{4x}{4}=\frac{-8}{4}$$
$$x=\mathbf{-2}$$

Check:

L.H.S. $=3(-4-6)+2(-2+3)$

$=3(-10)+2(1)=-30+2=-28$

R.H.S. $=4(-2-5)=4(-7)=-28$

Hence the solution $x=-2$ is correct.

Further problems similar to 7 and 8, on simple equations involving brackets may be found in Section 5.3 (Problems 22–35), page 125.

Worked problems on simple equations involving fractions

Problem 9. Solve $\frac{4}{x} = \frac{2}{7}$

This is achieved by multiplying each term of the equation by the lowest common multiple (L.C.M.) of the denominators. If

$$\frac{4}{x} = \frac{2}{7}$$

the L.C.M. of the denominators, i.e. the lowest algebraic expression that both x and 7 will divide into, is $7x$.

Hence $7x\left[\frac{4}{x}\right] = 7x\left[\frac{2}{7}\right]$

In the L.H.S. the x can be 'cancelled'. In the R.H.S. the 7 can be 'cancelled'.

Hence $7(4) = x(2)$

i.e. $28 = 2x$

or $x = \mathbf{14}$

Check:

L.H.S. $= \frac{4}{14} = \frac{2}{7}$

R.H.S. $= \frac{2}{7}$

Hence the solution $x = 14$ is correct.

Problem 10. Solve $\frac{2x}{5} + \frac{4}{7} = 3 - \frac{2x}{3}$

The L.C.M. of the denominators is 105. Thus

$$105\left(\frac{2x}{5}\right) + 105\left(\frac{4}{7}\right) = 105(3) - 105\left(\frac{2x}{3}\right)$$

Terms may then be 'cancelled' where appropriate. Thus

$$21(2x) + 15(4) = 105(3) - 35(2x)$$

Therefore $42x + 60 = 315 - 70x$

$$42x + 70x = 315 - 60$$

$$112x = 255$$

$$x = \mathbf{\frac{255}{112} = 2\frac{31}{112}}$$

Obviously it is more arithmetically difficult to check such an answer as this, but it should still be done.

Check:

$$\text{L.H.S.} = \frac{2}{5}\left[\frac{255}{112}\right] + \frac{4}{7}$$

$$= \frac{51}{56} + \frac{4}{7} = \frac{51 + 32}{56} = \frac{83}{56}$$

$$\text{R.H.S.} = 3 - \frac{2}{3}\left[\frac{255}{112}\right]$$

$$= 3 - \frac{85}{56} = \frac{168 - 85}{56} = \frac{83}{56}$$

Hence the solution $x = 2\frac{31}{112}$ is correct.

Problem 11. Solve $\frac{4}{x - 2} = \frac{5}{3x + 4}$

The L.C.M. of the denominators is $(x - 2)(3x + 4)$. Thus

$$(x - 2)(3x + 4)\left[\frac{4}{x - 2}\right] = (x - 2)(3x + 4)\left[\frac{5}{3x + 4}\right]$$

The next step is to cancel where appropriate. Hence

$$\cancel{(x - 2)}(3x + 4)\left[\frac{4}{\cancel{x - 2}}\right] = (x - 2)\cancel{(3x + 4)}\left[\frac{5}{\cancel{3x + 4}}\right]$$

Therefore $(3x + 4)4 = (x - 2)5$

$$12x + 16 = 5x - 10$$

$$12x - 5x = -10 - 16$$

$$7x = -26$$

$$x = \frac{-26}{7}$$

$$x = -3\tfrac{5}{7}$$

Check:

$$\text{L.H.S.} = \frac{4}{-3\frac{5}{7} - 2} = \frac{4}{-5\frac{5}{7}} = \frac{4}{-\frac{40}{7}} = -\frac{4(7)}{40}$$

$$= -\frac{28}{40} = -\frac{7}{10}$$

$$\text{R.H.S.} = \frac{5}{3\left[\frac{-26}{7}\right] + 4} = \frac{5}{\frac{-78}{7} + 4} = \frac{5}{\frac{-78 + 28}{7}}$$

$$= \frac{5}{\frac{-50}{7}} = \frac{-5(7)}{50} = \frac{-35}{50} = \frac{-7}{10}$$

Hence the solution $x = -3\frac{5}{7}$ is correct.

Further problems similar to 9 to 11, on simple equations involving fractions may be found in Section 5.3 (Problems 36–70), page 125.

Worked problems involving squares and square roots

Equations involving squares and square roots are not 'simple' equations as such; however, for completeness, they have been included here since they occur so often.

Problem 12. Solve $\sqrt{x} = 5$

Whenever square root signs are involved with the unknown, both sides of the equation must be squared.

Hence $(\sqrt{x})^2 = (5)^2$

Therefore $x = \mathbf{25}$

Check:

L.H.S. $= \sqrt{25} = 5$

R.H.S. $= 5$

Hence the solution $x = 25$ is correct.

Problem 13. Solve $2\sqrt{x} = 6$

It is often easier to arrange the equation so that the term involving the square root of the unknown is on its own.

Hence $\dfrac{2\sqrt{x}}{2} = \dfrac{6}{2}$

Therefore $\sqrt{x} = 3$

Squaring both sides gives $x = \mathbf{9}$

Check:

L.H.S. $= 2\sqrt{9} = 2(3) = 6$

R.H.S. $= 6$

Hence the solution $x = 9$ is correct.

Problem 14. Solve $\dfrac{\sqrt{x} + 1}{\sqrt{x}} = 4$

To remove the fraction, each term must be multiplied by $\sqrt{x}$.

Hence $\sqrt{x}\left[\dfrac{\sqrt{x}+1}{\sqrt{x}}\right] = \sqrt{x}(4)$

Therefore

$$\begin{aligned}
\sqrt{x} + 1 &= 4\sqrt{x} \\
1 &= 4\sqrt{x} - \sqrt{x} \\
1 &= 3\sqrt{x} \\
\frac{1}{3} &= \frac{3\sqrt{x}}{3} \\
\frac{1}{3} &= \sqrt{x}
\end{aligned}$$

Squaring both sides gives $\left[\dfrac{1}{3}\right]^2 = x$

Therefore $x = \mathbf{\dfrac{1}{9}}$

Check:

$$\text{L.H.S.} = \frac{\sqrt{\left[\frac{1}{9}\right]} + 1}{\sqrt{\left[\frac{1}{9}\right]}} = \frac{\frac{1}{3} + 1}{\frac{1}{3}} = \frac{\frac{4}{3}}{\frac{1}{3}} = 4$$

R.H.S. $= 4$

Hence the solution $x = \dfrac{1}{9}$ is correct.

Problem 15. Solve $x^2 = 16$

Whenever a square of the unknown is involved, the square root of both sides is taken

Hence $\sqrt{x^2} = \sqrt{16}$

Therefore $x = 4$

However, $x = -4$ is also a solution of the equation because $(-4) \times (-4) = +16$. Therefore it should be noted that whenever the square root of a number is found there are always **two** answers, one positive and the other negative.

The solution of this problem is written as $\boldsymbol{x = \pm 4}$

Problem 16. Solve $\dfrac{2}{9} = \dfrac{16}{3x^2}$

The L.C.M. of the denominator is $9x^2$.

Hence $9x^2\left[\dfrac{2}{9}\right] = 9x^2\left[\dfrac{16}{3x^2}\right]$

Therefore
$$2x^2 = 48$$
$$x^2 = 24$$
$$x = \sqrt{24}$$
$$\boldsymbol{x = \pm 4.899}$$

Check:

L.H.S. $= \dfrac{2}{9}$

R.H.S. $= \dfrac{16}{3(24)} = \dfrac{2}{3(3)} = \dfrac{2}{9}$

Hence the solution $x = \pm 4.899$ is correct.

Further problems similar to 12 to 16, on equations involving squares and square roots, may be found in Section 5.3 (Problems 71–81), page 127.

Summary of important points in the solution of simple equations

1. The equality of an equation must be maintained at every step.
2. Fractions are normally undesirable and are therefore removed. This is accomplished by multiplying each term by the lowest common multiple of the denominators.
3. Brackets are normally undesirable and are therefore removed at the earliest opportunity by simply 'multiplying out'.
4. The unknown terms must be grouped on one side of the equation and all other terms on the other side.

5. When the square root of the unknown is involved it is best to isolate that term on one side of the equation, and then square both sides.
6. When the square of the unknown is involved it is best to isolate that term on one side of the equation, and then take the square root of both sides. When taking square roots there are always **two** answers, one positive and the other negative.
7. All solutions must be checked.
8. Do not try to take more than one step at a time.

5.2 Practical problems

There are many different practical applications where the solutions of simple linear equations are needed. Various worked problems are shown below.

From now on it will be assumed that you will check all solutions by substitution into the original equation.

Worked problems on practical applications of simple equations

Problem 1. A copper wire of length, $l = 2$ km has a resistance of $4\ \Omega$ and a resistivity, ρ, of $17.2 \times 10^{-6}\ \Omega$ mm. Find the cross-sectional area, a, of the wire given that

$$R = \frac{\rho l}{a}$$

$$\text{Therefore } 4 = \frac{(17.2 \times 10^{-6})(2\,000 \times 10^{3})}{a}$$

$$= \frac{34.4}{a}$$

$$4a = 34.4$$

$$a = \frac{34.4}{4} = 8.6 \text{ mm}^2$$

Hence the cross-sectional area of the wire, a, is **8.6 mm²**

Problem 2. The temperature coefficient of resistance, α, may be calculated from the formula $R_t = R_0(1 + \alpha t)$. Find α, given $R_t = 0.726$, $R_0 = 0.6$, and $t = 50$

As $R_t = R_0(1 + \alpha t)$ then

$$0.726 = 0.6(1 + \alpha(50))$$

$$0.726 = 0.6 + \alpha(50)(0.6)$$

$$0.726 - 0.6 = 30\alpha$$
$$0.126 = 30\alpha$$
$$\alpha = \frac{0.126}{30}$$
$$\alpha = \mathbf{0.0042}$$

Problem 3. When three resistors in an electrical circuit are connected in parallel the total resistance, R_T, is found from the formula

$$\frac{1}{R_T} = \frac{1}{R_1} + \frac{1}{R_2} + \frac{1}{R_3}$$

Find the total resistance when resistors of 4 Ω, 6 Ω and 12 Ω are connected in parallel.

Therefore $\frac{1}{R_T} = \frac{1}{4} + \frac{1}{6} + \frac{1}{12}$

The L.C.M. of the denominators is $12R_T$

Therefore $12R_T\left[\frac{1}{R_T}\right] = 12R_T\left[\frac{1}{4}\right] + 12R_T\left[\frac{1}{6}\right] + 12R_T\left[\frac{1}{12}\right]$

$$12 = 3R_T + 2R_T + R_T$$
$$12 = 6R_T$$
$$\frac{12}{6} = \frac{6R_T}{6}$$
$$\mathbf{R_T = 2\ \Omega}$$

Alternatively, the fractions on the R.H.S. could be added first, as shown below:

$$\frac{1}{R_T} = \frac{1}{4} + \frac{1}{6} + \frac{1}{12} = \frac{3+2+1}{12}$$

$$\frac{1}{R_T} = \frac{6}{12} = \frac{1}{2}$$

Taking the reciprocal of both sides gives $R_T = 2\ \Omega$

Problem 4. A formula used to calculate power, P, in a d.c. electrical circuit is $P = \frac{V^2}{R}$ where V is the supply voltage and R is the resistance of the circuit. If the resistance is 6.25 Ω and the power measured is 100 watts find the supply voltage.

$$P = \frac{V^2}{R}$$

Therefore $100 = \dfrac{V^2}{6.25}$

$$6.25(100) = 6.25\left[\frac{V^2}{6.25}\right]$$

$$625 = V^2$$

$$V = \sqrt{625}$$

$$V = \mathbf{\pm 25\ volts}$$

Problem 5. A rectangular box with square ends has its length 10 cm greater than its breadth and the total length of its edges is 1.52 m. Find the width of the box.

Let x cm = breadth = width of the box. The box is shown in Fig. 5.1.

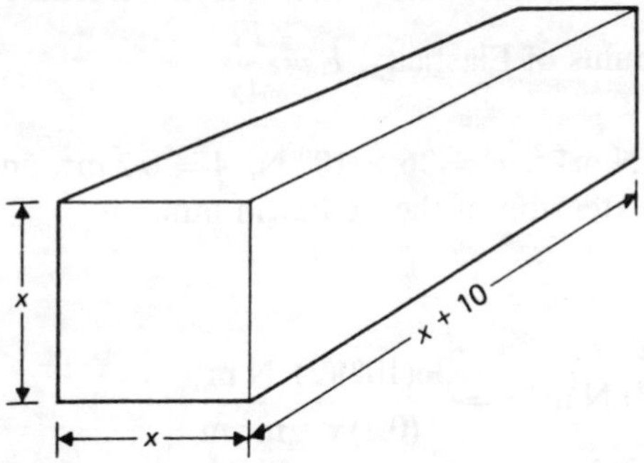

Fig. 5.1

Then the length of the box is $(x + 10)$ cm and the lengths of the edges are $2(4x) + 4(x + 10)$ cm.

Therefore $152 \text{ cm} = (8x + 4x + 40) \text{ cm}$

$$152 - 40 = 12x$$

$$112 = 12x$$

$$x = 9\tfrac{1}{3} \text{ cm}$$

Hence the width of the box = $\mathbf{9\tfrac{1}{3}}$ **cm**

Problem 6. A number of people agree to pay equally for the use of a hire-tool, and each pays £15. If there had been two more people in the party each would have paid £10. Find how many people there were in the party and the cost of the hire-tool.

Let the number of people in the party be x. Therefore the cost of the hire-tool for x people at £15 each is £$15x$.

Two more people in the party would make $(x + 2)$ people and it would now cost £10 each. So the cost of hire-tool for $(x + 2)$ people at £10 each is £10 $(x + 2)$.

Therefore $15x = 10(x + 2)$

$$15x = 10x + 20$$
$$15x - 10x = 20$$
$$5x = 20$$
$$x = \frac{20}{5} = 4$$

Hence there are **4 people in the party.**
Cost of hire-tool = **£15x = £15(4) = £60**

Problem 7. The extension, x m, of a mild steel tie bar of length l m and cross-sectional area A m^2 when carrying a load F N is given by the Modulus of Elasticity, $E = \frac{Fl}{Ax}$

If $E = 200 \times 10^9$ N m^{-2}, $F = 36 \times 10^6$ N, $A = 0.2$ m^2, and $l = 2$ m, find the extension of the tie bar in mm.

$$E = \frac{Fl}{Ax}$$

Therefore $200(10^9)\text{ N m}^{-2} = \frac{36(10^6)(2)}{(0.2)x} \frac{\text{N m}}{\text{m}^2\text{ m}}$

(Note: the unit x is in metres.)
Therefore $200(10^9)(0.2)x = 36(10^6)(2)$

$$x = \frac{36(10^6)(2)}{200(10^9)0.2}\text{ m}$$
$$x = 0.001\,8\text{ m}$$
$$x = (0.001\,8)(1\,000)\text{ mm}$$
$$x = 1.8\text{ mm}$$

Hence the extension of the tie bar is **1.8 mm**

Problem 8. A formula for distance, s metres, travelled in time t seconds is given by $s = ut + \frac{1}{2}at^2$ where u is the initial velocity in m/s and a is the acceleration in m/s^2. Find the acceleration of a body if it travels 150 m in 5 s with an initial velocity of 5 m/s

$s = ut + \frac{1}{2}at^2$
$s = 150$, $u = 5$, and $t = 5$

$$\text{Therefore}\qquad 150 = (5)(5) + \tfrac{1}{2}a(5)^2$$

$$150 = 25 + \frac{25}{2}a$$

$$150 - 25 = \frac{25}{2}a$$

$$125 = \frac{25}{2}a$$

$$125(2) = 25a$$

$$250 = 25a$$

$$\frac{250}{25} = a$$

$$a = 10$$

Hence the acceleration of the body is **10** m/s^2

Problem 9. A formula relating initial and final states of pressures p, volumes, V, and absolute temperatures, T, of an ideal gas, is

$$\frac{p_1V_1}{T_1} = \frac{p_2V_2}{T_2}$$

If $p_1 = 2\,063 \times 10^3$, $p_2 = 101.3 \times 10^3$, $T_1 = 323$, $T_2 = 273$ and $V_1 = 0.5$, find V_2.

$$\text{Therefore}\quad \frac{2\,063(10^3)0.5}{323} = \frac{101.3(10^3)V_2}{273}$$

$$(273)(323)\left[\frac{2\,063(10^3)0.5}{323}\right] = (273)(323)\left[\frac{101.3 \times 10^3}{273}\right]V_2$$

Cancelling gives: $273(2\,063)10^3(0.5) = 323(101.3)10^3\,V_2$

Dividing both sides by $(323)(101.3)10^3$ gives

$$\frac{(273)(2\,063)10^3(0.5)}{(323)(101.3)(10^3)} = \frac{(323)(101.3)(10^3)V_2}{(323)(101.3)10^3}$$

$$V_2 = \frac{273(2\,063)10^3 0.5}{323(101.3)10^3}$$

Hence $V_2 =$ **8.6**

Problem 10. A welder is paid £6 per hour for a basic 40 hour week, but overtime is paid for at one-and-a-half times this rate. Find how many hours he has to work in a week to earn £321

Basic rate per hour = £6

Rate for overtime per hour = 1.5(£6) = £9

Let the number of overtime hours worked = x
Therefore $40(6) + x(9) = 321$

$$240 + 9x = 321$$
$$9x = 321 - 240$$
$$9x = 81$$
$$x = \frac{81}{9} = 9$$

Thus 9 hours overtime would have to be worked to earn £321 per week. Hence total number of hours worked is **49**.

Problem 11. When two resistors R_1 and R_2 are connected in parallel the voltage across R_2 may be found from

$$V_2 = \left[\frac{R_1 R_2}{R_1 + R_2}\right] I$$

where I is the total circuit current. Find R_2 given $I = 14A$, $R_1 = 5\ \Omega$ and $V_2 = 20$ V.

$$V_2 = \left[\frac{R_1 R_2}{R_1 + R_2}\right] I$$

Therefore $$20 = \left[\frac{5R_2}{5 + R_2}\right] 14$$

$$20(5 + R_2) = (5R_2)14$$
$$100 + 20R_2 = 70R_2$$
$$100 = 70R_2 - 20R_2$$
$$100 = 50R_2$$
$$R_2 = \frac{100}{50} = 2\ \Omega$$

Hence $R_2 = 2\ \Omega$

Further problems similar to 1 to 11 on the practical type of simple equations may be found in the following Section (5.3) (Problems 82–110), page 128.

5.3 Further problems

Simple equations

Solve the following equations:

1. $a + 3 = 5$ [2]
2. $x - 2 = 9$ [11]
3. $3s = 12$ [4]

4. $7 - q = 4$ [3]
5. $2t + 2 = 1$ [$-\frac{1}{2}$]
6. $\frac{r}{3} = 5$ [15]
7. $3x + 4 = 8$ [$\frac{4}{3}$]
8. $7p + 12 = 5$ [-1]
9. $3b - 4b + 2b = 7$ [7]
10. $14 - 3c = 8$ [2]
11. $3u - 2 = 7u - 6$ [1]
12. $6m + 11 = 25 - m$ [2]
13. $3p - 18 = 8p + 22$ [-8]
14. $0.4a = 1.6$ [4]
15. $2.6x - 0.4 = 0.9x + 1.3$ [1]
16. $2d - 1 - 3d = 4d - 21$ [4]
17. $2a - 4 = 3a - 2 - 5a$ [$\frac{1}{2}$]
18. $3x + 1 - 4x = 0$ [1]
19. $0 = 2 + 4b$ [$-\frac{1}{2}$]
20. $2r + 5 - 3r - 2 = 0$ [3]
21. $18x - 5 + x = 9x + 2 - 7$ [0]

Equations involving brackets

22. $3(x - 1) = 6$ [3]
23. $4(x + 2) = 14$ [$\frac{3}{2}$]
24. $3(e - 1) - 4(2e + 3) = 15$ [-6]
25. $5(g + 2) = 29 + 3(g - 5)$ [2]
26. $5(h - 3) = 2h$ [5]
27. $4(3 - 2x) - 5 = -3(x - 2)$ [$\frac{1}{5}$]
28. $3(2y - 1) - 2 = 17$ [$3\frac{2}{3}$]
29. $4(3x - 4) - (x + 1) = 10$ [$2\frac{5}{11}$]
30. $0 = 2(3a - 5) - 5$ [$2\frac{1}{2}$]
31. $0 = 2(b - 2) - 3(3b + 1)$ [-1]
32. $5(2x + 3) = 6(x + 2) - 2(x - 4)$ [$\frac{5}{6}$]
33. $20 + 8(3c - 2) = (c - 1) + 3(2c - 4)$ [-1]
34. $16 - (p + 2) = 3(p - 7) + 10$ [$6\frac{1}{4}$]
35. $8 + 5(d - 1) - 6(d - 3) = 3(4 - d)$ [$-4\frac{1}{2}$]

Equations involving fractions

36. $\frac{1}{3}a + 2 = 5$ [9]
37. $\frac{3}{4}x + 2 = \frac{1}{3}x + \frac{1}{2}$ [$-3\frac{3}{5}$]
38. $\frac{1}{3}(2a - 1) + 4 = \frac{1}{2}$ [$-4\frac{3}{4}$]
39. $3 - \frac{1}{3}(2y + 1) = \frac{3}{4}$ [$2\frac{7}{8}$]

40. $\frac{1}{2}(2f-3)+\frac{1}{3}(f-4)=0$ $[2\frac{1}{8}]$
41. $0=\frac{2}{3}(2m-1)-\frac{3}{4}(m+5)$ $[7\frac{4}{7}]$
42. $\frac{1}{3}(3a-2)-\frac{1}{4}(5a+1)+\frac{1}{5}(2a-7)=0$ $[15\frac{4}{9}]$
43. $-\frac{1}{9}(3x-4)+\frac{1}{5}(4x-1)=\frac{1}{15}-\frac{1}{3}(x+5)$ $[-2\frac{11}{36}]$
44. $\dfrac{t}{3}-\dfrac{t}{5}=2$ $[15]$
45. $\dfrac{V}{5}+\dfrac{V}{3}=\dfrac{47}{30}-\dfrac{V}{4}$ $[2]$
46. $\dfrac{m}{3}+3-\dfrac{m}{6}=1-\dfrac{m}{3}$ $[-4]$
47. $\dfrac{5}{x}=10$ $[\frac{1}{2}]$
48. $2t+\frac{1}{2}=\frac{4}{5}t-\frac{3}{10}$ $[-\frac{2}{3}]$
49. $\dfrac{3}{w}=\dfrac{9}{5}$ $[1\frac{2}{3}]$
50. $\dfrac{5x}{9}-\dfrac{2x}{5}=\dfrac{1}{15}$ $[\frac{3}{7}]$
51. $\dfrac{1}{3a}+\dfrac{1}{4a}=\dfrac{3}{20}$ $[3\frac{8}{9}]$
52. $\dfrac{c+3}{4}=2+\dfrac{c-3}{5}$ $[13]$
53. $\dfrac{3}{2}+\dfrac{3y}{20}=\dfrac{-(y-6)}{12}+\dfrac{2y}{15}$ $[-10]$
54. $\dfrac{5-d}{4}=\dfrac{d}{5}+\dfrac{7}{20}$ $[2]$
55. $4-a=\dfrac{2a+6}{5}$ $[2]$
56. $\dfrac{f-2}{f-3}=4$ $[3\frac{1}{3}]$
57. $\dfrac{3}{y-2}=\dfrac{4}{y+4}$ $[20]$
58. $\dfrac{2}{a-5}=\dfrac{3}{a-1}$ $[13]$
59. $\dfrac{4}{3b+2}=\dfrac{7}{5(b-3)}$ $[-74]$
60. $\dfrac{3}{4}-\dfrac{1}{2a-1}=\frac{1}{2}$ $[2\frac{1}{2}]$

61. $\dfrac{1}{3x-2}+\dfrac{1}{5x+3}=0$ $[-\frac{1}{8}]$

62. $\dfrac{2b}{2b-3}=\dfrac{3b-1}{3b+2}$ $[\frac{1}{5}]$

63. $\dfrac{1}{3(y-2)}-\dfrac{2}{5(y+3)}=0$ [27]

64. $\dfrac{3x+23}{3x+12}=\dfrac{4}{3}$ [7]

65. $(c-3)(2c+6)=2c(c-18)$ $[\frac{1}{2}]$

66. $\dfrac{1-a}{a+1}=9$ $[-\frac{4}{5}]$

67. $\dfrac{x-5}{x-2}=\frac{3}{4}$ [14]

68. $\dfrac{a}{4}-\dfrac{a+3}{2}=\dfrac{a+6}{5}$ [−6]

69. $\dfrac{x}{6}-\dfrac{3(x+4)}{2}=2(x-3)$ [0]

70. $\dfrac{2f+1}{3}-2=\dfrac{2f-3}{4}-\dfrac{1-f}{5}$ $[-21\frac{1}{2}]$

Equations involving squares and square roots

71. $4\sqrt{a}=12$ [9]
72. $2\sqrt{x}=\frac{5}{3}$ $[\frac{25}{36}]$

73. $6=\sqrt{\left(\dfrac{x}{2}\right)}-1$ [98]

74. $4=\sqrt{\dfrac{3}{x}}+2$ $[\frac{3}{4}]$

75. $\dfrac{2\sqrt{b}}{1-\sqrt{b}}=4$ $[\frac{4}{9}]$

76. $8=4\sqrt{\left(\dfrac{a}{2}-1\right)}$ [10]

77. $24=\dfrac{6}{a^2}$ $[\pm\frac{1}{2}]$

78. $\dfrac{4}{3}=\sqrt{\left[\dfrac{x+2}{x-2}\right]}$ $[7\frac{1}{7}]$

79. $16 = \frac{e^2}{9}$ [±12]

80. $\frac{8}{x} = \frac{2x}{25}$ [±10]

81. $\frac{9}{2} = 4 + \frac{2}{x^2}$ [±2]

Practical problems

82. A formula used for calculating resistance of an electrical cable is $R = \frac{\rho l}{a}$. Given $R = 0.112$, $l = 100$, $a = 25$, find the value of ρ. [0.028]

83. Ohm's Law states that $V = IR$. Find R when $I = 0.5$ and $V = 200$. [400]

84. If force, F N, mass, m kg and acceleration, a m/s^2 are linked by the formula $F = ma$ find the acceleration when a force of 2 kN is applied to a mass of 1 000 kg. [2 m/s^2]

85. Kinetic energy $= \frac{1}{2}mv^2$ where m is the mass in kg and v is the velocity in m/s. Find the mass of a body projected vertically upwards with velocity 20 m/s if the kinetic energy is 1 000 joules. [5 kg]

86. Two variables x and y are related by $y = ax + \frac{b}{x}$, where a and b are constants. If $y = 13$, $x = 3$ and $a = 3\frac{1}{3}$ find b. [9]

87. If $PV = mRT$ find R given $P = 100$, $T = 298$, $V = 2.00$ and $m = 0.322$ [2.084]

88. When 3 resistors, R_1, R_2 and R_3 in an electrical circuit are connected in parallel, the total resistance, R_T, is found from the formula

$$\frac{1}{R_T} = \frac{1}{R_1} + \frac{1}{R_2} + \frac{1}{R_3}$$

(*a*) Find the total resistance when $R_1 = 5$, $R_2 = 10$, and $R_3 = 30$ [3]

(*b*) Find R_2 given $R_T = 1$, $R_1 = 2$ and $R_3 = 6$ [3]

89. If $\frac{1}{R_T} = \frac{1}{R_1} + \frac{1}{R_2} + \frac{1}{R_3} + \frac{1}{R_4}$ find R_4 given $R_1 = 5$, $R_2 = 4.5$, $R_3 = 3.7$ and $R_T = 1.12$. [5]

90. Given that $\frac{p}{p+1} = \frac{p+1}{p+5}$ find the numerical value of $\frac{2p+1}{2p+6}$ [$\frac{1}{4}$]

91. Four sparking plugs and a fan belt cost £10.50. If the fan belt costs £1.75 more than a sparking plug find the cost of each component. [£1.75, £3.50]

92. If $R = \dfrac{R_1 R_2}{R_1 + R_2}$ find R_2 given that $R_1 = 3$ and $R = 2$ [6]

93. Ohm's Law can be represented by $I = \dfrac{V}{R}$ where I is the current in amperes, V is the voltage in volts and R is the resistance in ohms.

(*a*) A 25 ohm resistor is connected across a voltage of 12.5 V. Find the current flowing. [0.5 A]

(*b*) A current of 5 amperes is flowing in a circuit of 50 ohms. What is the voltage? [250 V]

(*c*) An electric soldering iron takes a current of 0.25 A at 240 V. Find the resistance of the element. [960 ohms]

94. If $R_2 = R_1(1 + \alpha t)$ find α given $R_1 = 3.5$, $R_2 = 3.65$ and $t = 42.8$ [0.001]

95. If $\dfrac{P_1 V_1}{T_1} = \dfrac{P_2 V_2}{T_2}$ find T_2 given that $P_1 = 2\,000 \times 10^3$, $V_1 = 5$, $T_1 = 300$, $P_2 = 10^6$, and $V_2 = 15$ [450]

96. If $s = ut + \frac{1}{2}ft^2$ find u given that $s = 500$, $t = 4$ and $f = -15$ [155]

97. A rectangle has a length of 15 cm and a width w cm. When its width is reduced by 2 cm its area becomes 45 cm^2. Find the original width and area of the rectangle. [5 cm, 75 cm^2]

98. If $v^2 = u^2 + 2as$ find u given $v = 25$, $a = -50$ and $s = 2.75$ [30]

99. A rectangular box with square ends has its length 3 m greater than its breadth and the total length of its edges is 21 m. Find the width of the box. [$\frac{3}{4}$ m]

100. If $I = \dfrac{V - E}{R + r}$ find r given $I = 0.5$, $R = 3$, $V = 10$ and $E = 8$ [1]

101. Two joiners and four mates earn £592 between them for fitting a kitchen. If a joiner earns £38 more than a mate find their wages for the job. [£124, £86]

102. If $t = 2\pi\sqrt{\dfrac{w}{Sg}}$ find the value of S given $t = 0.319\,6$, $w = 4$ and $g = 32.2$ [48]

103. If the area of a circle is given by $A = \pi r^2$ find r when the area is 15.0 cm^2. [2.185 cm]

104. A rectangular metal plate is 20 cm long. A strip 4 cm wide is cut from one end and a second strip of 2 cm is cut from the other. The remainder weighs 168 g. Find the width of the plate if 1 cm^2 of it weighs 1.5 g. [8 cm]

105. Evaluate R if $\dfrac{1}{R-1} = \dfrac{3}{1-t} - \dfrac{4}{1+t}$ and $t = 0.6$ [1.2]

106. The relation between the temperature on a Fahrenheit thermometer and that on a Celsius thermometer is expressed by the formula $F = \frac{9}{5}C + 32$. Express 27.5°C in Fahrenheit degrees. [81.5F]

107. The stress f in the material of a thick cylinder is given by the formula $\dfrac{D}{d} = \sqrt{\left(\dfrac{f+p}{f-p}\right)}$. Calculate the stress given $D = 19.5$, $d = 9.75$ and $p = 1\,500$ [2 500]

108. The perimeter of a rectangle is 14 cm. The length of a diagonal is given by $\sqrt{(w^2 + L^2)}$ where w and L are the width and length of the rectangle respectively. If the length is 1 cm greater than the width find the dimensions of the rectangle and the length of the diagonal.
[3 cm by 4 cm; 5 cm]

109. Sorel cement contains 36.5% magnesium chloride, 19.5% magnesium oxide and 44% water by mass. How much of the magnesium compounds is contained in a cement containing 48 g water? [magnesium compounds = 61.1 g]

110. An alloy contains 70% by weight of copper, the remainder being zinc. How much copper must be mixed with 40 kg of this alloy to give an alloy containing 80% copper? [20 kg]

Chapter 6

Simultaneous equations

6.1 Methods of solving simultaneous equations

It often happens that the solution of a problem involves the use of more than one unknown.

Consider the equation $2x + 3y = 11$. There are two unknowns involved, namely x and y.

If $2x + 3y = 11$
then $2x = 11 - 3y$
Therefore $x = \dfrac{11 - 3y}{2}$

This does not give the actual numerical value of x, but a result which involves the other unknown y.

If a value of y is known then a corresponding value of x can be found.

For example, if $y = 1$, $x = \dfrac{11 - 3(1)}{2}$

$$x = \frac{11 - 3}{2} = \frac{8}{2} = 4$$

Similarly, if $y = 2$, $x = \dfrac{11 - 3(2)}{2}$

$$= \frac{11 - 6}{2} = \frac{5}{2} = 2\tfrac{1}{2}$$

and if $y = 3$, $x = \dfrac{11 - 3(3)}{2}$

$$= \frac{11 - 9}{2} = \frac{2}{2} = 1$$

This process can be continued indefinitely, so that for every value of y there is a corresponding value of x, and there is an endless number of possibilities. If we need, for some reason, only **one** pair of values for x and y then some further information must be available in order that the correct pair can be found.

In many engineering/science applications there exist two or more unknown quantities, each of which has only one value: the value of one unknown depends on the value of the other unknown. In such cases it is found that one equation is insufficient to determine the value of each of the unknowns. If, as in the example $2x + 3y = 11$, there are two quantities x and y having one value each, then another equation is needed in order to find this unique pair of values.

Similarly, if a problem contains three quantities, each of which has only one value, then three equations are needed in order to evaluate the three quantities. If as stated

$$2x + 3y = 11 \qquad \text{[equation (1)]}$$

and **simultaneously**, the same quantities x and y also satisfy the equation:

$$4x + 2y = 10, \qquad \text{[equation (2)]}$$

then there is only one value of x and one value of y which satisfies both equations. Such equations are called **simultaneous equations**, the word 'simultaneous' meaning 'existing or taking place at the same time'.

Here we will deal only with simultaneous linear equations in two unknowns.

There are two methods of solving such equations.

Method 1: By substitution

Let $2x + 3y = 11$ (1)

and $4x + 2y = 10$ (2)

Then from equation (1), $x = \dfrac{11 - 3y}{2}$

Let this expression for x be substituted into equation (2). Thus

$$4\left[\frac{11 - 3y}{2}\right] + 2y = 10$$

This is now a simple equation in y and may be solved as shown in Chapter 5. Multiplying both sides of the equation by 2 gives

$$4(11 - 3y) + 4y = 20$$

Removing brackets gives

$$44 - 12y + 4y = 20$$

Rearranging gives

$$44 - 20 = 12y - 4y$$

$$24 = 8y$$

Hence $y = 3$

This value of y may be substituted into either equation (1) or equation (2). (The result should be the same in both cases.)

Substituting in equation (1) gives

$$2x + 3(3) = 11$$

Therefore $2x + 9 = 11$

$$2x = 11 - 9$$

$$2x = 2$$

Hence $\mathbf{x = 1}$

A check may be obtained by substituting in equation (2) for x and y. Thus

$$4(1) + 2(3) = 10$$

$$4 + 6 = 10$$

L.H.S. = R.H.S.

Therefore **the solution of the simultaneous equations $2x + 3y = 11$ and $4x + 2y = 10$ is $x = 1$ and $y = 3$.** This is the only pair of values that satisfies both equations.

Method 2: By elimination

Again

let $2x + 3y = 11$ (1)

and $4x + 2y = 10$ (2)

If equation (1) is multiplied throughout by 2 the resulting equation will be

$$4x + 6y = 22 \qquad (3)$$

The reason that equation (1) is multiplied by 2 is that the coefficient of x (i.e. the number multiplying x) in equation (2) and equation (3) is now the same. Sometimes it is necessary to multiply both equations by constants chosen so that the coefficients of x or y in each equation become the same.

Equation (2) can now be subtracted from equation (3). Thus

$$4x + 6y = 22 \quad (3)$$
$$4x + 2y = 10 \quad (2)$$

Subtracting $0 + 4y = 12$

Hence $4y = 12$

Therefore $y = 3$

This value of y may now be substituted in equation (1) or equation (2) exactly as in method 1 to find the value of x.

It will be found from experience that in many cases method 1, that of substitution, is unnecessarily cumbersome, so that method 2, the elimination procedure will be employed.

6.2 Procedure for solving any simultaneous linear equation in two unknowns by the method of elimination

The procedure will be outlined using the following problem.

Problem. Solve the simultaneous equations $3x + 4y = 5$ and $12 + 2x - 5y = 0$

1. The two equations should be written down with similar unknown terms under each other. It is usual to put the two unknown terms on the left-hand side of the equation with the constant term on the right-hand side.
   Thus $3x + 4y = 5$
   and $2x - 5y = -12$
2. The two equations are numbered. This merely makes later reference to particular equations easier. Thus
   $$3x + 4y = 5 \quad (1)$$
   $$2x - 5y = -12 \quad (2)$$
3. The equations must be multiplied in such a way that the coefficient of one unknown becomes the same in both equations, i.e. either the coefficients of x, or the coefficients of y, have to be made equal. The choice is arbitrary: it is made with convenience and ease as the prime considerations.

   If, in this particular problem, the x coefficients are to be made equal, then equation (1) would have to be multiplied by 2 and equation (2) would have to be multiplied by 3. Thus, multiplying equation (1) by 2 gives
   $$6x + 8y = 10$$
   and multiplying equation (2) by 3 gives:
   $$6x - 15y = -36$$

4. Any newly-formed equations should be numbered. Thus

$$6x + 8y = 10 \qquad (3)$$

$$6x - 15y = -36 \qquad (4)$$

5. By adding or subtracting newly-formed equations one unknown may be entirely eliminated, leaving a simple equation containing the other unknown.

If the sign of the common coefficients is the **same** in both equations (i.e. both positive or both negative) then one equation is **subtracted** from the other.

If the sign of the common coefficients is **different** in each equation (i.e. one coefficient is positive and the other negative) then the two equations are **added**.

In equations (3) and (4) the coefficients and sign of the x term are the same in both, i.e. $+6$. Thus equation (4) is subtracted from equation (3)

$$6x + 8y = 10 \qquad (3)$$

$$6x - 15y = -36 \qquad (4)$$

Subtracting $0 + 23y = 46$

It is very important to note that

$8y$ take away $-15y$ is the same as
$8y - -15y$, which is the same as
$8y + 15y = +23y$

Similarly 10 take away -36 is the same as
$10 - -36$, which is the same as
$10 + 36 = 46$

If mistakes are going to be made in solving simultaneous equations it is usually at this stage, where the rules of arithmetic must be strictly adhered to.

6. The new equation formed by the above subtraction contains one unknown only, i.e. y. It is therefore necessary to solve the simple equation

$$23y = 46$$

Dividing both sides of the equation by 23 gives

$$\mathbf{y = 2}$$

7. The solution $y = 2$ is substituted into either of the **original** equations, i.e. equation (1) or equation (2), which is then solved for the unknown, x. Once again the choice is arbitrary.

Substituting $y = 2$ in equation (1) gives

$$3x + 4(2) = 5$$

Therefore $3x + 8 = 5$

$$3x = 5 - 8$$

$$3x = -3$$

$$x = -1$$

8. It is essential to check that the pair of values found (i.e. $x = -1$ and $y = 2$) satisfy **both** of the **original** equations.

   It is important to note that stages 7 and 8, the substitution and checking stages, must be done in the original and not the derived equations.

   From above, $x = -1$ and $y = 2$ satisfies equation (1).

   Checking in equation (2)

   $$\text{L.H.S.} = 2x - 5y = 2(-1) - 5(2) = -2 - 10 = -12$$

   $$\text{R.H.S.} = -12$$

   Thus the solution of the simultaneous equations $3x + 4y = 5$ and $12 + 2x - 5y = 0$ is $x = -1$ and $y = 2$. This is the **only** pair of values that satisfies **both** equations.

In Chapter 8 it will be shown that equations such as the two used in the above problem are straight lines when plotted graphically. The above solution of the two equations represents the point where the two lines intersect, and two straight lines can intersect at only one point.

Worked problems on solving simultaneous equations

Problem 1. Solve $8x - 3y = 39$ (1)

$7x + 5y = -4$ (2)

When equation (1) is multiplied by 5 and equation (2) is multiplied by 3 the coefficient of y in each equation is numerically the same, i.e. 15. However, in this case, they are of opposite sign. Thus, equation (1) multiplied by 5 gives

$40x - 15y = 195$ (3)

and equation (2) multiplied by 3 gives

$21x + 15y = -12$ (4)

To eliminate the unknown y, equation (3) is added to equation (4). Thus

$61x = 183$

Dividing both sides of the equation by 61 gives $\boldsymbol{x = 3}$
Substituting in equation (1) gives

$$8(3) - 3y = 39$$

Therefore $24 - 3y = 39$

$$-3y = 39 - 24$$

$$-3y = 15$$

Dividing both sides by -3 gives

$$y = \frac{15}{-3}$$

$$\boldsymbol{y = -5}$$

Checking in equation (2)

L.H.S. $= 7(3) + 5(-5)$

$= 21 - 25$

$= -4 =$ R.H.S.

Thus the solution of the given equations is $\boldsymbol{x = 3}$ **and** $\boldsymbol{y = -5}$

Problem 2. Solve $4x - 18 = 3y$
$1 + x + 2y = 0$

Rearranging gives

$4x - 3y = 18$ (1)

$x + 2y = -1$ (2)

Multiplying equation (2) by 4 gives

$4x + 8y = -4$ (3)

Subtracting equation (3) from equation (1) gives

$$-11y = 22$$

Therefore $y = \dfrac{22}{-11}$

$$\boldsymbol{y = -2}$$

Substituting in equation (1) gives

$4x - 3(-2) = 18$

Therefore $4x + 6 = 18$

$$4x = 18 - 6$$

$$4x = 12$$

$$\boldsymbol{x = 3}$$

Checking in equation (2)

L.H.S. $= (3) + 2(-2)$

$= 3 - 4$

$= -1 =$ R.H.S.

Thus the solution is $\boldsymbol{x = 3, y = -2}$

Further problems on simple simultaneous equations as shown in worked Problems 1 and 2 may be found in Section 6.4 (Problems 1–15), page 150.

Worked problem on simultaneous equations involving fractions

Problem 3. Solve $\dfrac{a}{6} - \dfrac{b}{8} = 0$ (1)

$$\frac{a}{3} + \frac{b}{4} = 6 \quad (2)$$

It is usually easier to rid the equations of fractions before the procedure for solving simultaneous equations is attempted. This is achieved by multiplying each equation by the lowest common multiple of the denominators. Multiplying equation (1) by 24 gives

$$24\left[\frac{a}{6}\right] - 24\left[\frac{b}{8}\right] = 24(0)$$

Therefore $4a - 3b = 0$ (3)

Multiplying equation (2) by 12 gives

$$12\left[\frac{a}{3}\right] + 12\left[\frac{b}{4}\right] = 12(6)$$

Therefore $4a + 3b = 72$ (4)

The usual procedure for solution is now adopted using equations (3) and (4).

Therefore $4a - 3b = 0$ (3)

$$4a + 3b = 72 \quad (4)$$

By coincidence the coefficients of 'a' are the same in both equations. Also the coefficients of 'b' are the same although the sign is different. Adding equations (3) and (4) gives

$8a = 72$

$\boldsymbol{a = 9}$

Substituting in equation (3) gives

$$4(9) - 3b = 0$$

Therefore $36 - 3b = 0$

$$36 = 3b$$

$$\boldsymbol{b = 12}$$

Checking in equation (4) gives

L.H.S. $= 4(9) = 3(12)$

$= 36 + 36$

$= 72 =$ R.H.S.

Whenever fractions are involved in simultaneous equations a final check must be made in the original equations. Thus, in equation (1)

L.H.S. $= \frac{9}{6} - \frac{12}{8}$

$= 1\frac{1}{2} - 1\frac{1}{2}$

$= 0 =$ R.H.S.

And in equation (2)

L.H.S. $= \frac{9}{3} + \frac{12}{4}$

$= 3 + 3$

$= 6 =$ R.H.S.

Thus the solution is $\boldsymbol{a = 9, b = 12}$.

Further problems on simultaneous equations involving fractions as shown in worked Problem 3 may be found in Section 6.4 (Problems 16–23), page 151.

Worked problems on simultaneous equations involving decimal fractions

---

Problem 4. Solve

$1.2x - 1.8y = -21$ (1)

$2.5x + 0.6y = 65$ (2)

Many practical examples of simultaneous equations will contain decimal fractions. It is generally easier to rid the equations of their decimal fractions and in this Problem this is achieved by multiplying throughout by 10.

Thus, multiplying equation (1) by 10 gives

$$12x - 18y = -210 \quad (3)$$

and multiplying equation (2) by 10 gives

$$25x + 6y = 650 \quad (4)$$

Equations (3) and (4) do not contain decimal fractions, and are thus solved as follows:

Multiplying equation (4) by 3 gives

$$75x + 18y = 1\,950 \quad (5)$$

Adding equations (3) and (5) gives

$$87x = 1\,740$$

$$\text{Therefore } x = \frac{1\,740}{87}$$

$$\boldsymbol{x = 20}$$

Substituting in equation (1)

$$1.2(20) - 1.8y = -21$$

$$\text{Therefore } 24 - 1.8y = -21$$

$$24 + 21 = 1.8y$$

$$45 = 1.8y$$

$$y = \frac{45}{1.8}$$

$$\boldsymbol{y = 25}$$

The solution $x = 20$ and $y = 25$ must now be substituted into equation (2), to serve as a check.

Check:

In equation (2), L.H.S. $= 2.5x + 0.6y$

$= 2.5(20) + 0.6(25)$

$= 50 + 15$

$= 65 =$ R.H.S.

Thus the solution is $\boldsymbol{x = 20}$ **and** $\boldsymbol{y = 25}$

The following worked problem is included because it shows clearly that the use of a calculator is often invaluable. The procedure does not differ in any way from that previously adopted. Numbers can become large and cumbersome in practical applications and this is unavoidable.

Problem 5. Solve $0.460b + 0.150d = 1.990$ (1)

$1.21b - 0.536d = 4.304$ (2)

If equation (1) is multiplied by 100 and equation (2) by 1 000 the decimal fractions will be changed into integers. However, if a calculator is used, this changing of decimal fractions to integers is an unnecessary step.

If equation (1) is multiplied by 0.536 and equation (2) by 0.150 the coefficients of d will be the same, albeit of opposite sign. Thus

$$(0.536)(0.460)b + (0.536)(0.150)d = (0.536)(1.990) \quad (3)$$

$$(0.150)(1.21)b - (0.150)(0.536)d = (0.150)(4.304) \quad (4)$$

Therefore $0.246\,56b + 0.0804d = 1.066\,64$ (3)

$0.181\,50b - 0.0804d = 0.645\,60$ (4)

Adding equations (3) and (4) gives

$$0.428\,06b = 1.712\,24$$

Therefore $b = \dfrac{1.712\,24}{0.428\,06}$

$$\boldsymbol{b = 4}$$

Substituting in equation (1) gives

$$(0.46)4 + 0.150d = 1.99$$

Therefore $1.84 + 0.150d = 1.99$

$$0.150d = 1.99 - 1.84$$

$$0.150d = 0.15$$

$$\boldsymbol{d = 1}$$

Checking in equation (2)

$$\text{L.H.S.} = (1.21)(4) - (0.536)(1)$$

$$= 4.840 - 0.536$$

$$= 4.304 = \text{R.H.S.}$$

Thus the solution is $\boldsymbol{b = 4}$ **and** $\boldsymbol{d = 1}$

Further problems on simultaneous equations involving decimal fractions as shown in worked Problems 4 and 5 may be found in Section 6.4 (Problems 24–32), page 151.

Worked problems on simultaneous equations involving reciprocals

Problem 6. Solve $\frac{3}{a} + \frac{6}{b} = 4$ (1)

$$\frac{9}{a} - \frac{12}{b} = -3 \qquad (2)$$

When the reciprocals of the unknowns are included throughout, the following substitution should be made before continuing with the adopted procedure. Let

$\frac{1}{a} = x$ and $\frac{1}{b} = y$

Equations (1) and (2) then become

$3x + 6y = 4$ (3)

$9x - 12y = -3$ (4)

Multiplying equation (3) by 3 gives

$9x + 18y = 12$ (5)

Subtracting equation (4) from equation (5) gives

$30y = 15$

Therefore $y = \frac{15}{30}$

$\boldsymbol{y = \frac{1}{2}}$

Substituting in equation (3) gives

$3x + 6(\frac{1}{2}) = 4$

Therefore $3x + 3 = 4$

$3x = 4 - 3$

$3x = 1$

$\boldsymbol{x = \frac{1}{3}}$

But $x = \frac{1}{a}$ or $a = \frac{1}{x}$

Therefore $a = \frac{1}{\frac{1}{3}}$

$\boldsymbol{a = 3}$

Also $y = \frac{1}{b}$ or $b = \frac{1}{y}$

Therefore $b = \frac{1}{\frac{1}{2}}$

$\boldsymbol{b = 2}$

A check must be made in the original two equations. Substituting $a = 3$ and $b = 2$ in equation (1) gives

$$\text{L.H.S.} = \frac{3}{3} + \frac{6}{2}$$
$$= 1 + 3$$
$$= 4 = \text{R.H.S.}$$

Substituting in equation (2) gives

$$\text{L.H.S.} = \frac{9}{3} - \frac{12}{2}$$
$$= 3 - 6$$
$$= -3 = \text{R.H.S.}$$

Thus the solution is $\boldsymbol{a = 3}$ **and** $\boldsymbol{b = 2}$

Further problems on simultaneous equations involving reciprocals as shown in worked Problem 6 may be found in Section 6.4 (Problems 33–39), page 152.

Worked problems on simultaneous equations involving more difficult fractions

---

Problem 7. Solve $\frac{f-1}{3} + \frac{g+2}{2} = 3$ (1)

$$\frac{1-f}{6} + \frac{4-g}{2} = \frac{1}{2} \quad (2)$$

Before equations (1) and (2) can be simultaneously solved the fractions need to be removed. Thus, multiplying equation (1) by 6 gives

$$6\left[\frac{f-1}{3}\right] + 6\left[\frac{g+2}{2}\right] = 6(3)$$

Therefore $2(f-1) + 3(g+2) = 18$

$$2f - 2 + 3g + 6 = 18$$
$$2f + 3g = 18 + 2 - 6$$
$$2f + 3g = 14 \quad (3)$$

Multiplying equation (2) by 6 gives

$$6\left[\frac{1-f}{6}\right]+6\left[\frac{4-g}{2}\right]=6(\tfrac{1}{2})$$

Therefore $(1-f)+3(4-g)=3$

$$1-f+12-3g=3$$

$$-f\quad -3g=3-1-12$$

$$-f\quad -3g=-10$$

This equation can be multiplied throughout by -1 to give:

$f+3g=10$ (4)

Thus the initial problem containing fractions becomes simply the solution of the simultaneous equations

$2f+3g=14$ (3)

$f+3g=10$ (4)

Subtracting equation (4) from equation (3) gives

$\boldsymbol{f=4}$

Substituting in equation (3) gives

$$(2)(4)+3g=14$$

Therefore $8+3g=14$

$$3g=14-8$$

$$3g=6$$

$$\boldsymbol{g=2}$$

Checking in equation (4)

$$\text{L.H.S.}=4+3(2)$$

$$=4+6$$

$$=10=\text{R.H.S.}$$

For a complete check, the solutions $f=4$, $g=2$ should be substituted into equations (1) and (2). Thus, substituting in equation (1)

$$\text{L.H.S.}=\frac{4-1}{3}+\frac{2+2}{2}$$

$$=\frac{3}{3}+\frac{4}{2}$$

$$=1+2$$

$$=3=\text{R.H.S.}$$

Substituting in equation (2)

$$\text{L.H.S.} = \frac{1-4}{6} + \frac{4-2}{2}$$
$$= \frac{-3}{6} + \frac{2}{2}$$
$$= -\tfrac{1}{2} + 1$$
$$= \tfrac{1}{2} = \text{R.H.S.}$$

Thus, the solution is $\boldsymbol{f = 4}$ **and** $\boldsymbol{g = 2}$

Further problems on simultaneous equations involving more difficult fractions as shown in worked Problem 7 may be found in Section 6.4 (Problems 40–45), page 152.

6.3 Simultaneous equations involved in practical problems

In many practical problems it is necessary to firstly form two equations from given data, and then to solve using the procedure of Sections 6.1 and 6.2.

Worked problems on solving practical problems

The following worked problems are typical of some of the practical problems that may be encountered by scientists, technicians and engineers.

Problem 1. In a system of pulleys, the effort P required to raise a load W is given by $P = aW + b$, where a and b are constants. If $W = 50$ when $P = 10$ and $W = 80$ when $P = 13$ find the values of a and b.

Substituting $W = 50$ and $P = 10$ in the given equation gives

$$10 = 50a + b \qquad (1)$$

Substituting $W = 80$ and $P = 13$ in the given equation gives

$$13 = 80a + b \qquad (2)$$

Thus two equations have been produced from the given data. Subtracting equation (1) from equation (2) gives

$$3 = 30a$$

Therefore $a = \dfrac{3}{30}$

$$\boldsymbol{a = \frac{1}{10}}$$

Substituting in equation (1) gives

$$10 = 50\left[\frac{1}{10}\right] + b$$

Therefore $10 = 5 + b$

$10 - 5 = b$

$\boldsymbol{b = 5}$

Checking in equation (2)

$$\text{R.H.S.} = 80\left[\frac{1}{10}\right] + 5$$

$$= 8 + 5$$

$$= 13 = \text{L.H.S.}$$

Thus the solution is $\boldsymbol{a = \frac{1}{10}}$ **and** $\boldsymbol{b = 5}$

Problem 2. By using Kirchhoff's laws in a certain electrical network the currents I_1 and I_2 (in amperes) are connected by the equations

$$0.2I_1 + 1.6I_2 = 8.6 \quad (1)$$

$$1.8I_1 - 2.2I_2 = -5.6 \quad (2)$$

Find the currents I_1 and I_2.

Multiplying both equations by 10 changes the decimal fractions into integers. Hence,

$$2I_1 + 16I_2 = 86 \quad (3)$$

$$18I_1 - 22I_2 = -56 \quad (4)$$

Multiplying equation (3) by 9 gives

$$18I_1 + 144I_2 = 774 \quad (5)$$

Subtracting equation (4) from equation (5) gives

$$166I_2 = 830$$

Therefore $I_2 = \frac{830}{166}$

$\boldsymbol{I_2 = 5}$

Substituting in equation (3) gives

$$2I_1 + 16(5) = 86$$

Therefore $2I_1 + 80 = 86$

$$2I_1 = 86 - 80$$

$$2I_1 = 6$$

$$I_1 = \frac{6}{2}$$

$$\boldsymbol{I_1 = 3}$$

Checking in equation (1)

L.H.S. $= (0.2)(3) + (1.6)(5)$

$= 0.6 + 8.0$

$= 8.6 =$ R.H.S.

Checking in equation (2)

L.H.S. $= (1.8)(3) - (2.2)(5)$

$= 5.4 - 11$

$= -5.6 =$ R.H.S.

Thus the solution is $\boldsymbol{I_1 = 3}$ **A and** $\boldsymbol{I_2 = 5}$ **A**

Problem 3. The distance s m of a body travelling with constant acceleration a m s^{-2} in a straight line is given by $s = ut + \frac{1}{2}at^2$ where u m s^{-1} is the initial velocity and t s the time. Calculate the initial velocity and the acceleration given that $s = 36$ m at $t = 2$ s and $s = 124$ m when $t = 4$ s

Substituting $s = 36$ and $t = 2$ in the given equation gives

$$36 = 2u + \tfrac{1}{2}a(2)^2$$

Therefore $36 = 2u + 2a$ (1)

Substituting $s = 124$ and $t = 4$ in the given equation gives

$$124 = 4u + \tfrac{1}{2}a(4)^2$$

Therefore $124 = 4u + 8a$ (2)

Multiplying equation (1) by 2 gives

$$72 = 4u + 4a \qquad (3)$$

Subtracting equation (3) from equation (2) gives

$$52 = 4a$$

Therefore $a = \dfrac{52}{4}$

$$\boldsymbol{a = 13}$$

Substituting in equation (1) gives

$$36 = 2u + 2(13)$$

Therefore $36 = 2u + 26$

$$36 - 26 = 2u$$

$$10 = 2u$$

$$u = \frac{10}{2}$$

$$\boldsymbol{u = 5}$$

Checking in equation (2)

$$\text{R.H.S.} = 4(5) + 8(13)$$
$$= 20 + 104$$
$$= 124 = \text{L.H.S.}$$

Thus the solution is that the initial velocity is **5 m s^{-1}** and the acceleration is **13 m s^{-2}**

Problem 4. An engineering and five technicians on an oil rig together earn £14 208 per month, whilst three engineers and nine technicians together earn £28 800 per month. Find how much an engineer and how much a technician earn per month.

Let E represent the salary of an engineer and T that of a technician.

Thus $E + 5T = 14\,208$ (1)

and $3E + 9T = 28\,800$ (2)

Multiplying equation (1) by 3 gives

$$3E + 15T = 42\,624 \quad (3)$$

Subtracting equation (2) from equation (3) gives

$$6T = 13\,824$$

Therefore $T = \dfrac{13\,824}{6}$

$$T = 2\,304$$

Substituting in equation (1) gives

$$E + 5(2\,304) = 14\,208$$

Therefore $E + 11\,520 = 14\,208$

$$E = 14\,208 - 11\,520 = 2\,688$$

Checking in equation (2):

$$\text{L.H.S.} = 3(2\,688) + 9(2\,304)$$
$$= 8\,064 + 20\,736$$
$$= 28\,800 = \text{R.H.S.}$$

Thus the solution is that an **engineer earns £2 688 per month** and a **technician £2 304 per month.**

Problem 5. The resistance R ohms of a length of wire at t°C is given by the formula

$R = R_0(1 + \alpha t)$

where R_0 is the resistance at 0° C. Find α and R_0 if $R = 30$ ohms at 60°C and $R = 32$ ohms at 100° C.

Substituting $R = 30$ and $t = 60$ in the equation gives

$$30 = R_0(1 + 60\alpha) \qquad (1)$$

Substituting $R = 32$ and $t = 100$ in the equation gives

$$32 = R_0(1 + 100\alpha) \qquad (2)$$

If equation (1) is divided by equation (2) the resulting equation will involve only one unknown, α. Thus

$$\frac{30}{32} = \frac{R_0(1 + 60\alpha)}{R_0(1 + 100\alpha)}$$

Therefore $\dfrac{30}{32} = \dfrac{(1 + 60\alpha)}{(1 + 100\alpha)}$

The L.C.M. of the denominators is $32(1 + 100\alpha)$. Therefore, multiplying both sides of the equation by $32(1 + 100\alpha)$ gives

$$32(1 + 100\alpha)\frac{30}{32} = 32(1 + 100\alpha)\frac{(1 + 60\alpha)}{(1 + 100\alpha)}$$

Therefore $30(1 + 100\alpha) = 32(1 + 60\alpha)$

$$30 + 3\,000\alpha = 32 + 1\,920\alpha$$

$$3\,000\alpha - 1\,920\alpha = 32 - 30$$

$$1\,080\alpha = 2$$

$$\alpha = \frac{2}{1\,080}$$

$$\boldsymbol{\alpha = \frac{1}{540} \text{ or } 0.001\,85}$$

Substituting in equation (1) gives

$$30 = R_0\left(1 + \frac{1}{540}(60)\right)$$

Therefore $30 = R_0\left[1 + \dfrac{1}{9}\right]$

$$30 = \frac{10}{9}R_0$$

$$R_0 = \frac{9}{10}(30)$$

$$\boldsymbol{R_0 = 27}$$

Checking in equation (2)

$$\text{R.H.S.} = 27\left[1 + \frac{100}{540}\right]$$
$$= 27\left[1 + \frac{10}{54}\right]$$
$$= 27\left[\frac{64}{54}\right]$$
$$= 32 = \text{L.H.S.}$$

Thus the solution is $\boldsymbol{\alpha} = \mathbf{0.001\,85}$ and $\boldsymbol{R_0} = \mathbf{27}$ **ohms**

Further problems on simultaneous equations obtained from practical data as shown in worked Problems 1 to 5 may be found in the following Section (6.4) (Problems 46–60), page 153.

6.4 Further problems

Solve the following equations for the unknowns and verify the results.

Simple simultaneous equations

1. $x + y = 5$
$x - y = 2$
$[x = 3\frac{1}{2},\ y = 1\frac{1}{2}]$

2. $a - b = 8$
$a + b = 12$
$[a = 10,\ b = 2]$

3. $2s + 3t = 5$
$s + t = 2$
$[s = 1,\ t = 1]$

4. $d + e = 3$
$3d + 2e = 7$
$[d = 1,\ e = 2]$

5. $3g - 2h = 7$
$g + 2h = 5$
$[g = 3,\ h = 1]$

6. $x + 3y = 11$
$x + 2y = 8$
$[x = 2,\ y = 3]$

7. $4x - 3y = 18$
$x + 2y = -1$
$[x = 3,\ y = -2]$

8. $3m - 2n = -4.5$
$4m + 3n = 2.5$
$[m = -\frac{1}{2},\ n = 1\frac{1}{2}]$

9. $7a - 4b = 37$
$6a + 3b = 51$
$[a = 7,\ b = 3]$

10. $3x = 2y$
$4x + y = -11$
$[x = -2,\ y = -3]$

11. $4c = 2 - 5d$
$3d + c + 3 = 0$
$[c = 3,\ d = -2]$

12. $4x - 3y = 3$
$3x + 5y = 111$
$[x = 12,\ y = 15]$

13. $3a + 4b - 5 = 0$
$12 = 5b - 2a$
$[a = -1,\ b = 2]$

14. $6m - 19 = 3n$
$13 = 5m + 6n$
$[m = 3,\ n = -\frac{1}{3}]$

15. $4a - 6b + 2.5 = 0$
$7a - 5b + 0.25 = 0$
$[a = \frac{1}{2}, b = \frac{3}{4}]$

Simultaneous equations involving fractions

16. $\frac{a}{3} + \frac{b}{4} = 6$
$\frac{a}{6} - \frac{b}{8} = 0$
$[a = 9, b = 12]$

17. $s + t = 17$
$\frac{s}{5} - \frac{t}{7} = 1$
$[s = 10, t = 7]$

18. $\frac{a}{2} - 11 = -2b$
$\frac{3}{5}b = 9 - 3a$
$[a = 2, b = 5]$

19. $\frac{3}{2}p - 2q = \frac{1}{2}$
$p + \frac{3}{2}q = 6$
$[p = 3, q = 2]$

20. $\frac{2}{5}x - \frac{y}{4} = \frac{17}{20}$
$\frac{x}{4} + \frac{2}{7}y = \frac{13}{7}$
$[x = 4, y = 3]$

21. $\frac{c}{5} + \frac{d}{3} = \frac{43}{30}$
$\frac{c}{9} - \frac{d}{6} = -\frac{1}{12}$
$[c = 3, d = 2\frac{1}{2}]$

22. $y - \frac{x}{12} = 1$
$x + \frac{y}{4} = \frac{25}{2}$
$[x = 12, y = 2]$

23. $\frac{a}{8} + \frac{5}{2} = b$
$13 - \frac{b}{3} - 3a = 0$
$[a = 4, b = 3]$

Simultaneous equations involving decimal fractions

24. $1.2a - 1.8b = -21$
$2.5a + 0.6b = 65$
$[a = 20, b = 25]$

25. $2.5x + 0.45 - 3y = 0$
$1.6x + 0.8y - 0.8 = 0$
$[x = 0.30, y = 0.40]$

26. $1.2p + q = 1.8$
$p - 1.2q = 3.94$
$[p = 2.50, q = -1.20]$

27. $0.5b - 1.2c = -13$
$0.8b + 0.3c = 12.5$
$[b = 10, c = 15]$

28. $6.9 + 1.7y = 0.9x$
$4.3 + 1.3x + 2.3y = 0$
$[x = 2.0, y = -3.0]$

29. $1.4x - 3.2y = 7.06$
$-2.1x + 6.7y = -12.87$
$[x = 2.30, y = -1.20]$

30. $-0.5f + 0.4g = 0.7$
$1.2f - 0.3g = 3.6$
$[f = 5.0, g = 8.0]$

31. $1.25d - 0.75e = 1$
$0.25d + 1.25e = 17$
$[d = 8.0, e = 12.0]$

32. $0.1p + 0.2q = -0.2$
$1.5p - 0.4q = 10.6$
$[p = 6.0, q = -4.0]$

Simultaneous equations involving reciprocals

33. $\frac{3}{x} - \frac{2}{y} = 0$
$\frac{1}{x} + \frac{4}{y} = 14$
$[x = \frac{1}{2}, y = \frac{1}{3}]$

34. $\frac{2}{a} - \frac{3}{b} = -8$
$\frac{4}{a} + \frac{5}{b} = 50$
$[a = \frac{1}{5}, b = \frac{1}{6}]$

35. $\frac{3}{r} - \frac{4}{s} = \frac{5}{6}$
$\frac{1}{r} + \frac{3}{s} = 1$
$[r = 2, s = 6]$

36. $\frac{1}{2a} + \frac{3}{5b} = 7$
$\frac{4}{a} + \frac{1}{2b} = 13$
$[a = \frac{1}{2}, b = \frac{1}{10}]$

37. $\frac{3}{h} - \frac{2}{k} = \frac{1}{2}$
$\frac{5}{h} + \frac{3}{k} = \frac{29}{12}$
$[h = 3, k = 4]$

38. $\frac{10}{a} - \frac{4}{b} = 3$
$\frac{6}{a} + \frac{8}{b} = 7$
$[a = 2, b = 2]$

39. $\frac{5}{m} + \frac{4}{n} = 1.3$
$\frac{3}{m} - \frac{7}{n} = -1.1$
$[m = 10, n = 5]$

Simultaneous equations involving more difficult fractions

40. $\frac{a-1}{3} + \frac{b+2}{2} = 3$
$\frac{1-a}{6} + \frac{4-b}{2} = \frac{1}{2}$
$[a = 4, b = 2]$

41. $\frac{f-2}{3} + \frac{k-1}{4} = \frac{13}{12}$
$\frac{2-f}{2} + \frac{3+k}{3} = \frac{11}{6}$
$[f = 3, k = 4]$

42. $\dfrac{4}{a-b}=\dfrac{16}{21}$

$\dfrac{3}{a+b}=\dfrac{4}{9}$

$[a = 6, b = \frac{3}{4}]$

43. $\dfrac{2x+1}{5}-\dfrac{1-4y}{2}=\dfrac{5}{2}$

$\dfrac{1-3x}{7}+\dfrac{2y-3}{5}+\dfrac{32}{35}=0$

$[x = 2, y = 1]$

44. If $5p-\dfrac{2}{q}=11$ and $p+\dfrac{3}{q}=9$ find the value of $\dfrac{pq+1}{q}$ [5]

45. When the numerator of a certain fraction is increased by 3 and the denominator by 1, its value changes to $1\frac{1}{4}$; but when the numerator is increased by 4 and the denominator decreased by $1\frac{1}{2}$, it is equal to 4. Find the original fraction. $[\frac{2}{3}]$

Practical problems involving simultaneous equations

46. In an engineering process two variables x and y are related by $y = ax + \dfrac{b}{x}$ where a and b are constants. Find a and b if $y = 13$ when $x = 3$ and $y = 12$ when $x = 2$ $[a = 3, b = 12]$

47. When an effort E newtons is applied to a machine it is found that the resistance R newtons can be overcome and that E and R are connected by the formula: $E = a + bR$. An effort of 3.5 N overcomes a resistance of 5 N while an effort of 5.3 N overcomes a resistance of 8 N. Find a and b and the effort required to overcome a resistance of 10 N.
$[a = 0.50, b = 0.60, E = 6.5\text{ N}]$

48. When Kirchhoff's Laws are applied to a two mesh circuit the following equations are produced

$$3.0 = 0.10I_1 + (I_1 - I_2)$$
$$-2.0 = 0.05I_2 - (I_1 - I_2)$$

Find I_1 and I_2 $[I_1 = 7.42\text{ A}, I_2 = 5.16\text{ A}]$

49. 500 Cotswold bricks and 80 breeze blocks cost £182. Another order of 220 bricks and 50 breeze blocks cost £86. Find the cost of one breeze block and the cost, per hundred, of bricks.
[40p, £30]

50. The law connecting friction F and load L for an experiment to find the friction force between two surfaces is of the type $F = aL + b$ where a and b are constants. When $F = 6.0$ N, $L = 7.5$ N and when $F = 2.7$ N, $L = 2.0$ N. Find the values of a and b and the value of F when $L = 18$ N.
$[a = 0.60, b = 1.5, F = 12.3\text{ N}]$

51. The distance s metres of a vehicle travelling with constant acceleration a m s^{-2} in a straight line is given by: $s = ut + \frac{1}{2}at^2$ where u is the initial velocity and t the time. Calculate the initial velocity and the acceleration given that $s = 51.75$ m when $t = 3$ s and $s = 118.75$ m when $t = 5$ s.
[$u = 7.50$ m s^{-1}, $a = 6.50$ m s^{-2}]

52. The weekly wage bill for three instructors and thirty apprentices in a mechanical training workshop is £4 341. In an electrical training centre the wage bill is £2 664 for two instructors and eighteen apprentices. If the five instructors each receive the same pay, and the forty-eight apprentices each receive the same pay, calculate the weekly wage of an instructor and the weekly wage of an apprentice. [Instructor £297, apprentice £115]

53. $y = mx + c$ is the equation of a straight line of slope m and intercept on the y-axis of c. If a line passes through the point where $x = 3$ and $y = 2\frac{1}{2}$, and also through the point where $x = 5\frac{1}{2}$ and $y = 3\frac{3}{4}$, find the slope and intercept of the straight line. [$m = \frac{1}{2}$, $c = 1$]

54. The resistance R ohms of a length of wire at t°C is given by the formula $R = R_0(1 + \alpha t)$ where R_0 ohms is the resistance at 0°C and α is a constant.
(*a*) If $R = 12$ ohms at 20°C, find R at 60°C. Assume $\alpha = 0.004\,2$. Find also the resistance at 0°C.
(*b*) If $R = 25$ ohms at 50°C and $R = 26$ ohms at 90°C find α and R_0.
(*a*) [$R_{60} = 13.86$ ohms, $R_0 = 11.07$ ohms] (*b*) [$R_0 = 23.75$ ohms, $\alpha = \frac{1}{950}$ or 0.001 05]

55. Applying Kirchhoff's Laws to a two-mesh network gives the following equations:

$$14 = 0.2I_1 + 2I_1 + 8(I_1 - I_2)$$
$$0 = -8(I_1 - I_2) + 2I_2 + 10I_2$$

Find the values of I_1 and I_2. [$I_1 = 2$, $I_2 = 0.8$]

56. A garage is supplied with two orders for tyres. If 2 radial tyres and 3 crossply tyres cost £151.50, and 5 radials and 7 crossplies cost £365, find the cost of each type of tyre.
[Radial £34.50, crossply £27.50]

57. Equations connecting parallel resistances in an electrical circuit are

$$\frac{4}{R_1} + \frac{6}{R_2} + \frac{9}{R_3} = 6$$

$$\frac{15}{R_1} + \frac{11}{R_2} + \frac{2}{R_3} = 8\frac{1}{12}$$

If $R_2 = R_3$, find the values of R_1, R_2 and R_3.
[$R_1 = 4$, $R_2 = R_3 = 3$]

58. The compound C_yH_z reacts with oxygen in proportions obeyed by the equations

$$44y - 9z = 60$$

$$32y + 8z = 160$$

Determine y and z and hence the formula of C_yH_z [C_3H_8]

59. The molar heat capacity of a solid compound is given by the equation $c = a + bT$ where c is measured in J K^{-1} mole^{-1} and T in K. When c is 41, T is 100 K and when c is 108.5, T is 350 K. Find a and b. [$a = 14$, $b = 0.27$]

60. x g of cartridge brass (70% copper, 30% zinc, by mass) and y g of naval brass (62% copper, 37% zinc, 1% tin, by mass) when fused together give a new alloy containing 342.6 g of copper and 165.6 kg of zinc. Find x and y.
[$x = 330$, $y = 180$]

Chapter 7

Evaluation and transposition of formulae

7.1 Evaluation of formulae

The statement $I = \frac{V}{R}$ is said to be a **formula** for I in terms of V and R.

I, V and R are called **symbols.**

The single term on the left-hand side is called the **subject of the formula.**

Similarly, $A = 2\pi rh + 2\pi r^2$ is a formula for A in terms of r and h, and A is the subject of the formula.

Worked problems on the evaluation of formulae

In most practical branches of science and engineering the evaluation of formulae is essential. Below are some worked problems typical of the types of formulae met in day to day calculations. Use your calculator to check the answers.

The use of brackets in place of multiplication signs is employed throughout.

Problem 1. The surface area, A, of a hollow cone is given by the formula $A = \pi rl$. Find the surface area in square centimetres, when $r = 4.0$ cm and $l = 9.0$ cm.

Substituting the values for symbols in the formula $A = \pi rl$ gives:

$A = (\pi)(4.0)(9.0)\ \text{cm}^2$

$= 113\ \text{cm}^2$

Hence the surface area, A = **113 cm²**

Problem 2. In an electrical circuit the voltage V volts is given by $V = IR$. Find the voltage correct to 4 significant figures when $I = 7.240$ amperes and $R = 12.57$ ohms.

$V = IR$

Therefore $V = (7.240)(12.57)$ volts

$= 91.01$ volts

Hence the voltage, V = **91.01 volts**

Problem 3. A formula used for calculating velocity v m s^{-1} is given by $v = u + at$. If $u = 12.47$ m s^{-1}, $a = 5.46$ m s^{-2} and $t = 4.92$ s find v correct to 2 decimal places.

$$v = u + at$$

Therefore $v = 12.47 + (5.46)(4.92)$

$$= 12.47 + 26.86$$

$$= 39.33 \text{ m s}^{-1}$$

Hence the velocity, v = **39.33 m s^{-1}**

Problem 4. The area A m^2 of a circle is given by $A = \pi r^2$. Find the area correct to 2 decimal places given $r = 4.156$ m.

$$A = \pi r^2$$

Therefore $A = (\pi)(4.156)^2$ m^2

$$= 54.26 \text{ m}^2$$

Hence the area, A = **54.26 m^2**

Problem 5. Power P watts in an electrical circuit may be expressed by the formula $P = \dfrac{V^2}{R}$. Evaluate the power correct to 2 decimal places, given that $V = 24.62$ volts and $R = 45.21$ ohms.

$$P = \frac{V^2}{R}$$

Therefore $P = \dfrac{(24.62)^2}{45.21}$ watts

$$= \frac{606.14}{45.21} \text{ watts} = 13.41 \text{ watts}$$

Hence the power, P = **13.41 watts**

Problem 6. The volume V cm^3 of a right circular cone is given by the formula $V = \frac{1}{3}\pi r^2 h$. Given $r = 5.637$ cm and $h = 16.41$ cm find the volume in standard form correct to 3 significant figures.

$$V = \frac{1}{3}\pi r^2 h$$

$$\text{Therefore} \quad V = \frac{1}{3}(\pi)(5.637)^2(16.41)\ \text{cm}^3$$

$$= \frac{1}{3}(\pi)(31.78)(16.41)\ \text{cm}^3$$

$$= \frac{1\,638}{3}\ \text{cm}^3$$

$$= 546\ \text{cm}^3$$

Hence the volume $V = \mathbf{5.46 \times 10^2\ cm^3}$

Problem 7. If force F newtons is given by

$$F = \frac{Gm_1m_2}{d^2}$$

where m_1 and m_2 are masses, d their distance apart and G a constant, find the force given $G = 6.67 \times 10^{-11}\ \text{N m}^2\ \text{kg}^{-2}$, $m_1 = 8.43$ kg, $m_2 = 17.2$ kg and $d = 24.2$ m. Express the answer in standard form to 3 significant figures.

$$F = \frac{Gm_1m_2}{d^2}$$

$$= \frac{(6.67)(10^{-11})\ \text{N m}^2\ \text{kg}^{-2}(8.43)\ \text{kg}(17.2)\ \text{kg}}{(24.2)^2\ \text{m}^2}$$

$$= \frac{(6.67)(8.43)(17.2)(10^{-11})\ \text{N}}{585.6}$$

$$= \frac{(967.1)(10^{-11})}{585.6}\ \text{N}$$

$$= 1.651 \times 10^{-11}\ \text{N}$$

$$= 1.65 \times 10^{-11}\ \text{N to 3 significant figures}$$

Hence the force, $F = \mathbf{1.65 \times 10^{-11}\ N}$

Problem 8. The time t seconds of swing of a simple pendulum is given by the formula

$$t = 2\pi\sqrt{\left[\frac{l}{g}\right]}$$

Find the time, correct to 3 decimal places, given $l = 13.0$ m and $g = 9.81\ \text{m s}^{-2}$

$$t = 2\pi\sqrt{\left[\frac{l}{g}\right]}$$

$$t = (2)(\pi)\sqrt{\left[\frac{13.0 \text{ m}}{9.81 \text{ m s}^{-2}}\right]}$$
$$= (2)(\pi)\sqrt{(1.325 \text{ s}^2)}$$
$$= (2)(\pi)(1.151) \text{ s}$$
$$= 7.234 \text{ s}$$

Hence the time $t =$ **7.234 s**

Problem 9. A formula for resistance variation with temperature is $R = R_0(1 + \alpha t)$. Given $R_0 = 15.42$ ohms, $\alpha = 0.002\,70$ and $t = 78.4°\text{C}$ evaluate R ohms, correct to 2 decimal places.

$$R = R_0(1 + \alpha t)$$
$$= 15.42[1 + (0.002\,7)(78.4)] \text{ ohms}$$
$$= 15.52[1 + 0.211\,7] \text{ ohms}$$
$$= 15.42[1.211\,7] \text{ ohms}$$

Hence $R =$ **18.68 ohms**

Further problems on evaluation of formulae may be found in Section 7.3 (Problems 1–23), page 164.

7.2 Transposition of formulae

Formulae play a very important part in mathematics, science and engineering, as by them it is possible to give a concise, accurate and generalised statement of laws.

From Section 7.1 it can be seen that some formulae contain several symbols. Usually, one of the symbols is isolated on one side of the equation and is called the subject of the formula. Sometimes a symbol other than the subject is required to be calculated. In such circumstances it is usually easiest to rearrange the formula to make a new subject before numbers are substituted for symbols. This rearranging process is called **transposing the formula** or simply **transposition.** Transposition should be treated carefully because an error can result in a statement that is not true, with possibly serious effects in certain branches of industry.

Basically the rules used for transposition of formulae are the same as the rules used for the solution of simple equations (see Chapter 5).

Worked problems on transposition of formulae

In the following graded worked problems four types of transposition have been categorised (Case A to Case D); however, it will become evident that many practical examples will require the application of several of these categories in order to transpose a formula.

The objective, in transposition, is to obtain the required new subject on its own on the left-hand side (L.H.S.) of the equation. Therefore, as a first step in any transposition, the equation, if necessary, should be changed around so that the side containing the required new subject is on the left.

Case A. Symbols connected by a plus or minus sign

Problem 1. Transpose $a = b + c + d$ to make c the subject.

The first step is to change the equation around so that the new subject, i.e. c, is on the L.H.S. Thus

$b + c + d = a$

Subtracting $(b + d)$ from both sides of the equation gives:

$$(b + c + d) - (b + d) = a - (b + d)$$

Therefore $b + c + d - b - d \quad = a - b - d$

$$\mathbf{c = a - b - d}$$

It will be clear from a knowledge of the solution of simple equations that what has actually happened in the above example is that $+b$ has been moved from the L.H.S. to the R.H.S., with the necessary change of sign. Similarly $+d$ has been changed from the L.H.S. to the R.H.S. with a change of sign.

Problem 2. If $x + y = a - b + c$ express b as the subject.

$$a - b + c = x + y$$

Therefore $-b \quad = x + y - a - c$

Obtaining a negative subject instead of the required positive subject is not an uncommon occurrence. In such circumstances both sides of the equation should be multiplied by -1.

Hence $(-1)(-b) = (-1)(x + y - a - c)$

Therefore $\quad +b = -x - y + a + c$

The result of multiplying each side of the equation by -1 has simply changed all of the signs in the equation, i.e. the negative signs have become positive and vice versa.

Therefore $b = \mathbf{a + c - x - y}$

When a number of symbols having different signs are involved in a formula it is usual to express the positive symbols before the negative ones. That is, rather than $b = -x - y + a + c$, it is more conventional (although no more correct) to express the formula as $\mathbf{b = a + c - x - y}$

Case B. Formulae involving products

Problem 3. Transpose $V = IR$ to make I the subject.

$IR = V$

On the L.H.S. R is multiplying the required new subject, I. The opposite operation to multiplication is division. Hence, if both sides of the equation are divided by R, then

$$\frac{IR}{R} = \frac{V}{R}$$

Therefore $I = \dfrac{V}{R}$

Problem 4. If a body falls freely through a height h the velocity is given by the formula $V^2 = 2gh$. Express this formula with h as the subject.

$2gh = V^2$

Dividing both sides of the equation by $2g$ gives

$$\frac{2gh}{2g} = \frac{V^2}{2g}$$

Therefore $\boldsymbol{h = \dfrac{V^2}{2g}}$

Case C. Formulae containing fractions

Problem 5. If $K = \dfrac{M}{V}$, rearrange the equation to make M the subject.

$\dfrac{M}{V} = K$

On the L.H.S. the prospective new subject, M, is divided by V. The opposite operation to division is multiplication. Hence, if both sides of the equation are multiplied by **V** then

$V\left[\dfrac{M}{V}\right] = V(K)$

Therefore $\boldsymbol{M = VK}$

Note that, as in the solution of simple equations, to rid an equation of a fraction, both sides of the equation need to be multiplied by the lowest common multiple of the denominators, in the above case, V.

Problem 6. If $a = \frac{F}{m}$ rearrange the equation to make m the subject.

$\frac{F}{m} = a$

Multiplying both sides by m gives

$m\left[\frac{F}{m}\right] = m(a)$

Therefore $F = ma$

The required new subject is now on the R.H.S. Thus

$ma = F$

Dividing both sides by a gives

$$\frac{ma}{a} = \frac{F}{a}$$

Therefore $\boldsymbol{m = \frac{F}{a}}$

Problem 7. Transpose $R = \frac{\rho l}{a}$ to make (i) a and (ii) ρ the subject.

(i) $\frac{\rho l}{a} = R$

Multiplying both sides by a gives

$$a\left[\frac{\rho l}{a}\right] = a(R)$$

Therefore $\rho l = aR$

$$aR = \rho l$$

Dividing both sides by R gives

$$\frac{Ra}{R} = \frac{\rho l}{R}$$

Therefore $\boldsymbol{a = \frac{\rho l}{R}}$

(ii) $\frac{\rho l}{a} = R$

Multiplying both sides by a gives

$$a\left[\frac{\rho l}{a}\right] = a(R)$$

Therefore $\rho l = aR$

Dividing both sides by l gives

$$\frac{\rho l}{l} = \frac{aR}{l}$$

Therefore $\boldsymbol{\rho = \frac{aR}{l}}$

Problem 8. If $v = u + \dfrac{ft}{m}$ transpose the equation to make t the subject.

$$u + \frac{ft}{m} = v$$

From Case A: $\dfrac{ft}{m} = v - u$

From Case C: $m\left[\dfrac{ft}{m}\right] = m(v - u)$

Therefore $ft = m(v - u)$

From Case B: $\dfrac{ft}{f} = \dfrac{m(v - u)}{f}$

Therefore $\boldsymbol{t = \frac{m}{f}(v - u)}$

Case D. Formulae containing the required new subject in a bracket

---

Problem 9. Transpose the equation $s = \frac{1}{2}(u + v)t$ to make v the subject.

$\frac{1}{2}(u + v)t = s$

From Case C: $2(\frac{1}{2})(u + v)t = 2s$

Therefore $(u + v)t = 2s$

From Case B: $\dfrac{(u + v)t}{t} = \dfrac{2s}{t}$

Therefore $$u + v = \frac{2s}{t}$$

From Case A: $$v = \frac{2s}{t} - u$$

Note that the R.H.S. $$\frac{2s}{t} - u = \frac{2s - ut}{t}$$

Therefore $$v = \frac{2s - ut}{t}$$

Both of the above answers for v are correct, the latter being the neater version.

Problem 10. Transpose the equation $L_2 = L_1(1 + \alpha\theta)$ to make α the subject.

$L_1(1 + \alpha\theta) = L_2$

Whenever the new subject is contained within a bracket it is advisable to isolate the bracket on the L.H.S. Thus, dividing both sides of the equation by L_1 gives

$$\frac{L_1(1 + \alpha\theta)}{L_1} = \frac{L_2}{L_1}$$

Therefore $$1 + \alpha\theta = \frac{L_2}{L_1}$$

$$\alpha\theta = \frac{L_2}{L_1} - 1$$

$$\alpha\theta = \frac{L_2 - L_1}{L_1}$$

Dividing both sides by θ gives

$$\frac{\alpha\theta}{\theta} = \frac{L_2 - L_1}{L_1\theta}$$

Therefore $$\boldsymbol{\alpha = \frac{L_2 - L_1}{L_1\theta}}$$

Further problems on transposition of formulae may be found in the following Section (7.3) (Problems 24–59), page 166.

7.3 Further problems

Evaluation of formulae

1. The area A of a rectangle is given by the formula $A = lb$.

Evaluate the area when l is 14.21 cm and b is 7.46 cm. $[A = 106.0\ \text{cm}^2]$

2. The circumference C of a circle is given by the formula $C = 2\pi r$. Find the circumference given $r = 6.20$ cm. $[C = 38.96\ \text{cm}]$
3. The area A of a triangle may be evaluated by using the formula $A = \frac{1}{2}bh$. Find the area when $b = 7.0$ mm and $h = 12.5$ mm $\quad [A = 43.75\ \text{mm}^2]$
4. A formula used when dealing with gas laws is $R = \dfrac{pV}{T}$.

   Evaluate R when p is $2\,000\ \dfrac{\text{kN}}{\text{m}^2}$, V is 5 m^3 and T is 200 K.

   $\left[\dfrac{50\ \text{kN m}}{\text{K}} = \dfrac{50\ \text{kJ}}{\text{K}}\right]$
5. In the following formula evaluate X when $p = 751$, $a = 5$ and $t = 21$: $X = p\left[1 + \dfrac{at}{100}\right]$ $\quad [1\,540]$
6. The voltage drop, V, in an electrical circuit is given by the formula $V = E - IR$. Evaluate the voltage drop when $E = 4.81$ volts, $I = 0.680$ amperes and $R = 5.37$ ohms. $[V = 1.158\ \text{volts}]$
7. $a = \dfrac{uv}{u + v}$ Evaluate a when $u = 10.93$ and $v = 7.380$.

   $[a = 4.405]$
8. When a number of cells are connected together the current I amperes of the combination is given by the formula $I = \dfrac{nE}{R + nr}$ Evaluate the current when $n = 40$, $E = 2.3$ volts, $R = 2.9$ ohms and $r = 0.60$ ohms. $\quad [I = 3.42\ \text{amperes}]$
9. The power P watts of an electrical circuit is given by the formula $P = \dfrac{V^2}{R}$ Find the power when V is 14.8 volts and R is 19.5 ohms. $\quad [P = 11.23\ \text{watts}]$
10. If $W = 8.20$, $v = 10.0$ and $g = 9.81$ evaluate B given that $B = \dfrac{Wv^2}{2g}$ $\quad [B = 41.79]$
11. If $F = \frac{1}{2}m(v^2 - u^2)$ find F when $m = 17.0$, $v = 14.8$ and $u = 9.24$ $\quad [F = 1\,136]$
12. The time t seconds of oscillation for a simple pendulum is

given by $t = 2\pi\sqrt{\frac{l}{g}}$. Evaluate the time of oscillation when $l = 61.42$ m and $g = 9.81\,\frac{\text{m}}{\text{s}^2}$ [$t = 15.72$ s]

Evaluate the subject of each of the following formulae shown in Problems 13–23.

13. $S = 2\pi r^2 + 2\pi rh$ when $\pi = 3.14$, $r = 2.42$ and $h = 3.47$ [89.51]

14. $E = \frac{1}{2}LI^2$ when $L = 3.55$ and $I = 0.490$. [0.426]

15. $A = \frac{V}{100}\left[Q - \frac{mV^2}{g}\right]$ when $Q = 50.28$, $m = 1.7$, $V = 5.0$ and $g = 9.81$ [0.348]

16. $I = \frac{V}{\sqrt{(R^2 + X^2)}}$ when $V = 240$, $R = 10.0$ and $X = 15.0$ [13.31]

17. $a = \frac{T(1 - S)}{S(T + 1)}$ when $T = 2.31$ and $S = 2.62$ [−0.431 5]

18. $X = T\left(1 + \frac{b}{10}\right)^n$ when $T = b = n = \frac{3}{4}$. [0.791 8]

19. $A = \sqrt{s(s - a)(s - b)(s - c)}$ where $s = \frac{a + b + c}{2}$ and $a = 2.0$, $b = 3.0$ and $c = 4.0$ [2.91]

20. $y = \sqrt{\left(\frac{ak}{p} - \frac{b}{q}\right)}$ when $a = 0.72$, $k = 0.620$, $p = 0.003\,00$, $b = 41.36$ and $q = 0.480$ [7.914]

21. $Z = \sqrt{\left[R^2 + \left(wL - \frac{1}{wC}\right)^2\right]}$ when $R = 24$, $L = 0.30$, $w = 352$ and $C = 6.5 \times 10^{-5}$ [66.4]

22. $S = ut + \frac{1}{2}at^2$ when $u = 10.1$, $t = 5.20$ and $a = -3.80$. [1.144]

23. $a = \frac{T}{\sqrt{(ku^2 - lv^2)}}$ when $T = 17.42$, $k = 0.073\,0$, $u = 5.46$, $l = 0.089\,0$ and $v = 3.21$ [±15.52]

Transposition of formulae

Make the symbol indicated in round brackets the subject of each of the formulae shown in problems 24–54 and express each in its simplest form.

24. $k + l = m - n + p \quad (m)$

$[m = k + l + n - p]$

25. $a + 2b = c \quad (b)$

$\left[b = \dfrac{c - a}{2}\right]$

26. $2abc = d \quad (c)$

$\left[c = \dfrac{d}{2ab}\right]$

27. $c = 2\pi r \quad (r)$

$\left[r = \dfrac{c}{2\pi}\right]$

28. $5 - 2bc = p + q \quad (c)$

$\left[c = \dfrac{5 - p - q}{2b}\right]$

29. $y = mx + c \quad (m)$

$\left[m = \dfrac{y - c}{x}\right]$

30. $pV = c \quad (p)$

$\left[p = \dfrac{c}{V}\right]$

31. $A = \pi rl \quad (l)$

$\left[l = \dfrac{A}{\pi r}\right]$

32. $I = PRT \quad (R)$

$\left[R = \dfrac{I}{PT}\right]$

33. $I = \dfrac{V}{R} \quad (R)$

$\left[R = \dfrac{V}{I}\right]$

34. $P = \dfrac{RT}{V} \quad (V)$

$\left[V = \dfrac{RT}{P}\right]$

35. $S = \dfrac{a}{1 - r} \quad (r)$

$\left[r = \dfrac{S - a}{S}\right]$

36. $V = \dfrac{\pi}{12} d^2 h \quad (h)$

$\left[h = \dfrac{12V}{\pi d^2}\right]$

37. $a = \dfrac{wL^2}{4d} \quad (d)$

$\left[d = \dfrac{wL^2}{4a}\right]$

38. $F = \dfrac{9}{5}C + 32 \quad (C)$

$\left[C = \dfrac{5}{9}(F - 32)\right]$

39. $T = \dfrac{\lambda(x - C)}{C} \quad (x)$

$\left[x = \dfrac{C(T + \lambda)}{\lambda}\right]$

40. $Y = \dfrac{Fl}{Ax} \quad (l)$

$\left[l = \dfrac{Axy}{F}\right]$

41. $\dfrac{a + b}{b} = \dfrac{3p}{q} \quad (q)$

$\left[q = \dfrac{3bp}{a + b}\right]$

42. $A = 2\pi r(r + h) \quad (h)$

$\left[h = \dfrac{A}{2\pi r} - r\right]$

43. $A = B - 5.6C + 2 \quad (C)$

$\left[C = \dfrac{2 + B - A}{5.6}\right]$

44. $W = aq(x - t)$ (x)

$$\left[x = t + \frac{W}{aq}\right]$$

45. $P = \dfrac{N - a}{4b}$ (N)

$$[N = 4bP + a]$$

46. $B = \dfrac{8(D - d)}{L}$ (d)

$$\left[d = D - \frac{LB}{8}\right]$$

47. $I = \dfrac{bd^3}{12}$ (b)

$$\left[b = \frac{12I}{d^3}\right]$$

48. $y = \dfrac{Ml^2}{8EI}$ (I)

$$\left[I = \frac{Ml^2}{8Ey}\right]$$

49. $R_2 = R_1(1 + \alpha t)$ (t)

$$\left[t = \frac{R_2 - R_1}{R_1\alpha}\right]$$

50. $\dfrac{1}{R} = \dfrac{1}{R_1} + \dfrac{1}{R_2}$ (R_1)

$$\left[R_1 = \frac{RR_2}{R_2 - R}\right]$$

51. $I = \dfrac{E - e}{R + r}$ (r)

$$\left[r = \frac{E - e - IR}{I}\right]$$

52. $y^2 = 4a\left(x + \dfrac{c^2}{4a}\right)$ (x)

$$\left[x = \frac{y^2 - c^2}{4a}\right]$$

53. $S = \dfrac{n}{2}[2a + (n - 1)d]$ (d)

$$\left[d = \frac{2(S - an)}{n(n - 1)}\right]$$

54. $A = b\left(\dfrac{1}{1 - d} - r\right)$ (d) $\left[d = \dfrac{A + b(r - 1)}{A + br}\right]$

55. If P is the safe load which may be carried by a steel plate weakened by rivet holes then $P = f(b - nd)t$. Make f, the safe working stress in the steel and then n, the number of rivet holes, the subject of the formula.

$$\left[f = \frac{P}{(b - nd)t};\ n = \frac{fbt - p}{fdt}\right]$$

56. The modulus of elasticity of a structural material (E) is given by the formula $E = \dfrac{Wl}{Ax}$. Make x the subject of the formula.

$$\left[x = \frac{Wl}{EA}\right]$$

57. Van der Waals equation for the pressure of a real gas (p) is $\left(p + \dfrac{a}{v^2}\right)(V - b) = RT$. Make p the subject of the formula.

$$\left[p = \frac{RT}{V - b} - \frac{a}{v^2}\right]$$

58. The observed growth yield, Y, of an organism is given by $\dfrac{1}{Y} = \dfrac{m}{\mu} + \dfrac{1}{Y_G}$. Make Y_G, the true growth yield, the subject of the equation. $\left[Y_G = \dfrac{\mu Y}{\mu - mY}\right]$

59. The viscosity coefficient of a liquid (η) is given by the equation $\eta = \dfrac{\pi P r^4 t}{8vl}$. Make v the subject of the formula.

$\left[v = \dfrac{\pi P r^4 t}{8\eta l}\right]$

Chapter 8

Direct and inverse proportionality and straight line graphs

8.1 Direct and inverse proportionality

An expression such as $y = 2x$ contains two variables. For every value of x there is a corresponding value of y. The variable x is called the **independent variable** and **y** is called the **dependent variable**. The formula used for the circumference c of a circle is $c = 2\pi r$, where r is the radius of the circle. Changes in the radius cause a corresponding change in the circumference, thus r is the independent variable and c is the dependent variable.

Direct proportion

When an increase or decrease in an independent variable leads to an increase or decrease of the same proportion in the dependent variable then this is termed **direct proportion**. If $y = 2x$, then y is directly proportional to x, which may be written as $y \propto x$, where the sign '$\propto$' means 'is proportional to'. Another way of stating the relationship between y and x is $y = kx$, where k is a constant called the **coefficient of proportionality**. In this example, $k = 2$. Similarly, when $c = 2\pi r$ then c is directly proportional to r, i.e. $c \propto r$ and the coefficient of proportionality in this case is $k = 2\pi$.

Inverse proportion

When an increase in an independent variable leads to a decrease of the same proportion in the dependent variable (or vice versa) then this is termed **inverse proportion**. If y is inversely proportional to x then $y \propto \frac{1}{x}$ or $y = \frac{k}{x}$. Rearranging gives $k = xy$, that is, **for inverse proportionality the product of the variables is constant.**

Some practical examples of laws involving direct and inverse proportion in science include:

(*a*) **Hooke's law**, which states that within the elastic limit of a material, the strain ε produced is directly proportional to the stress σ producing it, i.e. $\varepsilon \propto \sigma$ or $\varepsilon = k\sigma$.

(*b*) **Boyle's law**, which states that for a gas at constant temperature, the volume V of a fixed mass of gas is inversely proportional to its absolute pressure p, i.e. $p \propto \frac{1}{V}$ or $p = \frac{k}{V}$ or $k = pV$.

(*c*) **Charles' law**, which states that for a given mass of gas at constant pressure the volume V is directly proportional to its thermodynamic temperature T, i.e. $V \propto T$ or $V = kT$.

(*d*) **Ohm's law**, which states that the current I flowing through a fixed resistor R is directly proportional to the applied voltage V, i.e. $I \propto V$ or $I = kV$.

Alternatively, for a fixed voltage V, $I \propto \frac{1}{R}$ or $I = \frac{k}{R}$

Worked problems on direct and inverse proportionality

Problem 1. If y is directly proportional to x and $y = 3.54$ when $x = 0.6$, determine (*a*) the coefficient of proportionality, (*b*) the value of y when x is 0.7, and (*c*) the value of x when y is 8.85.

(*a*) $y \propto x$ or $y = kx$

When $y = 3.54$ and $x = 0.6$ then $3.54 = k(0.6)$

Hence the coefficient of proportionality, $k = \frac{3.54}{0.6} = \mathbf{5.9}$

(*b*) $y = kx$, hence when $x = 0.7$, $y = (5.9)(0.7) = \mathbf{4.13}$

(*c*) $y = kx$, from which, $x = \frac{y}{k}$

Hence when $y = 8.85$ and $k = 5.9$, $x = \frac{8.85}{5.9} = \mathbf{1.5}$

Problem 2. Hooke's law states that stress σ is directly proportional to strain ε within the elastic limit of a material. For a mild steel specimen when the stress is 20×10^6 pascals the strain is 0.000 1. Determine
(*a*) the coefficient of proportionality and (*b*) the value of strain when the stress is 16×10^6 pascals.

(*a*) $\sigma \propto \varepsilon$ or $\sigma = k\varepsilon$, from which, $k = \frac{\sigma}{\varepsilon}$

Hence the coefficient of proportionality,

$$k = \frac{20 \times 10^6}{0.0001} = \mathbf{200 \times 10^9 \text{ pascals}}$$

The coefficient of proportionality k in this case is called the Young's modulus.

(*b*) Since $\sigma = k\varepsilon$ then $\varepsilon = \dfrac{\sigma}{k}$

Hence when $\sigma = 16 \times 10^6$, strain $\varepsilon = \dfrac{16 \times 10^6}{200 \times 10^9} = \mathbf{0.000\,08}$

Problem 3. Boyle's law states that at a given constant temperature the volume V of a fixed mass of gas is inversely proportional to its absolute pressure p. If a gas occupies a volume of 0.060 m^3 at a pressure of 1.80×10^6 pascals and at constant temperature, determine (*a*) the coefficient of proportionality, (*b*) the volume if the pressure is changed to 4.0×10^6 pascals and (*c*) the pressure when the volume is 0.045 m^3

(*a*) $V \propto \dfrac{1}{p}$, i.e. $V = \dfrac{k}{p}$ or $k = pV$

Hence the coefficient of proportionality, $k = (1.80 \times 10^6)(0.060)$
$= \mathbf{0.108 \times 10^6}$ **Pa m^3**

(*b*) Volume $V = \dfrac{k}{p} = \dfrac{0.108 \times 10^6}{4.0 \times 10^6} = \mathbf{0.027}$ **m^3**

(*c*) Pressure $p = \dfrac{k}{V} = \dfrac{0.108 \times 10^6}{0.045} = \mathbf{2.4 \times 10^6}$ **pascals**

Problem 4. The electrical resistance R of a piece of wire is inversely proportional to its cross-sectional area A. When $A = 8$ mm^2, $R = 4.2$ ohms. Determine (*a*) the coefficient of proportionality, (*b*) the cross-sectional area corresponding to a resistance of 3 ohms, and (*c*) the resistance when the cross-sectional area is 10 mm^2.

(*a*) $R \propto \dfrac{1}{A}$ or $R = \dfrac{k}{A}$ or $k = RA$

Hence when $R = 4.2$ and $A = 8$, the coefficient of proportionality, $k = (4.2)(8) = \mathbf{33.6}$

(*b*) Since $k = RA$ then $A = \dfrac{k}{R}$

When $R = 3$ ohms, the cross-sectional area

$A = \dfrac{33.6}{3} = \mathbf{11.2}$ **mm^2**

(*c*) $R = \dfrac{k}{A}$, hence when $A = 10$, $R = \dfrac{33.6}{10} = \mathbf{3.36}$ **ohms**

Further problems on direct and inverse proportionality may be found in Section 8.3 (Problems 1–8), page 192.

8.2 Straight line graphs

The usual method of drawing graphs is to use two **axes**, these being two lines drawn mutually at right angles. One line is drawn horizontally and is called the **horizontal axis**, and the other is drawn vertically and is called the **vertical axis**. The point where these axes cross is called the **origin**. The information to be depicted is given in two sets of data, say, set A and set B.

Figure 8.1 shows the conventions which have been adopted. The values from set A are called x-values and those from set B y-values. The **ordered pairs** (an x-value with its assigned value of y) are called **cartesian coordinates** and the system uses **cartesian** or **rectangular axes**.

A graph drawn on rectangular axes is very much a pictorial representation of information. It is usually drawn on squared paper showing how a quantity (denoted by y) depends on, or varies with a related quantity (denoted by x). Often the corresponding values of x and y are given in tabulated form and the graph is the best straight line joining the system of points plotted on the squared paper.

To analyse two sets of related data (or figures) meaningfully can often take a lot of time, but if the data can be transferred on to a graph an immediate visual analysis is provided.

A **scale** can take many forms. One of these is a set of marks graded at

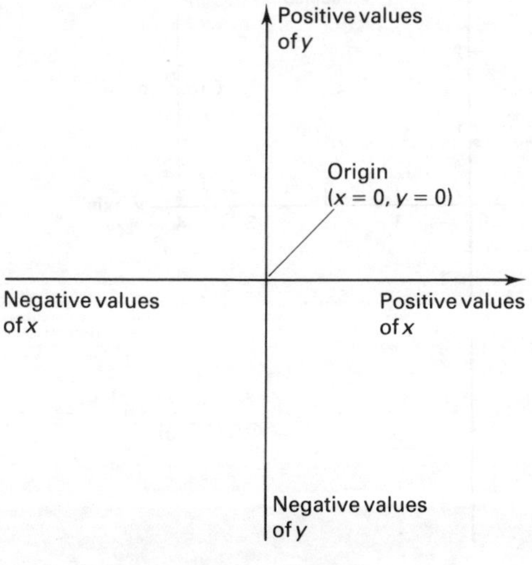

Fig. 8.1 Cartesian or rectangular axes

equal distances as on a ruler; that is, the distance from 0 to 1 is the same as the distance from 1 to 2 and so on. Such a scale is called **linear**.

The scale on each of the rectangular axes is chosen so that each space is equal to 1, 2, 4, 5, 10 or any other easily divisible unit. Scales where each space is equal to 3, 7, 11, 13 or similar units usually prove extremely awkward and should be avoided since these numbers cannot be readily divided into smaller parts. The scale chosen for the x-axis need not be the same as the scale chosen for the y-axis. Scales are usually chosen such that the graph occupies as much of the squared paper as possible, since the larger the graph the greater the accuracy.

The points on a graph are called **coordinates**. To specify a point on a graph two pieces of information are needed; the value of x, the horizontal component, and the value of y, the vertical component. If x is 4 units and the corresponding value of y is 3 units then the point or coordinate is termed (4, 3). Note that the x value is always specified first. In Fig. 8.2 rectangular axes x and y are shown. The coordinate A is shown at (4, 3) and marked by a cross. The coordinates B, shown at $(-2, 1)$, and C, shown at $(3, -3)$, are marked by a dot and circle. Coordinate D, shown at $(-3, -2)$, is marked by a dot only. All three methods may be used to denote coordinates on graphs.

The x distance of a point is called the **abscissa** and the y value the

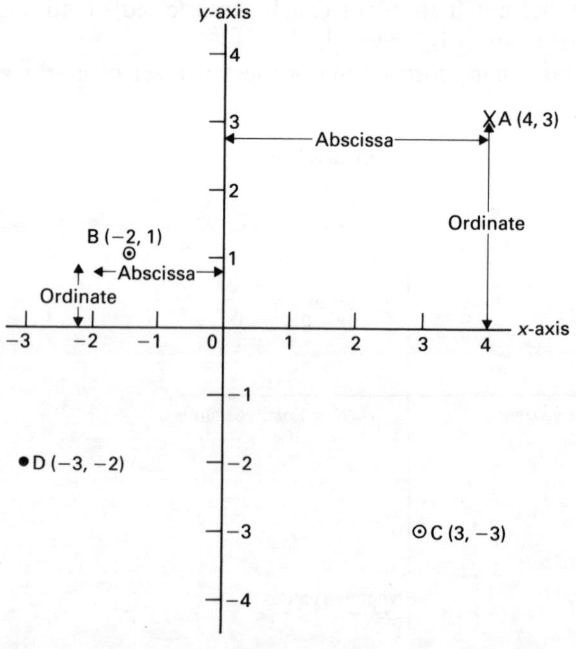

Fig. 8.2 Coordinates on rectangular axes

ordinate. Thus for the point A, the abscissa is 4 and the ordinate 3; for the point D, the abscissa is -3 and the ordinate -2.

Problem 1. The table below shows the temperature in degrees Celsius and the corresponding values in degrees Fahrenheit. Draw rectangular axes, choose a suitable scale and plot the graph of degrees Celsius (on the horizontal axis) against degrees Fahrenheit (on the vertical axis).

°C	10	25	40	65	80	100
°F	50	77	104	149	176	212

The coordinates (10, 50), (25, 77), (40, 104), and so on, are marked on the squared paper. If the coordinates are now joined, a straight line is produced as shown in Fig. 8.3.

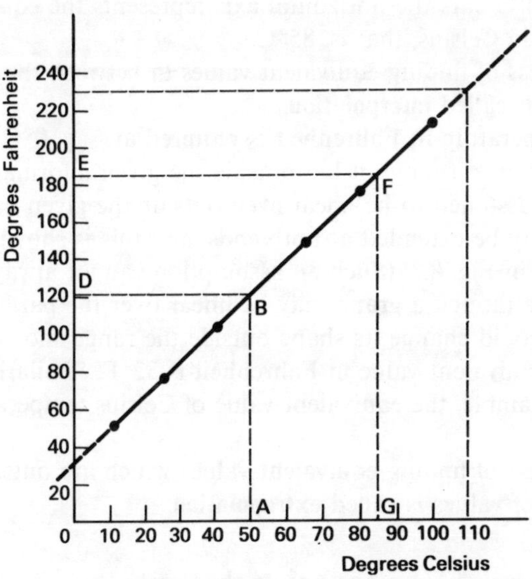

Fig. 8.3 Graph of degrees Celsius against degrees Fahrenheit

Since the resulting graph is a straight line, this shows that a linear relationship exists; that is, degrees Celsius are directly proportional to degrees Fahrenheit.

Interpolation and extrapolation

However, there is more information that can be obtained from the graph in Problem 1 above. For example, if the temperature in degrees Fahrenheit is required at 50° C this can be found from the graph. If the table given above was the only information available then all that could be said was that 50° C lies somewhere between 104° F and 149° F. However, the actual value may be found from the graph as follows:

1. The vertical broken line AB at 50° C cuts the graph at B.
2. The point where the horizontal line BD cuts the vertical axis indicates the corresponding temperature in degrees Fahrenheit, that is 122° F.

Hence 50° C ≙ 122° F. (The sign, ≙, means 'corresponds to'.)

Similarly the reverse process, that of finding an equivalent value of temperature in degrees Celsius to a given Fahrenheit temperature, can be read straight from the graph. Thus, to find the Celsius equivalent to 185° F, the horizontal line EF is drawn at 185° F which cuts the graph at F. Then the point where the vertical line FG cuts the horizontal axis represents the equivalent temperature in Celsius, that is, 85° C.

This process of finding equivalent values in between the given information is called **interpolation**.

If the temperature in Fahrenheit is required at, say, 0° C this can be found even though it lies outside the given information. If the graph is assumed to be linear even outside the given data then the graph may be extended at both ends, and this is shown by the broken lines in Fig. 8.3. (Such an assumption cannot always be made, since although a graph may be linear over the part being depicted, it could change its shape outside the range shown.) Thus at 0° C the equivalent value in Fahrenheit is 32° F. Similarly, at 230° F for example, the equivalent value of Celsius temperature is 110°C.

This process of finding equivalent values which are outside a given range of values is called **extrapolation**.

Axes where zero values are not at the origin

Whenever possible scales are chosen so that the zero value corresponds with the origin. Sometimes, however, this results in an accumulation of points in one corner of the graph paper. Whenever this happen it is best not to start the scales at zero. This is illustrated by Problem 2.

Problem 2. The resistance R ohms of a copper winding is measured at various temperatures t° C and the results are recorded in the following table:

R ohms	127.5	129.4	131.7	134.1	136
t°C	50	54	59	64	68

Plot a graph of resistance against temperature and find from it

(*a*) the temperature when the resistance is 130 ohms, and
(*b*) the resistance when the temperature is 66°C

If a graph is plotted with both the scales starting at zero at the origin, then the most appropriate scales would be as shown in Fig. 8.4

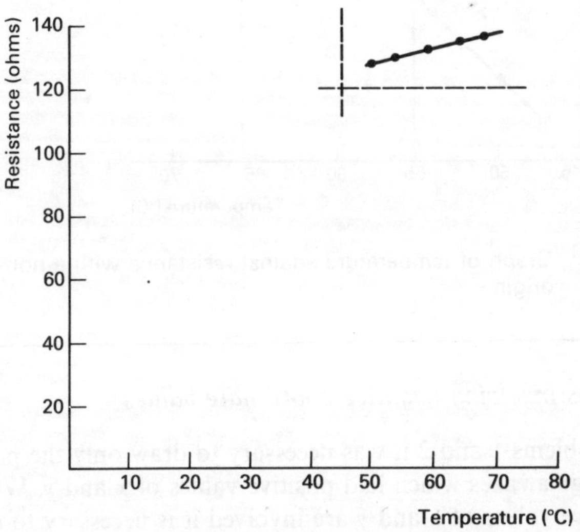

Fig. 8.4 Graph of temperature against resistance with cramped scale

It will be seen that all of the points fall in the top right-hand side of the graph paper, making interpolation difficult. A much more accurate graph is obtained if the x-axis (temperature) starts at 45°C and the y-axis (resistance) starts at 125 ohms. The axes corresponding to these values are shown by a broken line in Fig. 8.4 and are called **false axes** since the origin is not now at zero.

Choosing more appropriate scales the graph shown in Fig. 8.5 is produced. This is, in fact, a magnified version of the relevant part of the graph shown in Fig. 8.4

(*a*) **When the resistance is 130 ohms the equivalent value of temperature is 55°C**

(*b*) **When the temperature is 66°C the equivalent value of resistance is 135 ohms**

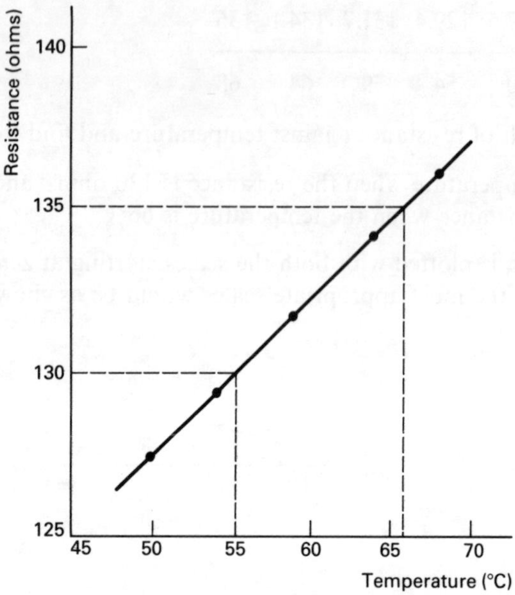

Fig. 8.5 Graph of temperature against resistance with a non-zero origin

---

Graphs involving negative coordinate values

In Problems 1 and 2 it was necessary to draw only the part of the rectangular axes which had positive values of x and y. When negative values of x and y are involved it is necessary to draw the four quadrants representing the complete rectangular axes. This is illustrated in Problem 3.

Problem 3. Draw a graph of $y = 3x + 2$ between the range of values $x = -3$ to $x = +5$. Hence find
(*a*) the value of y when $x = 2.5$ (*b*) the value of x when $y = -5.5$

Now $y = 3x + 2$ represents a simple equation and so for every value of x there is a corresponding value of y. For example, when $x = 0$, $y = 3(0) + 2 = 2$. Similarly, when $x = 1$, $y = 3(1) + 2 = 5$ and when $x = -2$, $y = 3(-2) + 2 = -6 + 2 = -4$, and so on. In this way a table of values may be drawn up as follows:

x	-3	-2	-1	0	1	2	3	4	5
y	-7	-4	-1	2	5	8	11	14	17

When only the equation is given, such a table, giving corresponding values of the two variables, is necessary.

The coordinates $(-3, -7)$, $(-2, -4)$, $(-1, -1)$, and so on, are plotted and joined together to produce the straight line as shown in Fig. 8.6

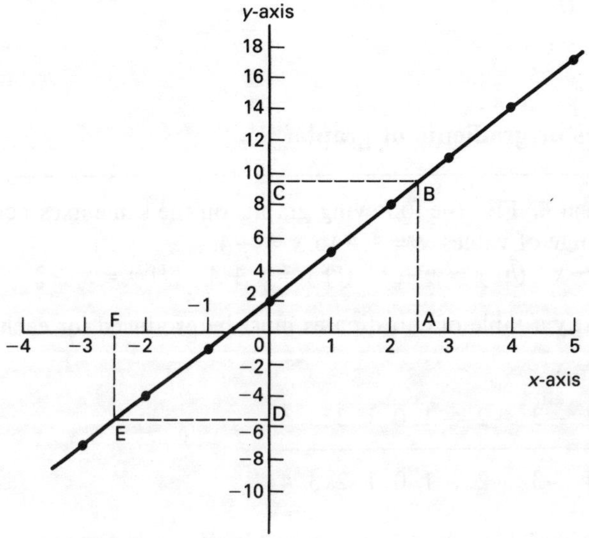

Fig. 8.6 Graph of $y = 3x + 2$

(*a*) **When $x = 2.5$, the value of y is 9.5**. This is obtained by drawing the vertical AB at $x = 2.5$ and reading off directly from the graph the y ordinate of BC, that is 9.5

(*b*) **When $y = -5.5$, the value of x is -2.5**. This is obtained by drawing the horizontal ED at $y = -5.5$ and reading off directly from the graph the x value corresponding to EF, that is, -2.5

The answers to (*a*) and (*b*) may be checked by calculation.

(*a*) $y = 3x + 2$
Therefore, when $x = 2.5$
$y = 3(2.5) + 2$
$y = 7.5 + 2$
$\mathbf{y = 9.5}$

(*b*) If $y = 3x + 2$
then $3x = y - 2$
and $x = \dfrac{y-2}{3}$

Therefore, when $y = -5.5$

$$x = \frac{-5.5 - 2}{3}$$

$$x = \frac{-7.5}{3}$$

$$x = -2.5$$

Slopes or gradients of graphs

Problem 4. Plot the following graphs on the same axes between the range of values $x = -4$ to $x = +4$:
(*a*) $y = x$ (*b*) $y = x + 1$ (*c*) $y = x + 4$ (*d*) $y = x - 2$

Firstly a table of coordinates must be produced for each graph.

(*a*) $y = x$

x	−4	−3	−2	−1	0	1	2	3	4
y	−4	−3	−2	−1	0	1	2	3	4

(*b*) $y = x + 1$

x	−4	−3	−2	−1	0	1	2	3	4
y	−3	−2	−1	0	1	2	3	4	5

(*c*) $y = x + 4$

x	−4	−3	−2	−1	0	1	2	3	4
y	0	1	2	3	4	5	6	7	8

(*d*) $y = x - 2$

x	−4	−3	−2	−1	0	1	2	3	4
y	−6	−5	−4	−3	−2	−1	0	1	2

The coordinates are plotted and joined for each graph. The result is shown in Fig. 8.7

It will be noticed that the straight lines shown are parallel to each other, that is, the **slope or gradient** is the same for each. The gradient of a straight line may be calculated as shown by the following example.

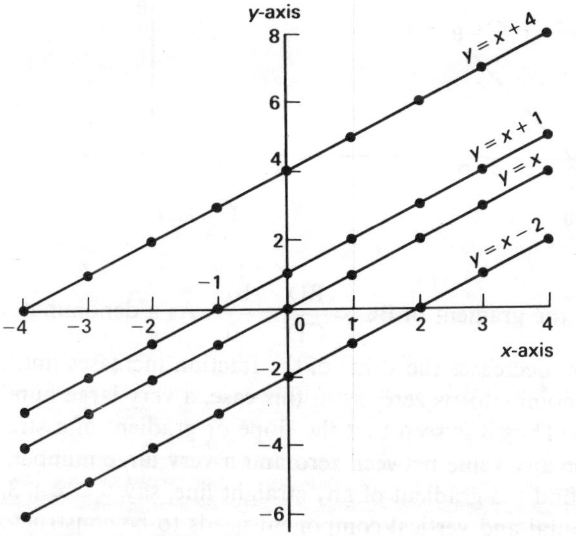

Fig. 8.7 Graph of $y = x$, $y = x + 1$, $y = x + 4$ and $y = x - 2$

Let OA represent a horizontal surface and BC a flat piece of metal hinged at C. This is illustrated in Fig. 8.8.
As BC lies flat on OA, BC is parallel to OA. The slope or gradient of BC in this position is zero, that is, BC has no gradient.

In Fig. 8.9, B has been raised.

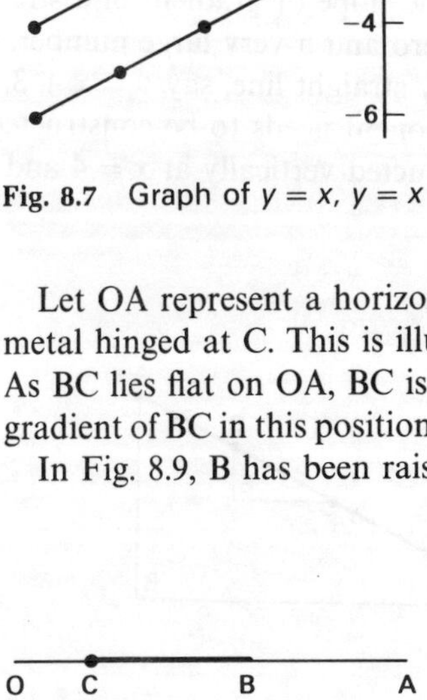

Fig. 8.8

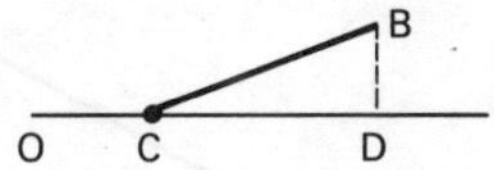

Fig. 8.9

If the broken line BD is vertical to OA then the **slope or gradient of BC is defined as the ratio $\frac{BD}{CD}$**. Hence if BD is 1 cm in length and CD is 10 cm in length then the gradient would be $\frac{1}{10}$. In other words, BC rises 1 unit vertically in 10 units horizontally.

In Fig. 8.10, B is raised until BD = CD.

Then the gradient of BC $= \frac{BD}{CD} = 1$. Thus when the vertical height BD equals the horizontal length CD then the gradient is 1.

In Fig. 8.11, B has been raised until BC is in a vertical position. Now BC = BD and CD = 0.

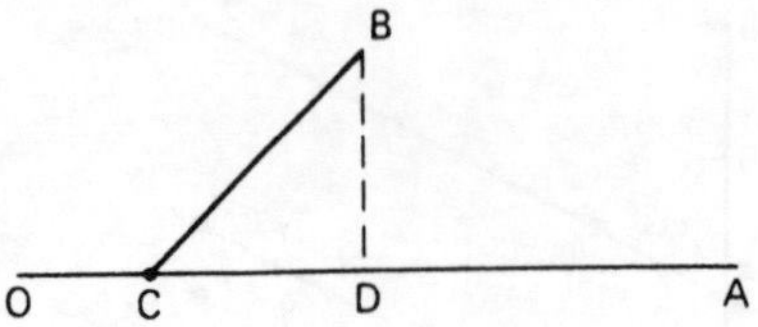

Fig. 8.10

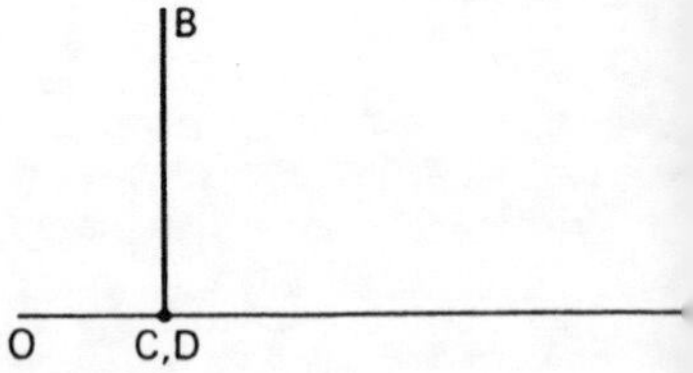

Fig. 8.11

Hence the gradient of BC $= \dfrac{BD}{CD} = \dfrac{BD}{0}$. As a denominator of a fraction decreases the value of the fraction increases until, when the denominator is zero, as in this case, a very large number results. Thus it is seen that the slope or gradient of a straight line may be any value between zero and a very large number.

To find the gradient of any straight line, say $y = x + 3$, a horizontal and vertical component needs to be constructed. Figure 8.12 shows AB constructed vertically at $x = 4$ and BC horizontally at $y = 3$.

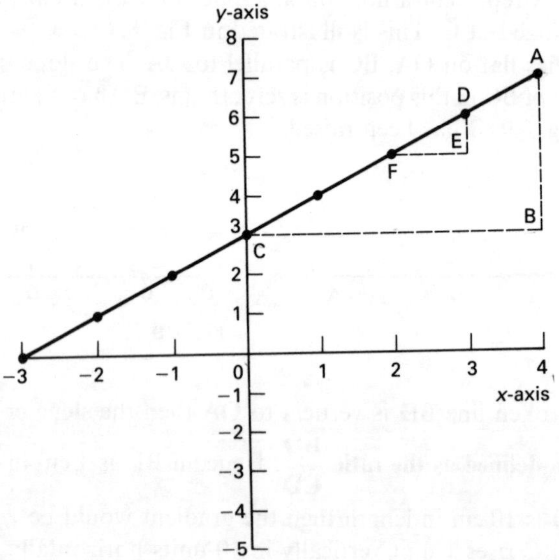

Fig. 8.12 Graph of $y = x + 3$

Thus the gradient of the straight line, that is, the gradient of AC is $\dfrac{AB}{BC}$.

Therefore its gradient is

$$\frac{AB}{BC} = \frac{7-3}{4-0} = \frac{4}{4} = 1$$

Thus the gradient of the straight line $y = x + 3$ is 1.

The positioning of AB and BC is unimportant. It would be equally correct in Fig. 8.12 to make the vertical component at DE and the horizontal component at EF. However, the greater the length involved the greater the accuracy. Then the gradient is

$$\frac{DE}{EF} = \frac{6-5}{3-2} = \frac{1}{1} = 1$$

The position of these components generally is best determined by trying to make the denominator an integer.

In Problem 4 the gradient of each of the straight lines can be shown to be unity.

Problem 5.

(*a*) Plot the following graphs on the same axes between the range of values $x = -3$ to $x = +3$ and find the gradient of each.

(*i*) $y = 2x$ (*ii*) $y = 2x + 3$ (*iii*) $y = 2x - 5$

(*b*) Plot the following graphs on the same axes between the range of values $x = -3$ to $x = +3$ and find the gradient of each.

(*i*) $y = 5x$ (*ii*) $y = 5x + 2$ (*iii*) $y = 5x - 6$

(*a*) A table of coordinates is firstly drawn up for each equation:

(*i*) $y = 2x$

x	−3	−2	−1	0	1	2	3
y	−6	−4	−2	0	2	4	6

(*ii*) $y = 2x + 3$

x	−3	−2	−1	0	1	2	3
y	−3	−1		1	3	5	7	9

(*iii*) $y = 2x - 5$

x	−3	−2	−1	0	1	2	3
y	−11	−9	−7	−5	−3	−1	1

Plotting and joining the coordinates, the graphs shown in Fig. 8.13 are produced.

The graphs are straight lines and are parallel to each other. If the gradient of the graph $y = 2x$ is found, this will also be the gradient of $y = 2x + 3$, and $y = 2x - 5$, since parallel lines have the same gradient.

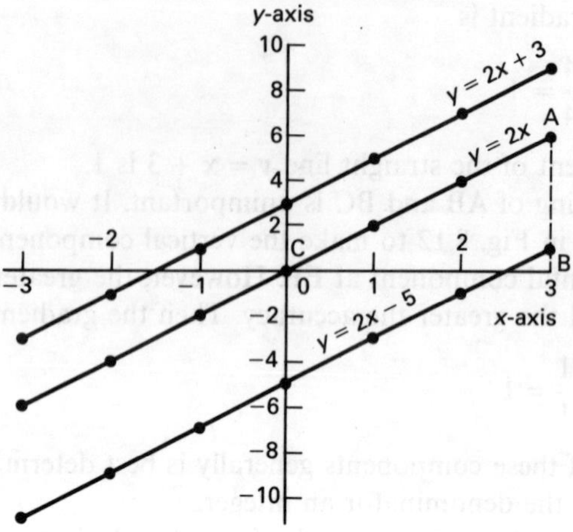

Fig. 8.13 Graphs of $y = 2x$, $y = 2x + 3$ and $y = 2x - 5$

Hence the gradient of AC $= \dfrac{AB}{BC} = \dfrac{6}{3} = 2$

Therefore **for each of the straight lines $y = 2x$, $y = 2x + 3$ and $y = 2x - 5$ the gradient is 2.**

(*b*) A table of coordinates is again drawn up for each equation:

(*i*) $y = 5x$

x	−3	−2	−1	0	1	2	3
y	−15	−10	−5	0	5	10	15

(*ii*) $y = 5x + 2$

x	−3	−2	−1	0	1	2	3
y	−13	−8	−3	2	7	12	17

(*iii*) $y = 5x - 6$

x	−3	−2	−1	0	1	2	3
y	−21	−16	−11	−6	−1	4	9

Plotting and joining the coordinates, the graphs shown in Fig. 8.14 are produced.

The graphs are straight lines and are parallel to each other.

If the gradient of the graph $y = 5x$ is found, this will also be the gradient of $y = 5x + 2$ and $y = 5x - 6$.

Hence the gradient of DF $= \dfrac{DE}{EF} = \dfrac{15}{3} = 5$

Therefore **for each of the straight lines $y = 5x$, $y = 5x + 2$, and $y = 5x - 6$ the gradient is 5**

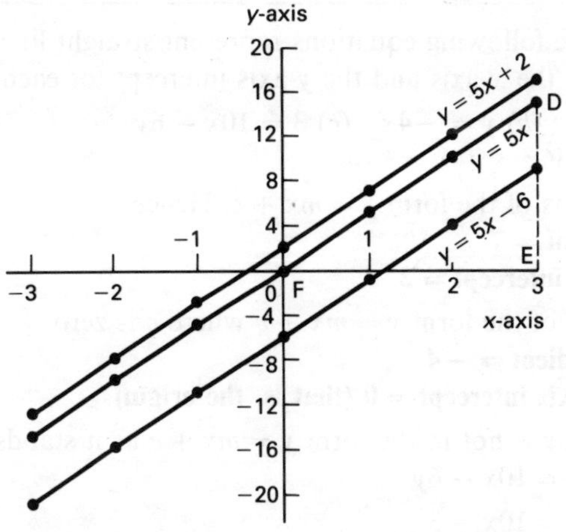

Fig. 8.14 Graphs of $y = 5x$, $y = 5x + 2$, $y = 5x - 6$

The equation of a straight line

Important properties relating to straight-line graphs can be deduced from problems 4 and 5.

When an equation is of the form $y = mx + c$ (where m and c are constants) then

(*i*) a graph of y against x produces a straight line;
(*ii*) m represents the slope or gradient of the line with the x-axis; and
(*iii*) c represents the value of y where the straight line cuts the y-axis. This is known as the intercept. (This is only true when the x scale is zero at the origin.)

When plotting a graph of the form $y = mx + c$, only two coordinates need be determined. When the coordinates are plotted a straight line is drawn between the two points. Normally, however, three coordinates are determined, the third one acting as a check.

If an equation with two variables can be arranged in the straight-line form $y = mx + c$ then the slope and the intercept with the y-axis can be readily found. An equation such as $y = 4$ represents a horizontal straight line intercepting the y-axis at 4. (That is, the line has no gradient with the x-axis.) An equation such as $x = 3$ represents a vertical straight line intercepting the x-axis at 3.

Problem 6. The following equations represent straight lines. Find the gradient to the x-axis and the y-axis intercept for each.

(*a*) $y = 7x + 2$ (*b*) $y = -4x$ (*c*) $3 = 10x - 6y$
(*d*) $7 - 63y = 36x$

(*a*) $y = 7x + 3$ is of the form $y = mx + c$. Hence
$m =$ **the gradient = 7**
$c =$ **the y-axis intercept = 2**

(*b*) $y = -4x$ is of the form $y = mx + c$ where c is zero.
Hence: **the gradient = −4**
the y-axis intercept = 0 (that is, the origin)

(*c*) $3 = 10x - 6y$ is not in the form $y = mx + c$ as it stands.
Therefore if $3 = 10x - 6y$
then $6y = 10x - 3$

$$y = \frac{10x - 3}{6}$$

$$y = \frac{10}{6}x - \frac{3}{6}$$

$$y = \frac{5}{3}x - \frac{1}{2}$$

This is now of the form $y = mx + c$
Hence: **the gradient = $\frac{5}{3}$**
the y-axis intercept = $-\frac{1}{2}$

(*d*) $7 - 63y = 36x$ is not of the form $y = mx + c$ as it stands.
Therefore if $7 - 63y = 36x$
then $7 - 36x = 63y$
or $63y = 7 - 36x$
Therefore $y = \dfrac{7 - 36x}{63}$

$$y = \frac{7}{63} - \frac{36x}{63}$$

$$y = \frac{1}{9} - \frac{4}{7}x$$

$$y = -\frac{4}{7}x + \frac{1}{9}$$

This is now of the form $y = mx + c$
Hence: **the gradient = $-\frac{4}{7}$**
the y-axis intercept = $\frac{1}{9}$

Problem 7. Plot a graph of $2y - 5x = 14$ between the range $x = -3$ and $x = +3$. Find from the graph:
(*a*) the slope to the *x*-axis (*b*) the value of *y* when *x* is $1\frac{1}{2}$
(*c*) the value of *x* when *y* is $-3\frac{1}{2}$
Check the answers by calculations.

If $2y - 5x = 14$

then $2y = 5x + 14$

Therefore $y = \frac{5}{2}x + 7$

$\mathbf{y = 2.5x + 7}$

A table of coordinates can now be produced:

x	−3	−2	−1	0	1	2	3
y	−0.5	2	4.5	7	9.5	12	14.5

Plotting and joining the coordinates the straight-line graph shown in Fig. 8.15 is produced.

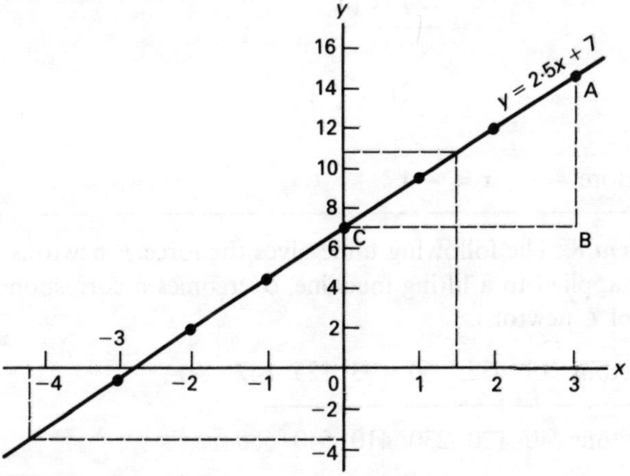

Fig. 8.15 Graph of $2y - 5x = 14$ (or $y = 2.5x + 7$)

(*a*) From the graph, the gradient or slope of AC is $\frac{AB}{BC}$

Hence **the slope of graph** $= \frac{14\frac{1}{2} - 7}{3 - 0} = \frac{7.5}{3} = \mathbf{2.5}$

This may be checked from the fact that $y = 2.5x + 7$ is of the form $y = mx + c$. Therefore the coefficient of *x* represents the gradient, that is, 2.5

(*b*) From the graph, when x is 1.5, the value of y is **10.8** (The decimal part can only be an estimation.)

By calculation: $y = 2.5x + 7$

therefore $y = (2.5)(1.5) + 7$

$= 3.75 + 7$

$y = \mathbf{10.75}$

(*c*) From the graph, when y is -3.5, the value of x is -4.2 (This may be read from the graph by extrapolating.)

By calculation: $y = 2.5x + 7$

therefore $y - 7 = 2.5x$

$$2y - 14 = 5x$$

$$5x = 2y - 14$$

$$x = \frac{2y - 14}{5}$$

When $y = -3.5$ $x = \dfrac{2(-3.5) - 14}{5}$

$$= \frac{-7 - 14}{5}$$

$$= \frac{-21}{5}$$

Therefore $\quad x = \mathbf{-4.2}$

Problem 8. The following table gives the force F newtons which, when applied to a lifting machine, overcomes a corresponding load of L newtons.

F newtons	19	35	50	93	125	147
L newtons	40	120	230	410	540	680

(*a*) Choose suitable scales and plot a graph with F representing the vertical axis and L the horizontal axis. Label the graph and its axes.
(*b*) Find the slope of the graph.
(*c*) From the graph determine
  (*i*) the force applied when the load is 325 N, and
  (*ii*) the load that a force of 45 N will overcome.
(*d*) If the graph were to continue in the same manner what value of force would be needed to overcome a 750 N load?
(*e*) State the equation of the graph.

(*a*) Plotting the coordinates (40, 19), (120, 35), (230, 50), and so on, the graph produced is the best straight line which can be drawn corresponding to these points, as shown in Fig. 8.16. The graph and each of its axes are labelled.
(*b*) To find the slope of the graph, horizontal and vertical components are constructed. Such components may be positioned anywhere along the length of the straight line. The horizontal component AB is chosen so that it is a number which makes any calculations easier, in this case 500. Thus the slope of graph = the slope of AC $= \frac{BC}{AB}$. Therefore the slope of the graph is

$$\frac{130 - 30}{600 - 100} = \frac{100}{500} = \frac{1}{5} \text{ or } 0.2$$

Hence the slope of the graph is **0.2**
(*c*) (*i*) When the load is 325 N the corresponding force is **75 N**
(*ii*) When the force is 45 N the corresponding load is **170 N**
Each of these can be seen from the broken lines shown in Fig. 8.16
(*d*) The graph is extended at its upper end and a load of 750 N corresponds to a force of **161 N**
(*e*) The equation of a straight line is $y = mx + c$ where y is plotted on the vertical axis and x on the horizontal axis. Also m represents the slope and c the y-axis intercept. Thus, in this case:

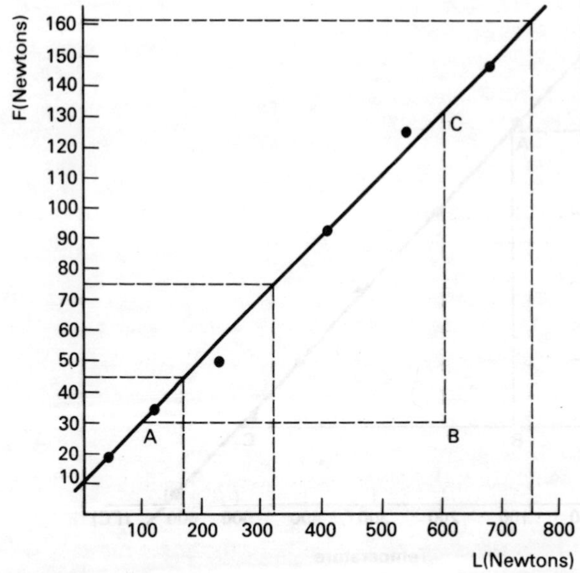

Fig. 8.16 Graph of force against load

Force F corresponds to y,
Load L corresponds to x,
m, the slope is 0.2,
c, the y-axis intercept is 11 (by extrapolation).
Hence the equation of the graph is $\boldsymbol{F = 0.2L + 11}$

Problem 9. Results obtained from determining the breaking stress σ of rolled copper at various temperatures, t, are as follows:

Stress, σ ($\times 10^4$ Pa)	8.28	7.99	7.70	7.48	7.30	7.03
temperature, y (°C)	90	205	320	410	480	590

(*a*) Plot a graph of stress (vertical) against temperature (horizontal), drawing the best straight line through the plotted coordinates.

Determine (*b*) the gradient of the graph, (*c*) the law of the graph, and (*d*) the value of stress when the temperature is 700°C if a linear relationship is assumed outside the given range of values.

(*a*) The coordinates (90, 8.28), (205, 7.99), and so on, are plotted as shown in Fig. 8.17 and the best straight line drawn through the points.

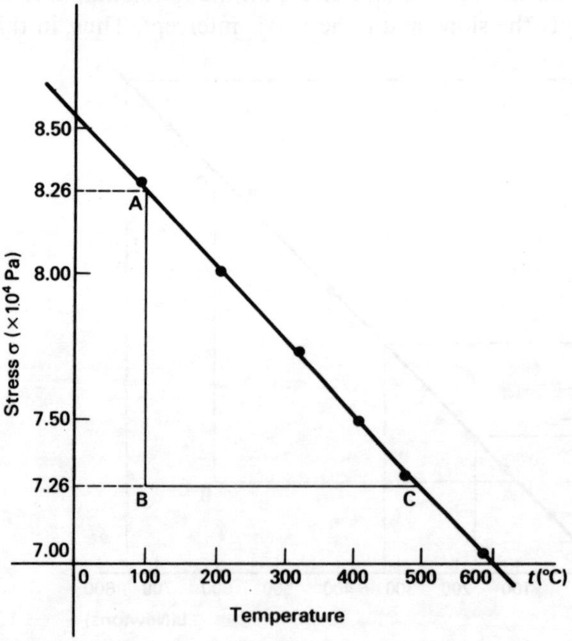

Fig. 8.17

(*b*) Gradient of the straight line $= \dfrac{AB}{BC} = \dfrac{(8.26 - 7.26)10^4}{100 - 500}$

$$= \mathbf{(-0.0025) \times 10^4}$$

(*c*) Intercept on the vertical axis $= 8.52 \times 10^4$
Hence the equation of the graph is $\sigma = \mathbf{(-0.0025t + 8.52)10^4}$

(*d*) When $t = 700°C$, the stress

$$\sigma = (-0.0025 \times 700 + 8.52)10^4$$

$$= (-1.75 + 8.52)10^4 = \mathbf{6.77 \times 10^4\ Pa}$$

Further problems on straight line graphs may be found in Section 8.3 (Problems 9–26) page 193.

Summary

The following general rules can be applied to any graph:

1. Give the graph a title, clearly explaining what is being illustrated.
2. The scales are chosen such that the graph occupies as much space as possible on the graph paper being used, consistent with simplicity of axes.
3. Scales are chosen so that interpolation is made as easy as possible. Usually, scales such as 1 cm = 1 unit, or 1 cm = 2 units or 1 cm = 10 units are used. Awkward scales such as 1 cm = 3 units or 1 cm = 7 units are not used.
4. The scales should not start at zero if starting at zero produces an accumulation of points within a small area of the graph paper.
5. The coordinates, or points, should be clearly marked. This may be done either by a cross, by a dot and circle or by a dot only.
6. A statement should be made along each axis explaining the numbers represented with their appropriate units.
7. Sufficient numbers should be written along each axis without cramping.
8. If the equation of a graph is of the form $y = mx + c$ the graph will always be a straight line, where m represents the slope or gradient and c the intercept with the y-axis, provided false zeros have not been introduced. Do not join each point in turn but draw the best straight line through the coordinates.
9. To find the gradient of a straight line graph a horizontal and a vertical component are constructed for any two points on the straight line. The gradient is then given by the value of the vertical component divided by the horizontal component.

8.3 Further problems

Direct and inverse proportionality

1. If y is directly proportional to x and $y = 45$ when $x = 7.2$, determine (*a*) the coefficient of proportionality and (*b*) the value of y when x is 10. (*a*) [6.25] (b) [62.5]
2. Charles' law states that for a given mass of gas at constant pressure the volume is directly proportional to its thermodynamic temperature. A gas occupies a volume of 4.5 dm^3 at 300 K. Determine, for constant pressure, (*a*) the coefficient of proportionality, (*b*) the volume at 380 K and (*c*) the temperature when the volume is 4.8 dm^3.
(*a*) [0.015] (*b*) [5.7 dm^3] (*c*) [320 K]
3. Hooke's law states that stress is directly proportional to strain within the elastic limit of a material. A spring has a coefficient of proportionality k of 150 N m^{-2}. (*a*) Determine the load required to extend it by 20 mm. (*b*) What extension corresponds to 13.5 N? (*a*) [3 N] (*b*) [90 mm]
4. 0.002 m^3 of aluminium has a mass of 5.3 kg. If mass is directly proportional to volume, determine (*a*) the mass of 0.001 2 m^3 of aluminium and (*b*) the volume of 8.48 kg of aluminium. (*a*) [3.18 kg] (*b*) [0.003 2 m^3]
5. Ohm's law states that the current flowing in a resistor is directly proportional to the applied voltage. When 40 volts are applied across a resistor the current flowing through the resistor is 3.2×10^{-3} amperes. Determine (*a*) the coefficient of proportionality, (*b*) the current when the voltage is 56 volts, and (c) the voltage when the current is 5.4×10^{-3} amperes. (*a*) [8×10^{-5}] (*b*) [4.48×10^{-3} A] (*c*) [67.5 V]
6. The distillation rate, R (in cm^3/min), of a given gas is inversely proportional to the temperature, T (K). If $R = 3$ when $T = 5$, (*a*) what is the rate when $T = 4$, and (*b*) at what temperature is $R = 5$? (*a*) [$3\frac{3}{4}$ cm^3/min] (*b*) [3 K]
7. Boyle's law states that for a gas at constant temperature, the volume of a fixed mass of gas is inversely proportional to its absolute pressure. If a gas occupies a volume of 2.25 m^3 at a pressure of 2×10^5 pascals determine for constant temperature (*a*) the coefficient of proportionality, (*b*) the volume when the pressure is 7.5×10^5 pascals, and (*c*) the pressure when the volume is 1.5 m^3. (*a*) [4.5×10^5] (*b*) [0.6 m^3] (*c*) [3×10^5 *Pa*]
8. The resistor R of a length of cable is inversely proportional to its cross-sectional area, a. When $a = 6$ mm^2, $R = 5.5$ ohms.

Determine (*a*) the coefficient of proportionality, (*b*) the cross-sectional area corresponding to a resistance of 2 ohms, and (*c*) the resistance when the cross-sectional area is 12 mm^2. (*a*) [33] (*b*) [16.5 mm^2] (*c*) [2.75 ohms]

Straight line graphs

9. Assuming graph paper measuring 20 cm by 20 cm is available, suggest suitable scales for the following ranges of values:

(*a*) Horizontal axis: 3 volts to 54 volts. [1 cm rep. 4 volts]
Vertical axis: 10 ohms to 180 ohms.
[1 cm rep. 10 ohms]

(*b*) Horizontal axis: 7 cm to 85 cm. [1 cm rep. 5 cm]
Vertical axis: 0.30 volts to 1.68 volts.
[1 cm rep. 0.1 volt]

(*c*) Horizontal axis: 210 N to 340 N.
[1 cm rep. 10 N (Note: not starting at zero)]
Vertical axis: 0.6 mm to 3.3 mm. [1 cm rep. 0.2 mm]

10. Corresponding values obtained experimentally for two quantities are:

x	-2	-1	0	1	2.5	4	5
y	-12	-7	-2	3	10.5	18	23

Using a horizontal scale for x of 1 cm = $\frac{1}{2}$ unit and a vertical scale for y of 1 cm = 2 units, draw a graph of x against y. Label the graph and each of its axes. By interpolation, find from the graph the value of y when x is 1.5. [5.5]

11. The equation of a line is $y = 3x + 2$. A table of corresponding values is produced and is shown below. Complete the table and plot a graph of y against x. Find the gradient of the graph.

x	-4	-3	-2	-1	0	1	2	3	4
y		-7			2		8		

[Gradient = 3]

12. Find the gradient and intercept on the y-axis for each of the following equations:
(*a*) $y = 3x + 4$ (*b*) $y = -2x + 3$ (*c*) $y = -5x - 1$
(*d*) $y = 6x$
(*a*) [3, 4] (*b*) [-2, 3] (*c*) [-5, -1] (*d*) [6, 0]

13. Find the gradient and the y-axis intercept for each of the following equations:

(*a*) $3y + 1 = 2x$ (*b*) $5x + 2y = -4$ (*c*) $4(2y - 3) = \frac{x}{2}$

(*d*) $3x - \frac{y}{2} = 5\frac{1}{2}$ (*e*) $y = 5$

Sketch the graph in each case showing the slope and *y*-axis intercept. (*a*) $[\frac{2}{3}, -\frac{1}{3}]$ (*b*) $[-2\frac{1}{2}, -2]$ (*c*) $[\frac{1}{16}, 1\frac{1}{2}]$ (*d*) $[6, -11]$ (*e*) $[0, 5]$

14. Draw on the same axes the graphs of $y = 3x - 1$ and $y + 2x = 4$. Find the coordinates of the point of intersection. [1, 2]

15. In an experiment on Charles' Law, the value of the volume of gas V m^3 was measured for various temperatures T°C. The results are tabulated below:

V m^3	26.68	27.16	27.56	28.08	28.52	29.00
T°C	10	20	30	40	50	60

Choose suitable scales and plot a graph of volume (vertically) against temperature (horizontally). Interpolate to find the volume at 35°C. Extrapolate to find the temperature when the volume is 30.0 m^3, assuming the relationship holds. [27.85 m^3, 81.7°C]

16. Plot the graphs of $y - 6.5 = 2x$ and $3x + 1 = -y$ and find the value of x and y at the point of intersection. Check the result obtained by solving the two simultaneous equations algebraically. [−1.5, 3.5]

17. In an experiment demonstrating Hooke's Law, the strain in a wire was measured for various stresses. The results are shown below:

Stress N/mm^2	10.8	21.6	33.3	37.8	45.9
Strain	0.000 12	0.000 24	0.000 37	0.000 42	0.000 51

Using scales of 1 cm = 5 N/mm^2 stress and 2 cm = 0.000 1 strain and plotting stress on the vertical axis, draw a graph of stress against strain. Find Young's Modulus of elasticity which is given by the gradient of the graph. [90 kN/mm^2]

18. The speed n rev/min of a motor changes when the voltage V across the armature is varied. The results are shown in the table below:

n rev/min	560	720	900	1 010	1 240	1 440
V volts	80	100	120	140	160	180

It is suspected that one of the readings taken of the speed is inaccurate. Plot a graph of speed (horizontally) against voltage (vertically) and find this value. Find also (*a*) the speed at a voltage of 132 V and (*b*) the voltage at a speed of 1 300 rev/min. [The 1 010 rev/min reading should be 1 088 rev/min (*a*) 1 018 rev/min (*b*) 164 V]

19. A test on a metal filament lamp gave the following values of resistance R ohms at various voltages V volts.

V volts	131		195	221	255	296	355
R ohms	78	95	121		162	190	230

Draw a graph of voltage (horizontally) against resistance (vertically) and find the missing values in the table. [$V = 156$ V at 95 ohms; $R = 139$ ohms at 221 V]

20. Tests carried out to determine the breaking stress S (N cm^{-2}) of rolled copper at different temperatures T (°C) gave the following results:

S N cm^{-2}	8.90	8.45	8.20	7.70	7.50	7.10
T °C	60	210	300	410	500	600

For the graph paper you have available, choose as big a scale as possible with stress on the vertical axis and temperature on the horizontal axis. Draw the best straight line through the plotted coordinates. Find the slope of the graph and the intercept with the y-axis. [$-\frac{1}{300}$, 9.1]

21. The velocity v m s^{-1} of a body at the end of a time interval t seconds was found by a series of experiments to be as tabulated below:

t seconds	1	3	6	8	11	12
v m s^{-1}	13.8	15.4	17.8	19.4	21.8	22.6

By plotting v vertically and t horizontally draw the graph of velocity against time. Find the equation of the graph. [$v = 0.8t + 13$]

22. The solubility of potassium bromide in g per 100 g of water varies with temperature in degrees Celsius as shown below.

Temperature (t)	10	20	30	40	50	60	70	80
Mass of potassium bromide (x)	2.97	3.23	3.50	3.77	4.04	4.30	4.57	4.83

Show that these results obey a law of the form $t = mx + c$ where m and c are constants. Determine m and c.
[$t = 38x - 103$]

23. The crushing strength of mortar in tonnes varies with the percentage of water used in its preparation as shown below.

Crushing strength (F)	1.51	1.33	1.15	0.97	0.79	0.61
% of water used (x)	8.0	10	12	14	16	18

Prove that these values obey the law $F = ax + b$ where a and b are constants. Determine a and b.
[$F = -0.090x + 2.23$]

24. The mass, m, of a steel joist varies with length, l, as follows:

mass (kg)	95.0	100	110	120
length (m)	3.480	3.670	4.030	4.400

Find the relationship between them.
[$m = 27.40l - 0.5500$]

25. The amount of light (L) absorbed by a coloured sulphide solution varies with the concentration (C in μg per 100 g of solution) as follows:

Amount of light (L)	0	0.09	0.18	0.27	0.36	0.45
Concentration (C)	0	2	4	6	8	10

Find the relationship between L and C. [$L = 0.045C$]

26. The molar heat capacity (c in J deg^{-1} mole^{-1}) varies with absolute temperature (T in K) as shown. (The results are for silicon carbide.)

Molar heat capacity (c)	38.77	41.29	43.81	46.33
Temperature (T)	100	300	400	700

How is c related to T? [$c = 0.0126T + 37.51$]

Chapter 9

Introduction to geometry

9.1 Definitions

A **line** is that which has length without breadth or thickness.
A **point** is that which has position but no magnitude.
Geometry is a part of mathematics in which the properties of points, lines, surfaces and solids are investigated.

9.2 Angles and their measurement

An **angle** is formed by the intersection of two **straight lines**. Let AOX, in Fig. 9.1 represent an angle. Consider the line OX to be fixed and the other line OA free to rotate about the point O. The point O is called the **vertex** of the angle, while OA and OX are called the **arms** of the angle.

By definition, **an angle is the amount of rotation between two straight lines**. Thus, in Fig. 9.1, OA has rotated from OX. The amount of rotation is known as angle AOX and is written as $\angle AOX$.

Angles may be measured in either **degrees** or **radians** (radian measure is dealt with in Chapter 11).

If the line OA lies along OX and is rotated until it lies along OX again, it is said to have made one **revolution**. If one revolution is divided into 360 equal parts then each part is called 1 degree. Thus

$$1 \text{ revolution} = 360 \text{ degrees (or 1 rev.} = 360°)$$

A degree is subdivided into 60 minutes (written as 60′) and a minute is

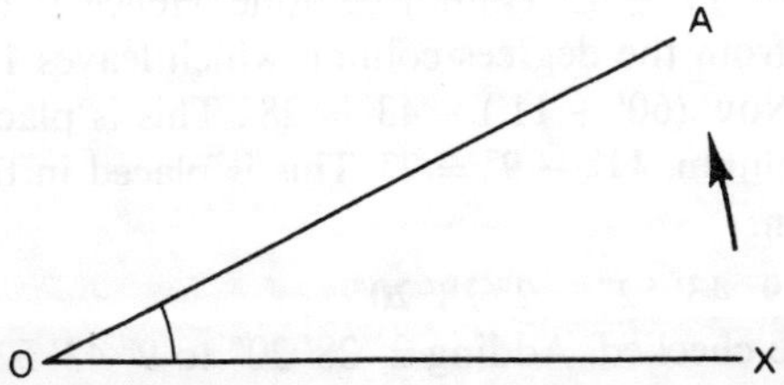

Fig. 9.1

subdivided into 60 seconds (written as 60″). Hence, summarising:

1 degree $= \frac{1}{360}$th of one revolution
1 minute $= \frac{1}{60}$th of a degree
1 second $= \frac{1}{60}$th of a minute

Worked problems on the measurement of angles

Problem 1. Add (*a*) 13° 52′ and 26° 29′ (*b*) 25° 37′ 51″ and 41° 29′ 16″.

(*a*)

13° 52′	
26° 29′	
40° 21′	
1°	

52′ + 29′ gives 81′. Since 60′ = 1°, 81′ = 1° 21′. Thus the 21′ is placed in the minutes column and 1° is carried in the degrees column. Then 13° + 26° + 1° (carried) = 40°. This is placed in the degrees column.

Thus 13° 52′ + 26° 29′ = **40° 21′**

(*b*)

25°	37′	51″
41°	29′	16″
67°	7′	7″
1°	1′	

51″ + 16″ gives 67″. Since 60″ = 1′, 67″ = 1′ 7″. Thus the 7″ is placed in the seconds column and 1′ is carried in the minutes column. Then 37′ + 29′ + 1′ (carried) = 67′. Since 60′ = 1°, 67′ = 1° 7′. Thus the 7′ is placed in the minute column and 1° is carried in the degrees column. Then 25° + 41° + 1° = 67°. This is placed in the degrees column.

Thus 25° 37′ 51″ + 41° 29′ 16″ = **67° 7′ 7″**

Problem 2. Subtract (*a*) 9° 43′ 52″ from 12° 12′ 12″ (*b*) 22° 45′ 39″ from 49° 28′ 15″

(*a*)

11°	11′	
~~12~~°	~~12~~°	12″
9°	43′	52″
2°	28′	20″

12″ − 52″ cannot be done. Hence 1′ or 60″ is 'borrowed' from the minutes column which leaves 11′ in that column. Now (60″ + 12″) − 52″ = 20″. This is placed in the seconds column. 11′ − 43′ cannot be done. Hence 1° or 60′ is 'borrowed' from the degrees column which leaves 11° in that column. Now (60′ + 11′) − 43′ = 28′. This is placed in the minutes column. 11° − 9° = 2°. This is placed in the degrees column.

Thus 12° 12′ 12″ − 9° 43′ 52″ = **2° 28′ 20″**

(This answer can be checked. Adding 2° 28′ 20″ to 9° 43′ 52″ should give 12° 12′ 12″.)

$$\begin{array}{l} 48^\circ\ 27' \\ (b)\ \ \not{4}\not{9}^\circ\ \not{2}\not{8}'\ 15'' \\ \underline{22^\circ\ 45'\ 39''} \\ \underline{26^\circ\ 42'\ 36''} \end{array}$$

15″ − 39″ cannot be done. Hence 1′ or 60″ is 'borrowed' from the minutes column which leaves 27′ in that column. Now (60″ + 15″) − 39″ = 36″. This is placed in the seconds column. 27′ − 45′ cannot be done. Hence 1° or 60′ is 'borrowed' from the degrees column which leaves 48° in that column. Now (60′ + 27′) − 45′ = 42′. This is placed in the minutes column. 48° − 22° = 26°. This is placed in the degrees column.

Thus 49° 28′ 15″ − 22° 45′ 39″ = **26° 42′ 36″**
(This may be checked by adding the bottom two lines of the subtraction sum. That is, 22° 45′ 39″ plus 26° 42′ 36″ should be equal to the top line, 49° 28′ 15″.)

Problem 3. Convert the following angles to degrees and decimals of a degree: (*a*) 23° 42′ (*b*) 84° 33′ 17″

(*a*) Since 1 minute = $\frac{1}{60}$th of a degree
42 minutes = $\frac{42}{60}$ of a degree
= 0.70 degree
Therefore 23° 42′ = **23.70°**

(*b*) Since 1 second = $\frac{1}{60}$th of a minute
17 seconds = $\frac{17}{60}$ minute
= 0.283 3 minute
Therefore 84° 33′ 17″ = 84° 33.283 3′
Since 1 minute = $\frac{1}{60}$th of a degree

$$33.283\,3 \text{ minutes} = \frac{33.283\,3}{60} \text{ degree}$$

= 0.554 7 degree correct to 4 decimal places.
Hence 84° 33′ 17″ = **84.554 7° correct to 4 decimal places.**

Problem 4. Convert 37.468° into degrees, minutes and seconds.

Since 1 degree = 60 minutes
0.468 degree = 0.468 × 60 minutes
= 28.08 minutes
Since 1 minute = 60 seconds
0.08 minute = 0.08 × 60 seconds
= 4.8 seconds
Hence 37.468° = **37° 28′ 5″, to the nearest second.**

Further problems on angular measurement may be found in Section 9.5 (Problems 1–4), page 206.

9.3 Acute, obtuse, reflex and right angles

There are four names given to angles, according to their size.

1. Any angle between 0° and 90° is called an **acute angle**. For example, 30° is an acute angle and is shown in Fig. 9.2.

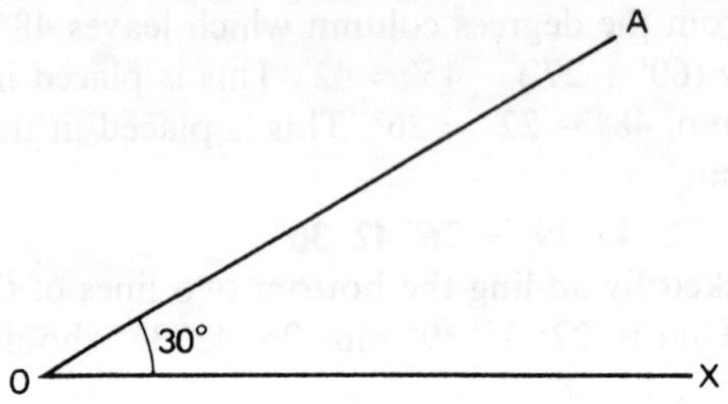

Fig. 9.2

2. An angle exactly equal to 90° is called a **right angle**. OA and OX in Fig. 9.3 are 90° apart and are said to be **perpendicular** to one another.

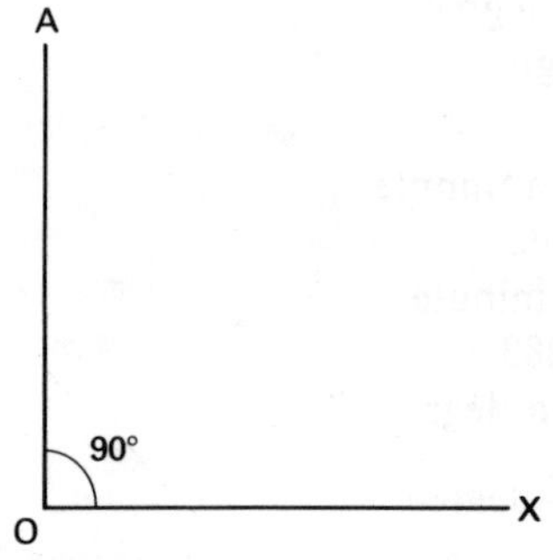

Fig. 9.3

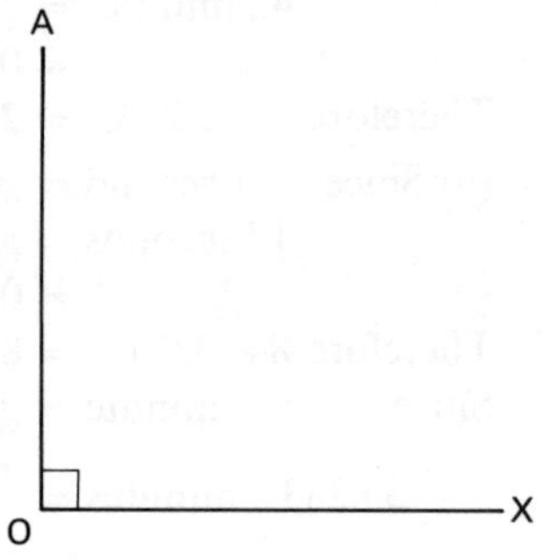

Fig. 9.4

There is a special sign used to denote a right angle and this is shown in Fig. 9.4. When this sign is used it is unnecessary to write the value of the angle as 90°, it is assumed to mean this.

Note that there are four right angles in one complete revolution.

3. Any angle between 90° and 180° is called an **obtuse angle**. For example 145° is an obtuse angle and is shown in Fig. 9.5.
4. Any angle greater than 180° and less than 360° is called a **reflex angle**. For example 230° is a reflex angle and is shown in Fig. 9.6.

9.4 Properties of angles and straight lines

1. An angle of 180° lies on a straight line. See Fig. 9.7.
2. If two angles add up to 90° they are called **complementary angles**. For

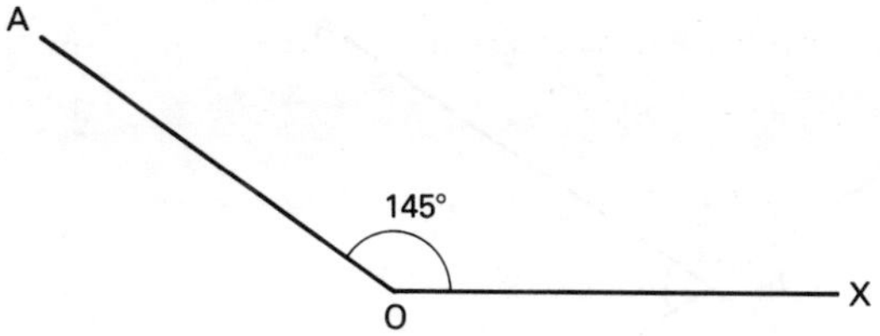

Fig. 9.5

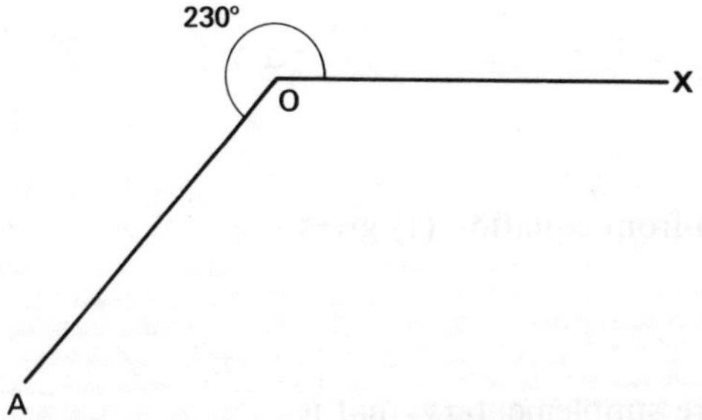

Fig. 9.6

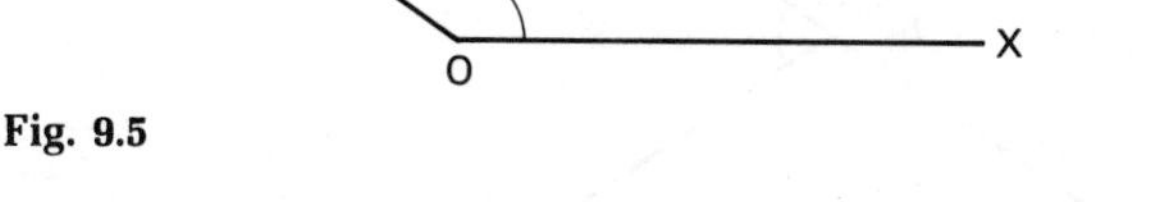

Fig. 9.7

example, 15° and 75° are complementary angles and each angle is called the complement of the other. Similarly, the complement of 27° is 63° and the complement of $12\frac{1}{2}°$ is $77\frac{1}{2}°$.

3. If two angles add up to 180° they are called **supplementary angles**. For example 76° and 104° are supplementary angles and each angle is called the supplement of the other. Similarly the supplement of 129° is 51° and the supplement of 14° 23′ is 165° 37′.
4. When two straight lines intersect, the vertically opposite angles are equal.

Proof: Let two straight lines RT and PQ intersect at O as shown in Fig. 9.8. Let the four angles produced be denoted by a, b, c and d where $\angle a = \angle \text{POR}$, $\angle b = \angle \text{POT}$, $\angle c = \angle \text{TOQ}$ and $\angle d = \angle \text{ROQ}$

Now $\angle a$ and $\angle b$ are supplementary, that is

$$\angle a + \angle b = 180 \qquad (1)$$

Also $\angle b$ and $\angle c$ are supplementary, that is

$$\angle b + \angle c = 180 \qquad (2)$$

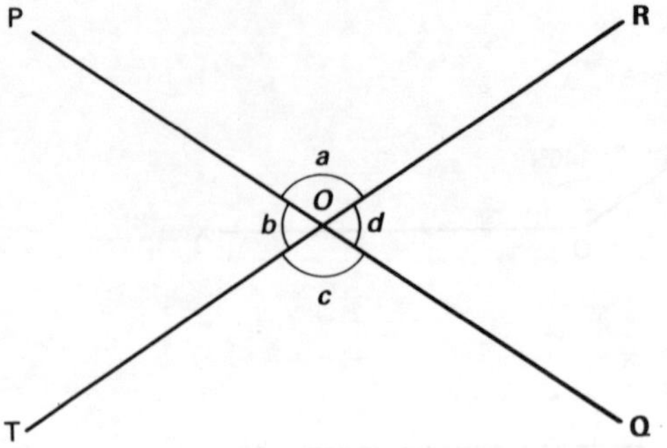

Fig. 9.8

Subtracting equation (2) from equation (1) gives

$$\angle a - \angle c = 0$$

Therefore $\angle a = \angle c$

Similarly $\angle a$ and $\angle d$ are supplementary, that is

$$\angle a + \angle d = 180 \qquad (3)$$

Subtracting equation (3) from equation (1) gives

$$\angle b - \angle d = 0$$

Therefore $\angle b = \angle d$

The pairs of angles, *a* and *c*, and *b* and *d* are called vertically opposite angles.

Thus when two straight lines intersect at a point the vertically opposite angles are equal.

---

5. **Parallel lines** may be defined as straight lines which are in the same plane (flat or level surface) and never meet, however far they may be produced in either direction. Therefore there can never be any angle between two parallel lines.

Let two parallel lines be represented by AB and CD in the plane of the paper as shown in Fig. 9.9. The arrows drawn on AB and CD are a way of denoting parallel lines.

A straight line which crosses two parallel lines is called a **transversal**. Let a transversal be represented by EF as shown in Fig. 9.9. Let *a*, *b*, *c*, *d*, *e*, *f*, *g* and *h* represent the angles formed at the intersections of the transversal with the parallel lines.

Since the lines AB and CD are parallel and the transversal EF is equally inclined to both lines, the following facts emerge and may be stated:

(*i*) $\angle a = \angle e$, $\angle b = \angle f$
$\angle c = \angle g$, $\angle d = \angle h$

Such pairs of angles are called **corresponding angles**.

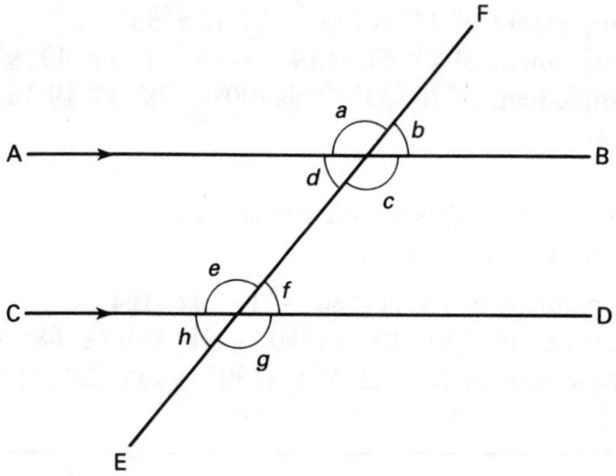

Fig. 9.9

(*ii*) $\angle d = \angle f$, $\angle c = \angle e$

Such pairs of angles are called **alternate angles.**

(*iii*) $\angle d + \angle e = 180°$
$\angle c + \angle f = 180°$

Such pairs of angles are called **interior angles.**

Thus the sum of the interior angles on the same side of the transversal is equal to 180°. That is, $\angle d$ and $\angle e$, and also $\angle c$ and $\angle f$, are supplementary.

If a transversal crosses two straight lines, and any one of the three conditions, (*i*), (*ii*) or (*iii*), holds good, then the two straight lines must be parallel.

Worked problems on types and properties of angles

Problem 1. State the general name given to the following angles: (*a*) 163° (*b*) 90° (*c*) 47° (*d*) 215°

(*a*) 163° lies between 90° and 180° and is therefore called an **obtuse angle.**
(*b*) 90° is called a **right angle.**
(*c*) 47° lies between 0° and 90° and is therefore called an **acute angle.**
(*d*) 215° is greater than 180° and less than 360° and is therefore called a **reflex angle.**

Problem 2. Find the angles complementary to (*a*) 37° (*b*) 49° 51′ (*c*) 78° 33′ 19″

(*a*) The complement of 37° is (90° – 37°) i.e. **53°**
(*b*) The complement of 49° 51′ is (90° – 49° 51′) i.e. **40° 9′**
(*c*) The complement of 78° 33′ 19″ is (90° – 78° 33′ 19″) i.e. **11° 26′ 41″**

Problem 3. Find the angles supplementary to
(*a*) 16° (*b*) 91° 10′ (*c*) 138° 23′ 5″

(*a*) The supplement of 16° is (180° – 16°) i.e. **164°**
(*b*) The supplement of 91° 10′ is (180° – 91° 10′) i.e. **88° 50′**
(*c*) The supplement of 138° 23′ 5″ is (180° – 138° 23′ 5″) i.e. **41° 36′ 55″**

Problem 4. Two straight lines MN and PQ intersect at O. If ∠MOP is 32° find ∠MOQ, ∠QON and ∠PON.

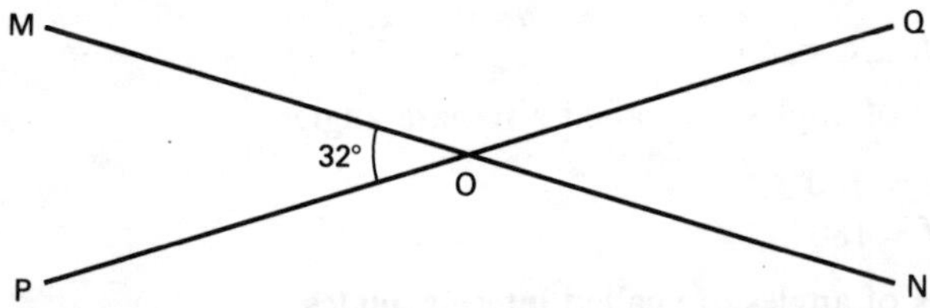

Fig. 9.10

In Fig. 9.10 ∠MOQ is supplementary to ∠MOP
Therefore ∠MOQ = (180° – 32°) = **148°**
When two straight lines intersect the vertically opposite angles are equal.
Therefore ∠QON = **32°**
∠PON = **148°**

Problem 5. Find the angle x in Fig. 9.11

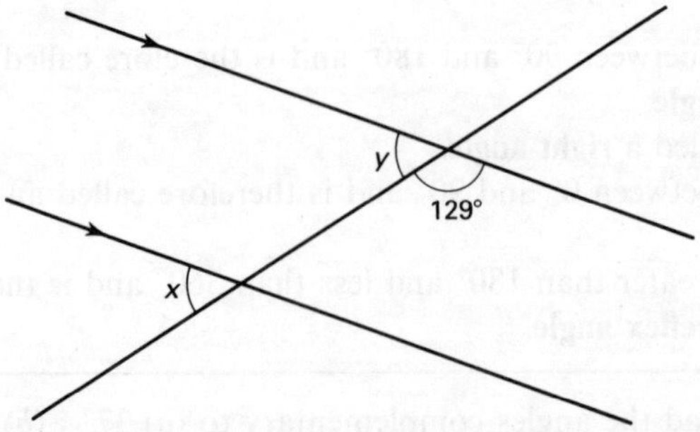

Fig. 9.11

Let $\angle y$ be the angle shown in Fig. 9.11, the supplement of 129°.
Therefore $\angle y = 180° - 129° = 51°$
But $\angle x = \angle y$ (corresponding angles between parallel lines)
Thus $\angle x =$ **51°**

Problem 6. Find the angle θ in Fig. 9.12

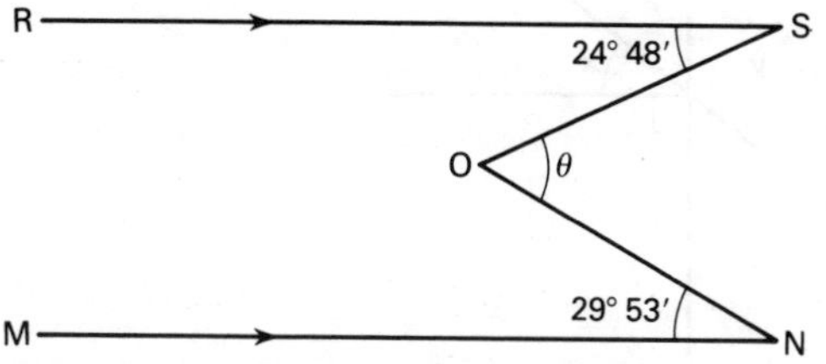

Fig. 9.12

Let a line AB be drawn through O such that AB is parallel to RS and MN as shown in Fig. 9.13

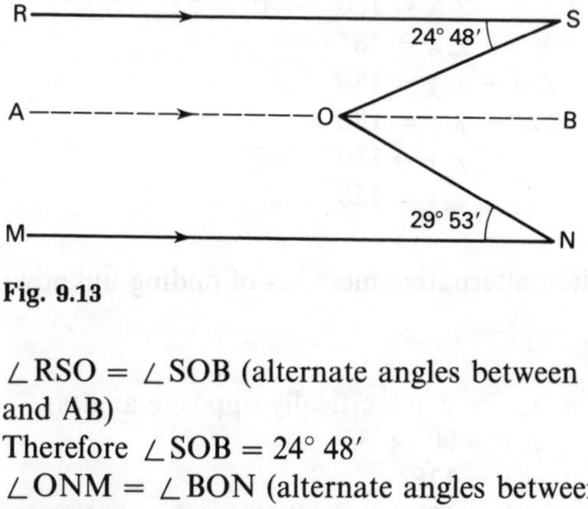

Fig. 9.13

$\angle$RSO = $\angle$SOB (alternate angles between parallel lines RS and AB)
Therefore $\angle SOB = 24° 48'$
$\angle$ONM = $\angle$BON (alternate angles between parallel lines AB and MN)
Therefore $\angle BON = 29° 53'$
From Figs 9.12 and 9.13, $\angle\theta = \angle SOB + \angle BON$
Therefore $\angle\theta = 24° 48' + 29° 53'$
$= 54° 41'$
Hence $\angle\theta =$ **54° 41′**

Problem 7. Find angles x and y in Fig. 9.14

Let a and b be the angles shown in Fig. 9.14
Therefore $\angle a = 52°$ (corresponding angles between parallel lines)
Also $\angle a + \angle x + 90° = 180°$ (angles on a straight line)

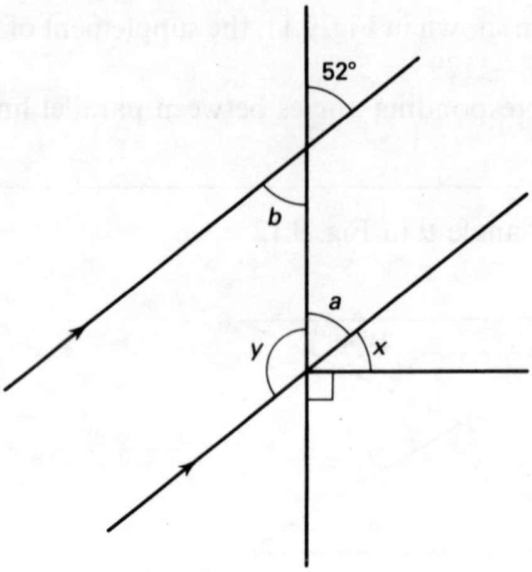

Fig. 9.14

$$\text{Therefore } 52° + \angle x + 90° = 180°$$
$$\angle x = 180° - 90° - 52°$$
$$\angle x = \mathbf{38}°$$

Similarly $\angle a + \angle y = 180°$

Therefore $52° + \angle y = 180°$

$$\angle y = 180° - 52°$$
$$\angle y = \mathbf{128}°$$

There are often alternative methods of finding unknown angles. For example:

$90° + \angle x = \angle y$ (vertically opposite angles)

Therefore $\angle y = 90° + 38°$

$$\angle y = \mathbf{128}°$$

or $\angle b = 52°$ (vertically opposite angles)

Therefore $\angle y + \angle b = 180°$ (interior angles on the same side of a transversal)

$$\angle y = 180° - \angle b$$
$$= 180° - 52°$$
$$\angle y = \mathbf{128}°$$

Further problems on types and properties of angles may be found in Section 9.5 (Problems 5–14), page 207.

9.5 Further problems

Angular measurement

1. Add together the following angles:
(*a*) 15° 32′ and 17° 19′ (*b*) 19° 23′ 42″ and 68° 4′ 33″
(*c*) 26° 46′ 11″ and 36° 38′ 51″
(*d*) 41° 10′ 58″, 32° 11′ 46″ and 11° 12′ 13″
(*e*) 63° 9′ 25″, 73° 19′ 37″ and 22° 22′ 22″
(*a*) [32° 51′] (*b*) [87° 28′ 15″] (*c*) [63° 25′ 2″]
(*d*) [84° 34′ 57″] (*e*) [158° 51′ 24″]
2. Subtract the following angles:
(*a*) 15° 11′ 18″ from 24° 33′ 27″ (*b*) 29° 13′ 29″ from 47° 8′ 15″
(*c*) 4° 44′ 53″ from 18° 39′ 48″ (*d*) 34° 30′ 10″ from 179° 23′ 8″
(*e*) 47° 53′ 46″ from 123° 0′ 0″
(*a*) [9° 22′ 9″] (*b*) [17° 54′ 46″] (*c*) [13° 54′ 55″]
(*d*) [144° 52′ 58″] (*e*) [75° 6′ 14″]
3. Convert the following angles to degrees and decimals of a degree, to 3 decimal places:
(*a*) 13° 13′ (*b*) 29° 12′ 35″ (*c*) 58° 58′ 48″ (*d*) 146° 32′ 43″
(*a*) [13.217] (*b*) [29.210] (*c*) [58.980] (*d*) [146.545]
4. Convert the following angles into degrees, minutes and seconds:
(*a*) 18.3° (*b*) 27.57° (*c*) 76.817° (*d*) 152.013°
(*a*) [18° 18′] (*b*) [27° 34′ 12″] (*c*) [76° 49′ 1″]
(*d*) [152° 0′ 47″]

Types and properties of angles

5. State the general name given to the following angles:
(*a*) 79° (*b*) 238° (*c*) 90° (*d*) 151°
(*a*) [Acute] (*b*) [Reflex] (*c*) [Right angle] (*d*) [Obtuse]
6. Find the angles complementary to the following:
(*a*) 19° (*b*) 35° 42′ (*c*) 83° 41′ 19″ (*d*) 14° 14′ 14″
(*a*) [71°] (*b*) [54° 18′] (*c*) [6° 18′ 41″] (*d*) [75° 45′ 46″]
7. Find the angles supplementary to the following:
(*a*) 101° (*b*) 32° (*c*) 49° 47′ 9″ (*d*) 178° 53′ 48″
(*a*) [79°] (*b*) [148°] (*c*) [130° 12′ 51″] (*d*) [1° 6′ 12″]
8. With reference to Fig. 9.15, what is the name given to the line AB? [transversal]
Give examples of each of the following:
(*a*) vertically opposite angles (*b*) supplementary angles
(*c*) corresponding angles (*d*) alternate angles.
(*a*) [1 and 3, or 2 and 4, or 5 and 7, or 6 and 8]

(*b*) [1 and 2, or 2 and 3, or 3 and 4, or 4 and 1, or 5 and 6, or 6 and 7, or 7 and 8, or 8 and 5, or 4 and 5, or 3 and 6]
(*c*) [1 and 5, or 4 and 8, or 2 and 6, or 3 and 7]
(*d*) [4 and 6, or 3 and 5]

9. If, in Fig. 9.15, $\angle 1 = 60°$ find all the other numbered angles.
[$\angle 3 = \angle 5 = \angle 7 = 60°$ $\angle 2 = \angle 4 = \angle 6 = \angle 8 = 120°$]

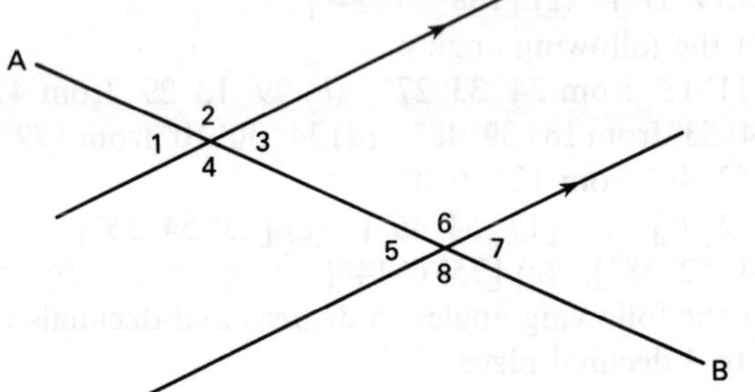

Fig. 9.15

10. In Fig. 9.16, find $\angle x$. [48° 27′]

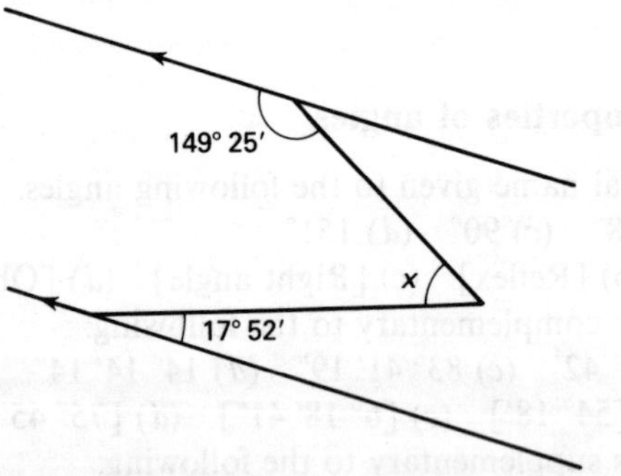

Fig. 9.16

11. XYZ is a straight line and RY is a straight line such that $\angle$RYZ = 126°. If $\angle$RYX is bisected by a straight line YS, find $\angle$XYS. [27°]

12. In Fig. 9.17, find angles *x*, *y* and *z*.

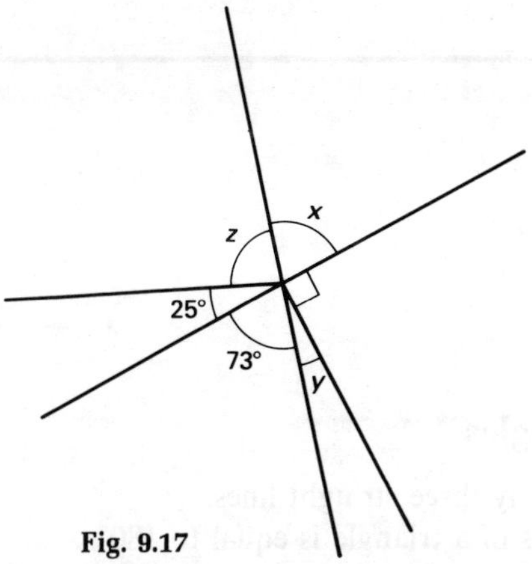

Fig. 9.17 $[x = 73° \quad y = 17° \quad z = 82°]$

13. Which of the following pairs of lines are parallel in Fig. 9.18? [(*d*)]

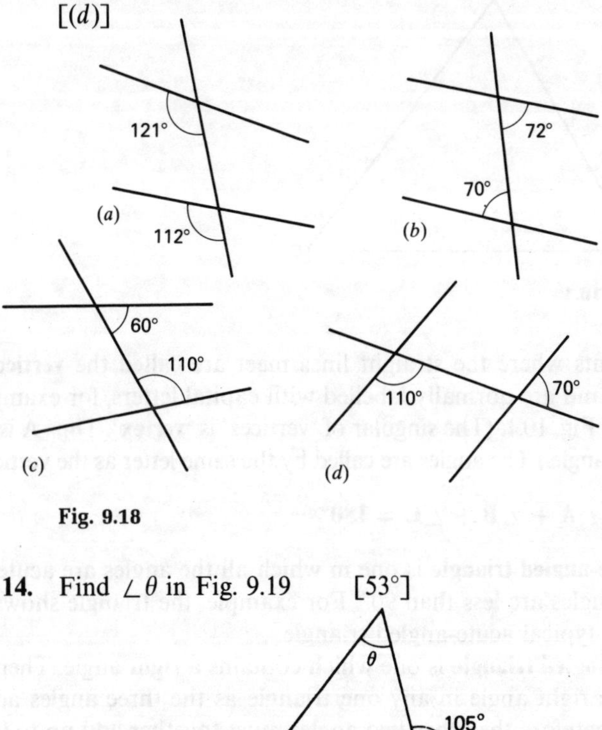

Fig. 9.18

14. Find $\angle \theta$ in Fig. 9.19 [53°]

Fig. 9.19

Chapter 10

Triangles

10.1 Properties of triangles

A triangle is a figure enclosed by three straight lines.

1. the sum of the three angles of a triangle is equal to 180°.

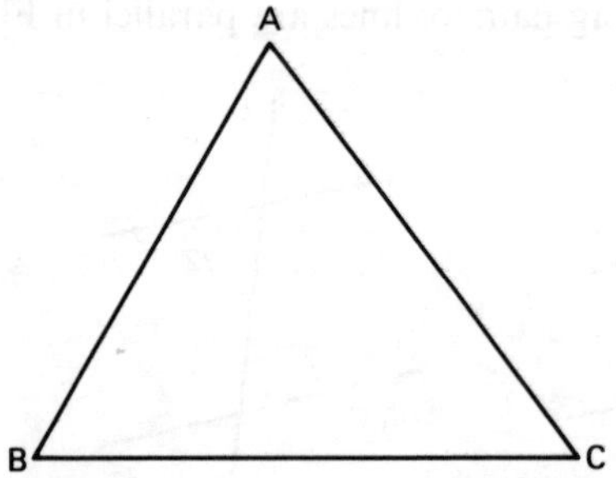

Fig. 10.1

The points where the straight lines meet are called the **vertices** of the triangle and are normally labelled with capital letters, for example, A, B and C of Fig. 10.1. (The singular of 'vertices' is **'vertex'**. Thus A is a vertex of the triangle.) The angles are called by the same letter as the vertices. Thus

$$\angle \mathbf{A} + \angle \mathbf{B} + \angle \mathbf{C} = \mathbf{180°}$$

2. An **acute-angled triangle** is one in which all the angles are acute, that is, all the angles are less than 90°. For example, the triangle shown in Fig. 10.2 is a typical acute-angled triangle.
3. A **right-angled triangle** is one which contains a right angle. There can be only one right angle in any one triangle as the three angles add up to 180°. (Therefore the other two angles must together add up to 90°.) The triangle shown in Fig. 10.3 is a typical right-angled triangle.
   The side DF, that is, the side opposite the right angle, is called the **hypotenuse**.
4. An **obtuse-angled triangle** is one which contains one obtuse angle, that is, one angle which lies between 90° and 180°. There can be only one such

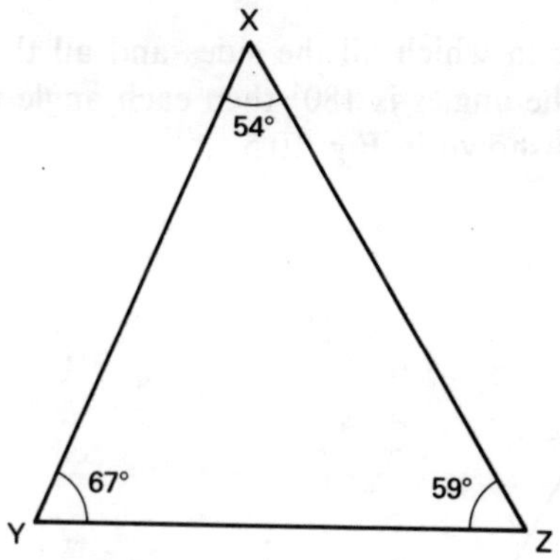

Fig. 10.2

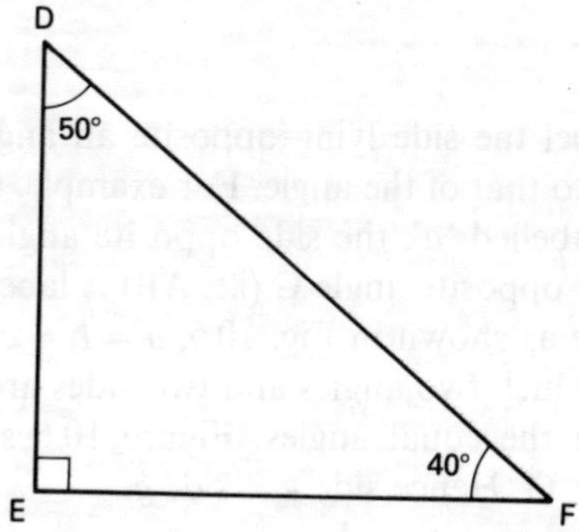

Fig. 10.3

angle in any one triangle as the three angles add up to 180°. (Therefore the other two angles must together add up to less than 90°.) Figure 10.4 shows a typical obtuse-angled triangle.

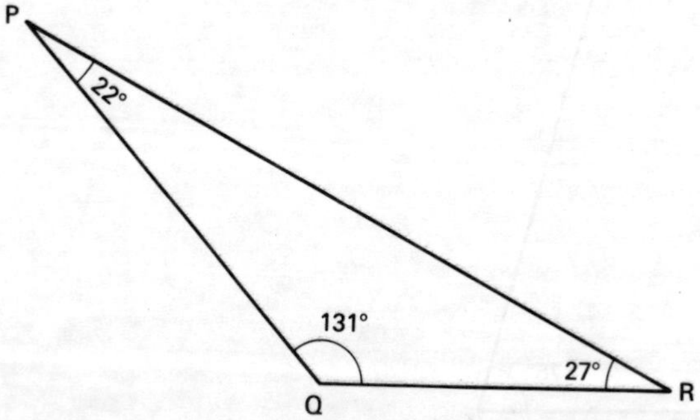

Fig. 10.4

5. An **equilateral triangle** is one in which all the sides and all the angles are equal. Since the sum of the angles is 180° then each angle must be 60°. An equilateral triangle is shown in Fig. 10.5.

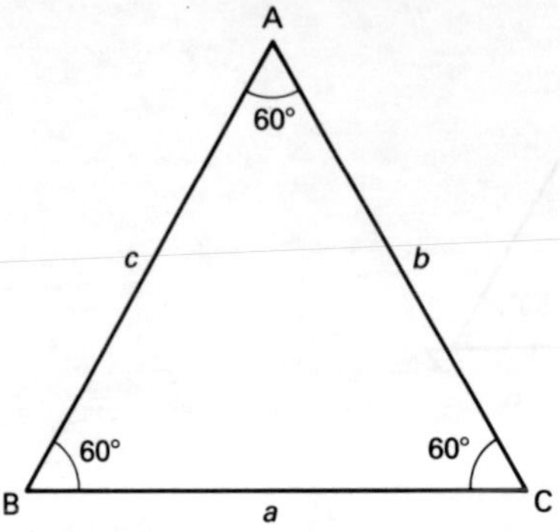

Fig. 10.5

It is normal convention to label the side lying opposite an angle with the small letter corresponding to that of the angle. For example, the side opposite angle A (i.e. BC) is labelled 'a', the side opposite angle B (i.e. AC) is labelled 'b' and the side opposite angle C (i.e. AB) is labelled 'c'. Thus, in an equilateral triangle as shown in Fig. 10.5, $a = b = c$.

6. An **isosceles triangle** is one in which two angles and two sides are equal, the equal sides lying opposite the equal angles. Figure 10.6 shows a typical triangle where $\angle E = \angle G$. Hence side e = side g.

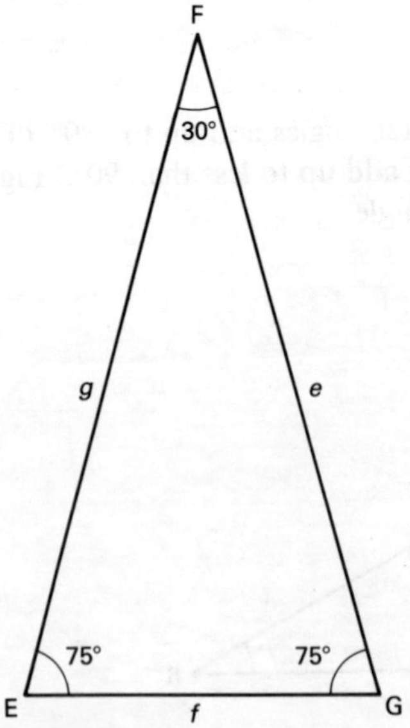

Fig. 10.6

In the isosceles triangle of Fig. 10.6, if a perpendicular to EG is drawn from F, then the angle F will be bisected (that is, cut equally into two parts). The base of the triangle EG will also be bisected. This is shown in Fig. 10.7.

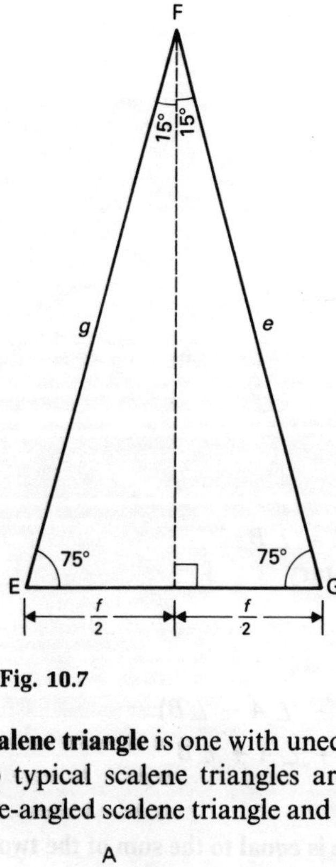

Fig. 10.7

7. A **scalene triangle** is one with unequal angles and therefore unequal sides. Two typical scalene triangles are shown in Fig. 10.8 where (*a*) is an acute-angled scalene triangle and (*b*) is an obtuse-angled scalene triangle.

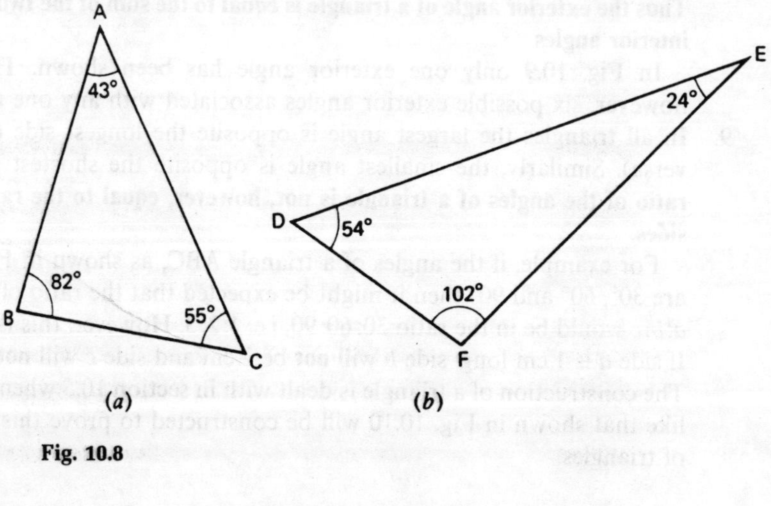

Fig. 10.8

8. The three angles contained within a triangle are called interior angles. In Fig. 10.9, $\angle A$, $\angle B$ and $\angle C$ are called **interior** angles. If the side BC is produced (i.e. extended) as shown, then the angle X is known as an

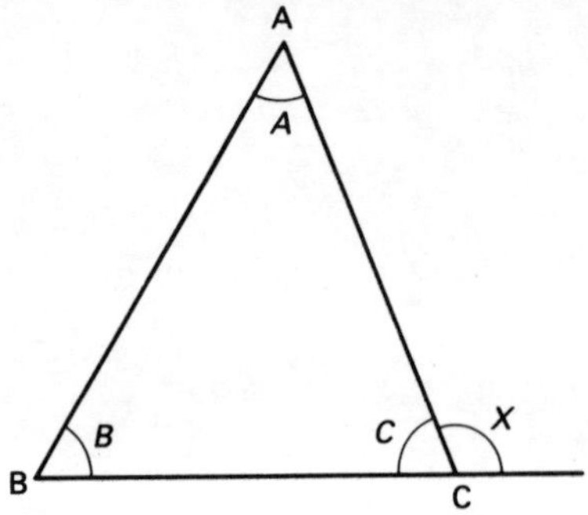

Fig. 10.9

exterior angle of the triangle.

If $\angle A + \angle B + \angle C = 180°$

then $\angle C = 180° - \angle A - \angle B$

Also, $\angle C$ and $\angle X$ are supplementary

i.e. $\angle C + \angle X = 180°$

$$\angle X = 180° - \angle C$$
$$= 180° - (180° - \angle A - \angle B)$$
$$= 180° - 180° + \angle A + \angle B$$

Therefore $\angle X = \angle A + \angle B$

Thus the exterior angle of a triangle is equal to the sum of the two opposite interior angles.

In Fig. 10.9 only one exterior angle has been shown. There are, however, six possible exterior angles associated with any one triangle.

9. In all triangles the largest angle is opposite the longest side (and vice versa). Similarly, the smallest angle is opposite the shortest side. **The ratio of the angles of a triangle is not, however, equal to the ratio of the sides.**

For example, if the angles of a triangle ABC, as shown in Fig. 10.10, are 30°, 60° and 90° then it might be expected that the ratio of the sides a:b:c would be in the ratio 30:60:90, i.e. 1:2:3. However, this is **not** true. If side a is 1 cm long, side b will **not** be 2 cm and side c will **not** be 3 cm. The construction of a triangle is dealt with in section 10.5 when triangles like that shown in Fig. 10.10 will be constructed to prove this property of triangles.

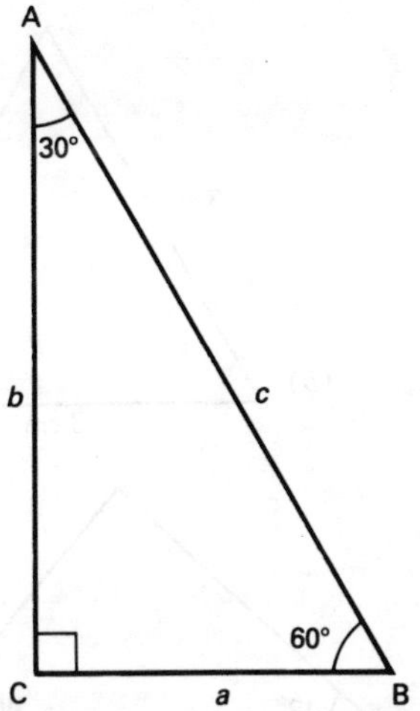

Fig. 10.10

10. In any triangle the sum of the lengths of any two of the sides is always greater than the length of the third side. In Fig. 10.11, y is the longest side since it lies opposite the largest angle.
Always $x + z > y$ (The sign $>$ means 'is greater than')

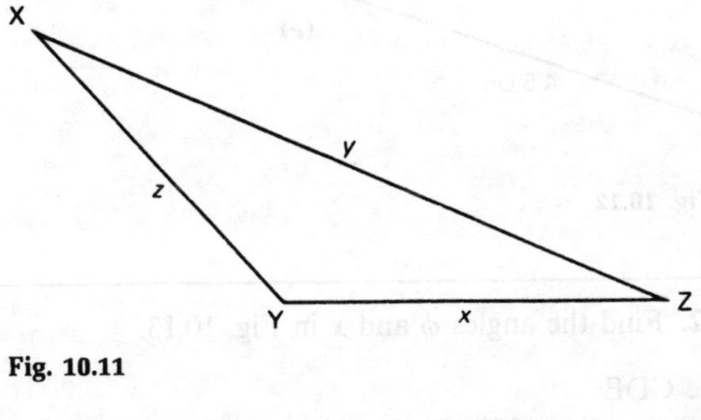

Fig. 10.11

Note that $x + y + z$ is called the **perimeter** of the triangle.

Worked problems on the properties of triangles

Problem 1. Name the types of triangles shown in Fig. 10.12.
(*a*) Obtuse-angled scalene triangle (*b*) Equilateral triangle
(*c*) Acute-angled scalene triangle (*d*) Right-angled triangle
(*e*) Isosceles triangle

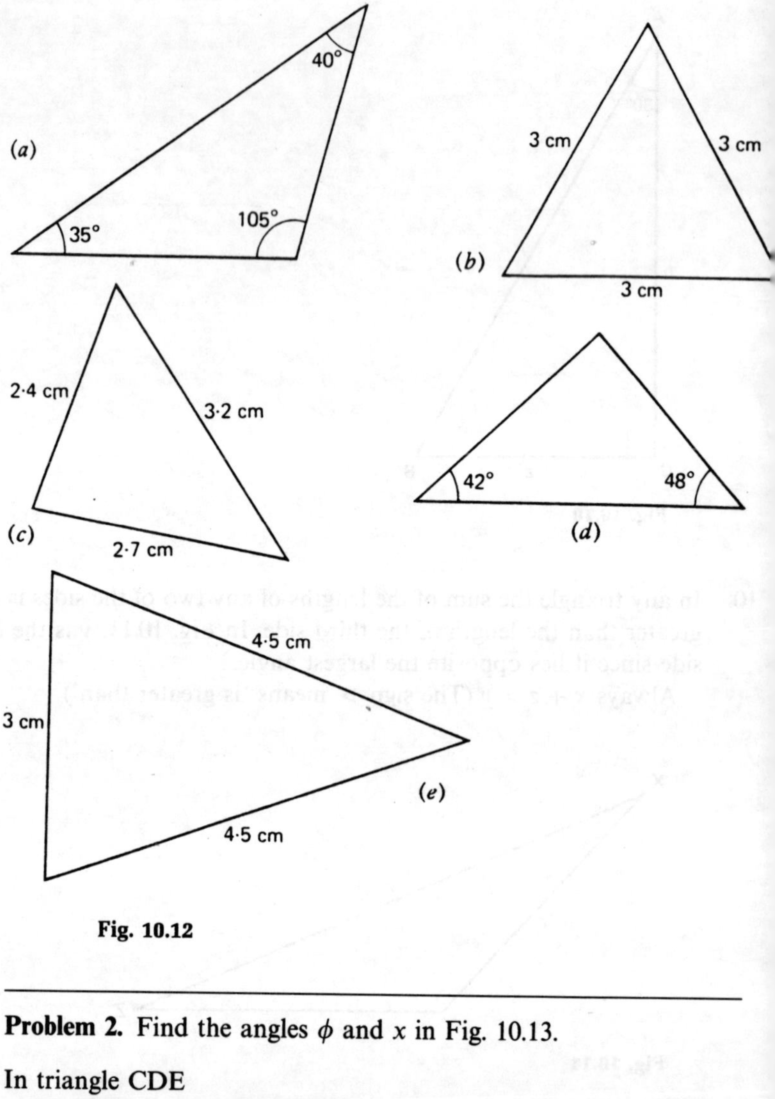

Fig. 10.12

Problem 2. Find the angles ϕ and x in Fig. 10.13.

In triangle CDE

$\angle C + \angle D + \angle E = 180°$ (Angles in a triangle add up to 180°)

Therefore $\angle DCE = 180° - 90° - 58°$

$= 32°$

$\angle BCA = \angle DCE = 32°$ (vertically opposite angles).

$\angle\phi = \angle BAC + \angle ACB$ (exterior angle of a triangle is equal to the sum of the two opposite interior angles).

$\angle\phi = 19° + 32°$

$\angle\phi = \mathbf{51°}$

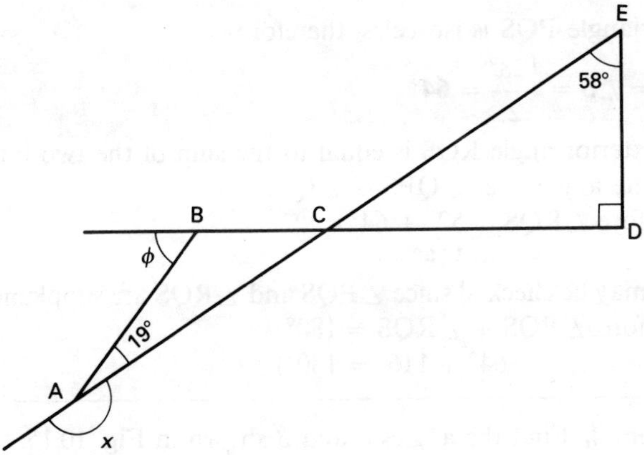

Fig. 10.13

$\angle x$ and 19° are supplementary.

Thus $\quad \angle x = 180° - 19°$

Therefore $\angle x =$ **161°**

Problem 3. PQS is an isosceles triangle in which the unequal angle QPS is 52° and PQ is extended to R (see Fig. 10.14). Find $\angle$RQS.

Since the three interior angles of a triangle add up to 180°

$$52° + \angle Q + \angle S = 180°$$

Therefore $\angle Q + \angle S = 180° - 52°$

$$= 128°$$

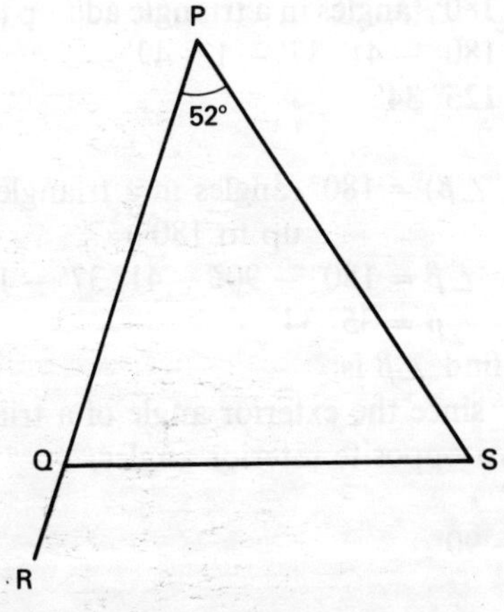

Fig. 10.14

But triangle PQS is isosceles, therefore

$$\angle Q = \angle S = \frac{128°}{2} = \mathbf{64°}$$

The exterior angle RQS is equal to the sum of the two interior opposite angles, i.e. $\angle QPS + \angle QSP$.

Therefore $\angle RQS = 52° + 64°$

$= \mathbf{116°}$

(This may be checked since $\angle PQS$ and $\angle RQS$ are supplementary.

Therefore $\angle PQS + \angle RQS = 180°$

$64° + 116° = 180°$)

Problem 4. Find the angles α and β shown in Fig. 10.15.

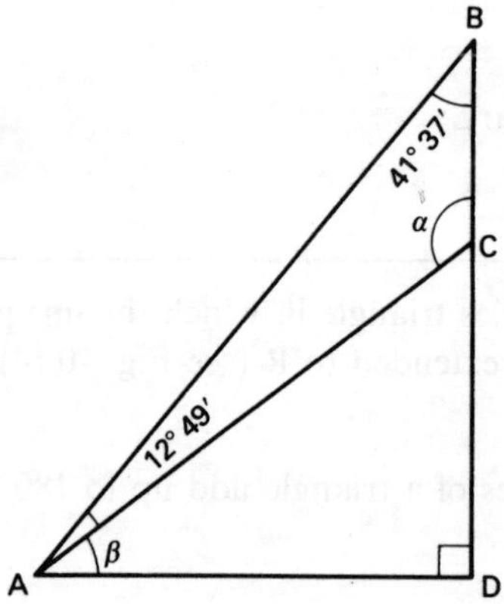

Fig. 10.15

In triangle ABC:

$41° \, 37' + 12° \, 49' + \angle\alpha = 180°$ (angles in a triangle add up to 180°)

Therefore $\angle\alpha = 180° - 41° \, 37' - 12° \, 49'$

$\angle\alpha = \mathbf{125° \, 34'}$

In triangle ABD:

$90° + 41° \, 37' + (12° \, 49' + \angle\beta) = 180°$ (angles in a triangle add up to 180°)

Therefore $\angle\beta = 180° - 90° - 41° \, 37' - 12° \, 49'$

$\angle\beta = \mathbf{35° \, 34'}$

An alternative method to find $\angle\beta$ is:

$\angle\alpha = \angle\beta + 90°$ since the exterior angle of a triangle is equal to the sum of the two opposite interior angles.

Therefore $\angle\beta = \angle\alpha - 90°$

$= 125° \, 34' - 90°$

$\angle\beta = \mathbf{35° \, 34'}$

Problem 5. Find the angles *a, b, c, d* and *e* in Fig. 10.16.

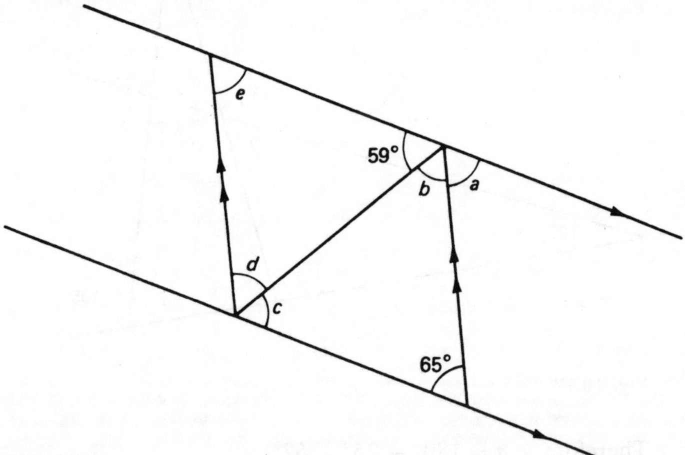

Fig. 10.16

When two pairs of parallel lines are depicted on the same figure, single arrows can be used to denote one pair of parallel lines and double arrows used to denote the other pair of parallel lines, as shown in Fig. 10.16.

$\angle \mathbf{c} = \mathbf{59}°$ (alternate angles between parallel lines)
$\angle \mathbf{a} = \mathbf{65}°$ (alternate angles between parallel lines)
$\angle b + 59° + 65° = 180°$ (angles in a triangle add up to 180° or angles on a straight line add up to 180°)
Therefore $\angle b = 180° - 59° - 65°$
$\angle \mathbf{b} = \mathbf{56}°$
$\angle b = \angle \mathbf{d} = \mathbf{56}°$ (alternate angles between parallel lines marked by double arrows)
$\angle e + 59° + 56° = 180°$ (angles in a triangle add up to 180°)
Therefore $\angle e = 180° - 59° - 56°$
$\angle \mathbf{e} = \mathbf{65}°$

Check:
$\angle e = \angle a = 65°$ (corresponding angles between parallel lines)

Problem 6. In Fig. 10.17 find all the unknown angles.

$\angle \mathbf{d} = \mathbf{123}°$ (vertically opposite angles)
$\angle e$ is the supplement of 123°
Therefore $\angle e = 180° - 123°$
$\angle \mathbf{e} = \mathbf{57}°$
$\angle \mathbf{e} = \angle \mathbf{b} = \mathbf{57}°$ (vertically opposite angles)
$\angle a + \angle b + 25° = 180°$ (angles in a triangle add up to 180°)

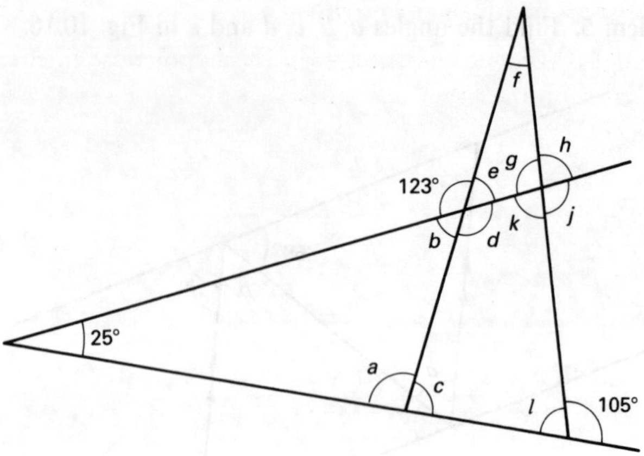

Fig. 10.17

Therefore $\angle a = 180° - 25° - 57°$

$\angle \boldsymbol{a} = \mathbf{98}°$

$\angle a$ and $\angle c$ are supplementary.

Therefore $\angle c = 180° - 98°$

$\angle \boldsymbol{c} = \mathbf{82}°$

$\angle l$ and 105° are supplementary.

Therefore $\angle l = 180° - 105°$

$\angle \boldsymbol{l} = \mathbf{75}°$

$\angle c + \angle f + \angle l = 180°$ (angles in a triangle add up to 180°)

Therefore $\angle f = 180° - 82° - 75°$

$\angle \boldsymbol{f} = \mathbf{23}°$

$\angle e + \angle f + \angle g = 180°$ (angles in a triangle add up to 180°)

Therefore $\angle g = 180° - 57° - 23°$

$\angle \boldsymbol{g} = \mathbf{100}°$

$\angle g = \angle \boldsymbol{j} = \mathbf{100}°$ (vertically opposite angles)

$\angle h$ is the supplement of 100°, therefore

$\angle h = 180° - 100°$

$\angle \boldsymbol{h} = \mathbf{80}°$

$\angle h = \angle \boldsymbol{k} = \mathbf{80}°$ (vertically opposite angles)

Check:

$25° + \angle k + \angle l = 180°$ (angles in a triangle add up to 180°)

$25° + 80° + 75° = 180°$

Further problems on the properties of triangles may be found in Section 10.6 (Problems 1–10), page 245.

10.2 The theorem of Pythagoras

In any right-angled triangle the square on the hypotenuse is equal to the sum of the squares on the other two sides.

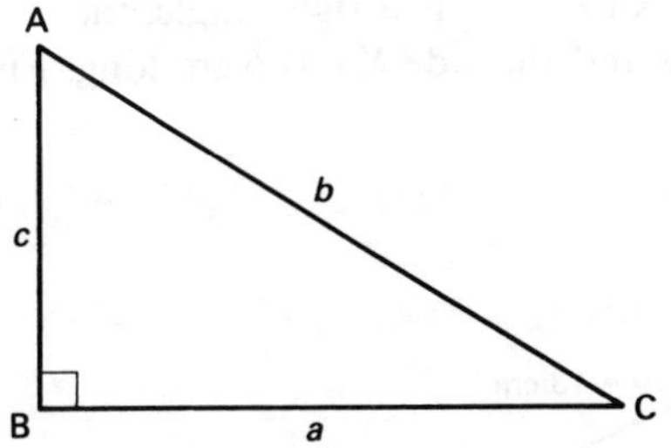

Fig. 10.18

For the triangle ABC, shown in Fig. 10.18, b is the hypotenuse. (The hypotenuse is always the side opposite the right angle.) Therefore, from above $(AC)^2 = (BC)^2 + (AB)^2$

that is $b^2 = a^2 + c^2$

Worked problems on the theorem of Pythagoras

Problem 1. In a triangle PQR, $\angle Q = 90°$, PQ = 3 cm, and RQ = 4 cm. Find the length of PR.

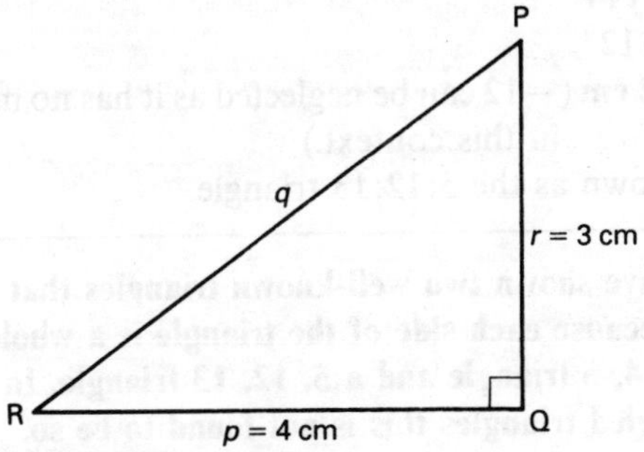

Fig. 10.19

The triangle PQR is shown in Fig. 10.19. By Pythagoras's theorem:

$$q^2 = p^2 + r^2$$

Therefore $q^2 = (4)^2 + (3)^2$

$$= 16 + 9$$

$$q^2 = 25$$

$$q = \sqrt{25} = \pm 5$$

The value -5 is meaningless in this context and is thus ignored.
Therefore PR = **5 cm**
Triangle PQR is known as the 3:4:5 triangle.

Problem 2. In a triangle XYZ, $\angle$Y is a right angle, the hypotenuse is 13 cm long and the side XY is 5 cm long. Find the length of the side YZ.

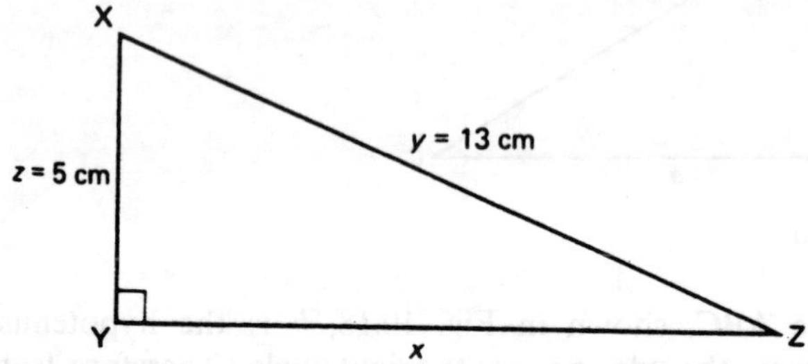

Fig. 10.20

The triangle XYZ is shown in Fig. 10.20. By Pythagoras's theorem:

$$y^2 = x^2 + z^2$$

Therefore $(13)^2 = x^2 + (5)^2$

$$169 = x^2 + 25$$
$$x^2 = 169 - 25$$
$$= 144$$
$$x = \sqrt{144}$$
$$= \pm 12$$

YZ = **12 cm** (-12 can be neglected as it has no meaning in this context.)

Triangle XYZ is known as the 5:12:13 triangle.

Problems 1 and 2 have shown two well-known triangles that are of particular interest because each side of the triangle is a whole number, that is a 3, 4, 5 triangle and a 5, 12, 13 triangle. In the majority of right angled triangles this is not found to be so.

Problem 3. In the triangle ABC shown in Fig. 10.21 find the length of the side b

By Pythagoras's theorem:

$$b^2 = a^2 + c^2$$
$$= (6.2)^2 + (4.1)^2$$
$$= 38.44 + 16.81$$
$$= 55.25$$

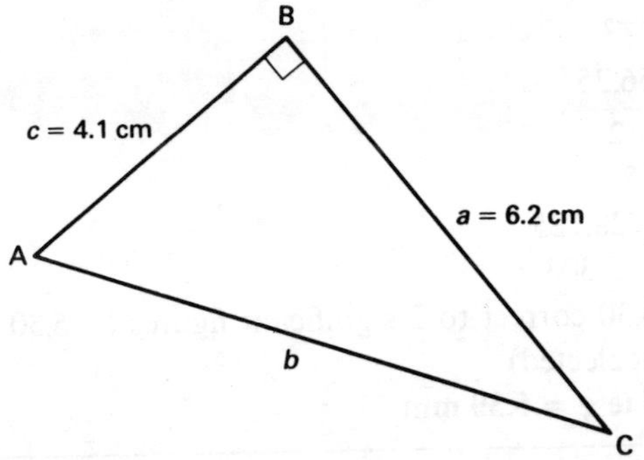

Fig. 10.21

Therefore $b = \sqrt{55.25}$
$= \pm 7.43$ (-7.43 can be neglected)
Hence the side b is **7.43 cm**

Problem 4. Triangle EFG is isosceles and $\angle$E is a right angle. If the hypotenuse is 7.5 mm find the length of the other two sides of the triangle.

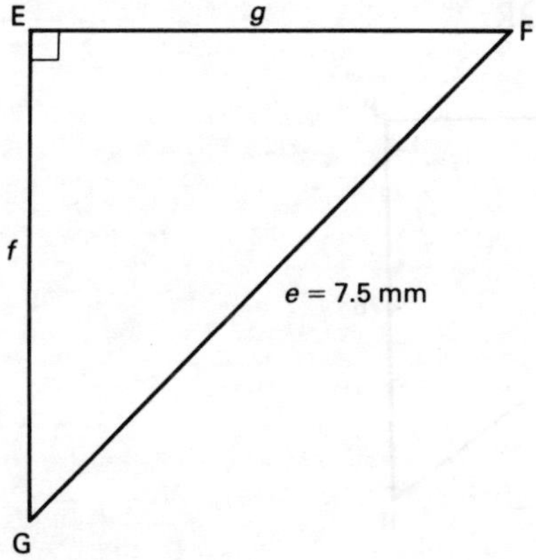

Fig. 10.22

The triangle EFG is shown in Fig. 10.22. By Pythagoras's theorem:
$e^2 = f^2 + g^2$
However if triangle EFG is isosceles then $\angle G = \angle F = 45°$ and side f = side g.
Therefore $e^2 = f^2 + f^2$
$e^2 = 2f^2$
$(7.5)^2 = 2f^2$

$$56.25 = 2f^2$$
$$f^2 = \frac{56.25}{2}$$
$$= 28.125$$
$$f = \sqrt{28.125}$$
$$= \pm 5.303$$
$= 5.30$ correct to 3 significant figures (-5.30 can be neglected)

Hence **side f = side g = 5.30 mm**

Problem 5. Two ships leave a port together. One travels due south at 26 knots and the other due west at 35 knots. If 1 knot = 1 nautical mile per hour, find how far apart the two ships are after 3 hours.

After 3 hours the first ship would have travelled a distance of $3 \times 26 = 78$ nautical miles due south.

Similarly, after 3 hours, the second ship would have travelled a distance of $3 \times 35 = 105$ nautical miles due west.

If, in Fig. 10.23, PR represents 78 nautical miles and PQ represents 105 nautical miles, then the distance apart after 3 hours is given by the distance QR.

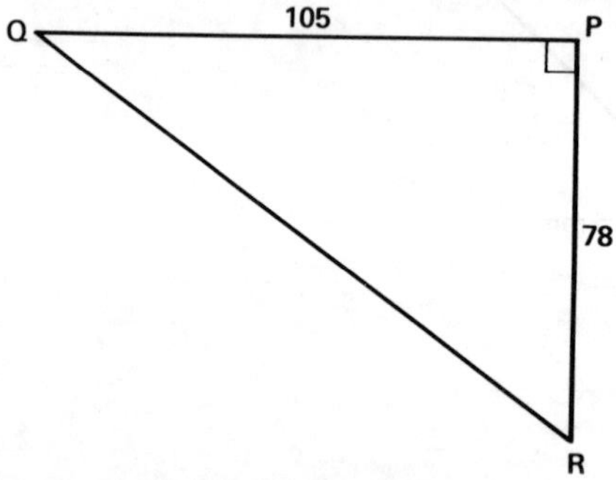

Fig. 10.23

By Pythagoras's theorem:

$$(QR)^2 = (PQ)^2 + (PR)^2$$

Therefore $QR^2 = (105)^2 + (78)^2$
$$= 11\,025 + 6\,084$$
$$= 17\,109$$
$$QR = \sqrt{17\,109}$$
$$= \pm 131$$
$= 131$ (-131 can be neglected)

Hence the distance apart after 3 hours is **131 nautical miles.**

Further problems on the theorem of Pythagoras may be found in Section 10.6 (Problems 11–21), page 248.

10.3 Congruent triangles

Two triangles are said to be **congruent** if they are equal in all respects, that is, three angles and three sides in one triangle are equal to three angles and three sides in the other triangle. Their areas must consequently be the same. Six facts determine a triangle completely – the three sides and the three angles.

However, to show that two triangles are congruent it is only necessary to show that three particular facts are equal in both triangles. There are only four possible combinations of facts and these are shown below:

1. Two triangles are congruent if the three sides of one are equal to the three sides of the other. (That is, side, side, side, which is sometimes written as *S.S.S.* for short.)

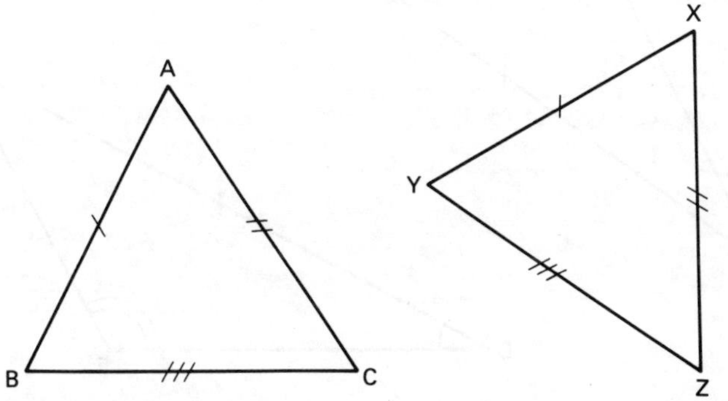

Fig. 10.24

   Thus, in Fig. 10.24, if AB = XY, AC = XZ and BC = YZ then triangle ABC and triangle XYZ are congruent.

2. Two triangles are congruent if they have two sides of the one equal to two sides of the other, and if the angles included by these sides are equal. (That is, side, angle, side, which is sometimes written as *S.A.S.* for short.)

   If, in Fig. 10.25, PQ = XY, QR = YZ and ∠Q = ∠Y then triangle PQR and triangle XYZ are congruent.

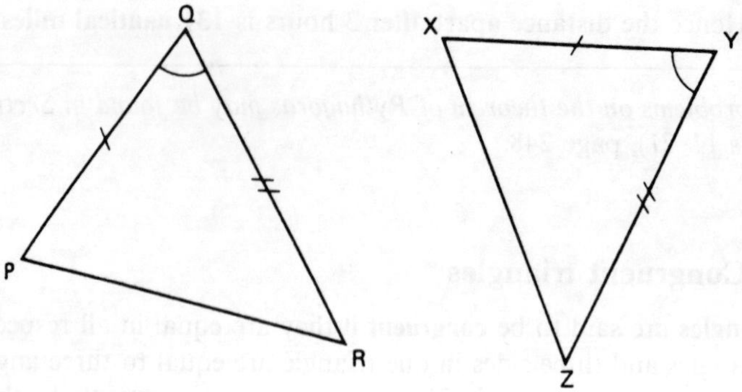

Fig. 10.25

3. Two triangles are congruent if two angles of the one are equal to two angles of the other, and any side of the first is equal to the corresponding side of the other. (That is, angle, side, angle, which is sometimes written as *A.S.A.* for short.)

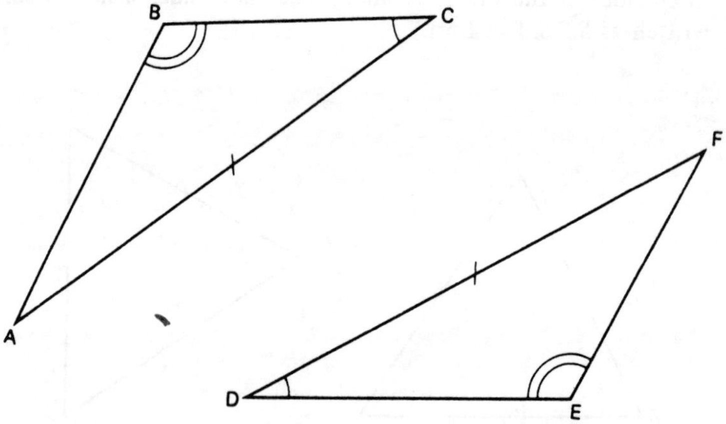

Fig. 10.26

If, in Fig. 10.26, AC = DF, ∠C = ∠D and ∠B = ∠E then triangle ABC and triangle FED are congruent.

Note that if two angles of one triangle are equal to two angles of the other triangle then the third angles in the two triangles are also equal since the sum of the interior angles of a triangle is 180°.

4. Two right-angled triangles are congruent if their hypotenuses are equal, and if one other side of one is equal to the corresponding side of the other. (That is, right angle, hypotenuse, side, which is sometimes written as *R.H.S.* for short.)

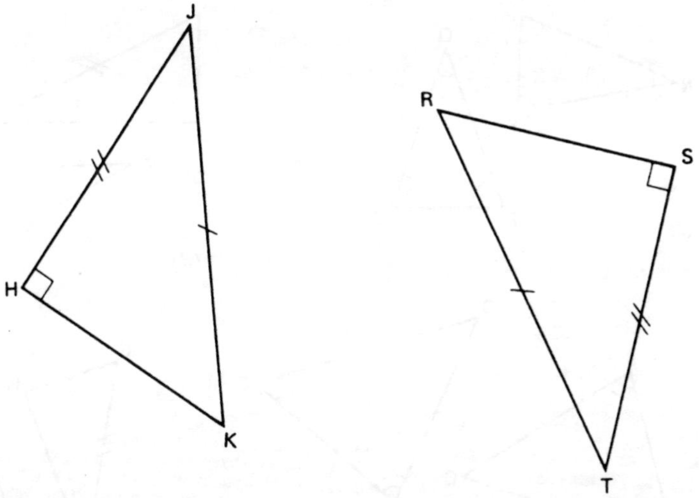

Fig. 10.27

If, in Fig. 10.27, JK = RT, HJ = ST and $\angle H = \angle S = 90°$, then triangle HJK and triangle STR are congruent.

When stating congruency between two triangles it is necessary to state the triangles in their correct sequence. Referring to Fig. 10.27, for example, it is true to say that triangles HJK and STR are congruent since H corresponds to S, J corresponds to T and K corresponds to R. It is untrue to say that triangles HJK and RST are congruent because of their incorrect sequence.

Worked problems on congruent triangles

Problem 1. State which of the following pairs of triangles are congruent and name their correct sequence.

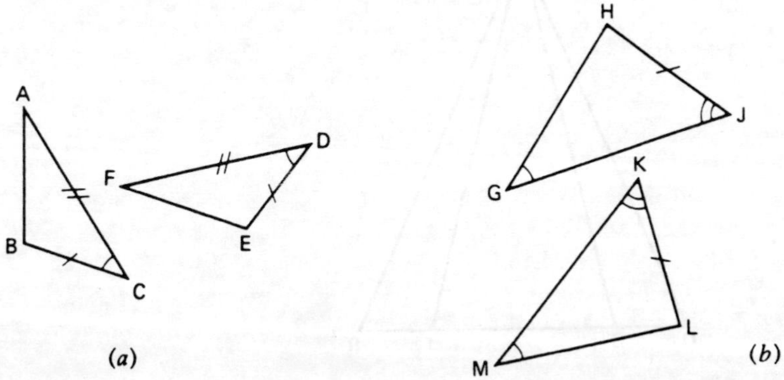

Fig. 10.28

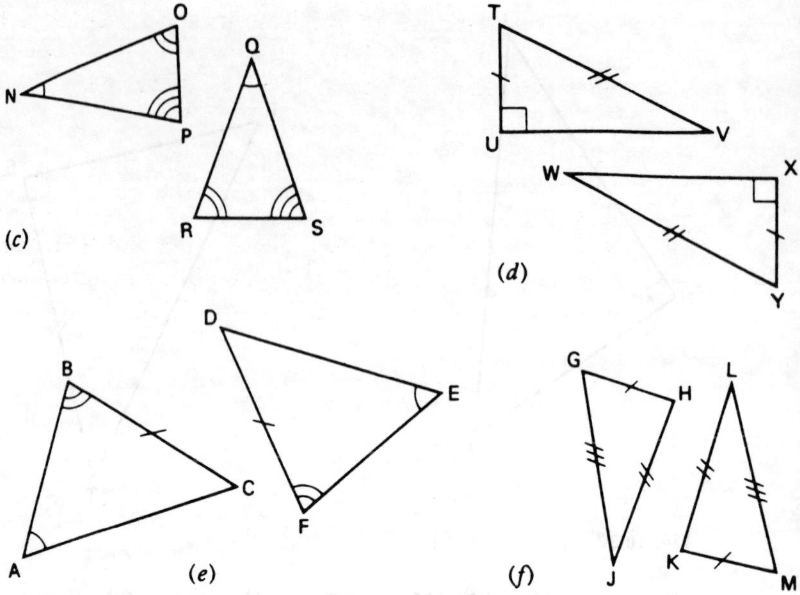

Fig. 10.28 (*continued*)

(*a*) Congruent. ACB, FDE (*S.A.S.*)
(*b*) Congruent. GHJ, MLK (*A.S.A.*)
(*c*) Not necessarily congruent. (*d*) Congruent. TUV, YXW (*R.H.S.*)
(*e*) Congruent. ABC, EFD (*A.S.A.*)
(*f*) Congruent. GHJ, MKL (*S.S.S.*)

Problem 2. If, in Fig. 10.29, PQ = PR and QS = TR show that triangle PST is isosceles.

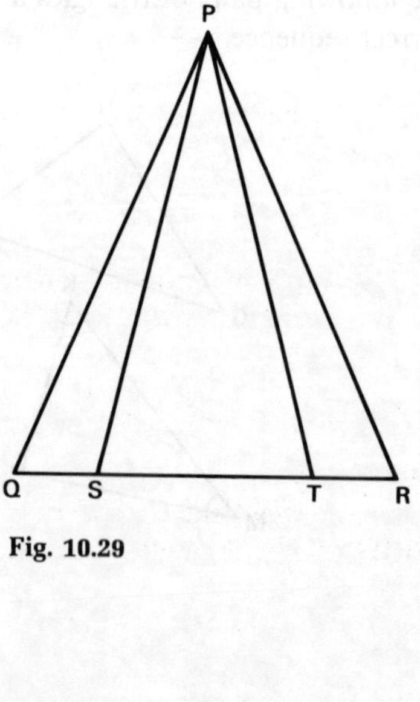

Fig. 10.29

If PQ = PR then triangle PQR is isosceles and ∠PQR = ∠PRQ.
If QS = TR then triangle PSQ and triangle PTR are congruent since PQ = PR, QS = TR and ∠PQS = ∠PRT. (S.A.S.)
Hence PS = PT
Therefore triangle PST is isosceles.

Problem 3. Figure 10.30 shows an isosceles triangle ABC with E the mid point of AB. Prove that triangles ADE and BFE are congruent, and that triangles CDE and CFE are congruent. Find the values of angles CAE and CDE.

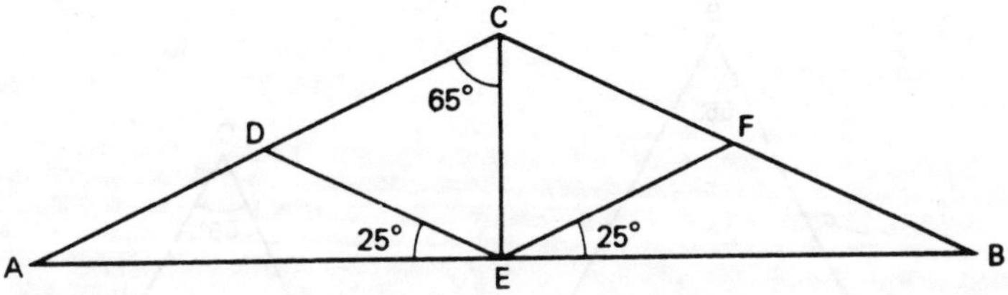

Fig. 10.30

Since triangle ABC is isosceles, AC = CB.
If E is the mid point of AB then triangles ACE and CEB are congruent (S.S.S.) and CE bisects the angle ACB and CE is perpendicular to AB, that is, ∠AEC = 90°
Therefore ∠CED = 90° − 25° = 65°
Similarly ∠CEF = 65°
In triangle CDE, ∠CDE + 65° + 65° = 180° (angles in a triangle add up to 180°)
Therefore ∠CDE = 180° − 65° − 65°
= 50°
Similarly ∠CFE = 50°
∠ADE is supplementary to ∠EDC, hence
∠ADE = 180° − 50°
= 130°
Similarly ∠BFE = 130°
Therefore ∠DAE + 130° + 25° = 180° (angles in a triangle add up to 180°)
∠DAE = 180° − 130° − 25°
= 25°
Similarly ∠FBE = 25°
Hence **∠CAE = 25°** and **∠CDE = 50°**
Triangles ADE and BFE are congruent since ∠DAE = ∠FBE, AE = BE and ∠DEA = ∠FEB (A.S.A.). Triangles CDE and CFE

are congruent since $\angle DCE = \angle FCE$, CE is common to both triangles and $\angle CED = \angle CEF$

Further problems on congruent triangles may be found in Section 10.6 (Problems 22 and 23), page 250.

10.4 Similar triangles

Two triangles are said to be **similar** if the angles of one triangle are equal to the angles of the other triangle.

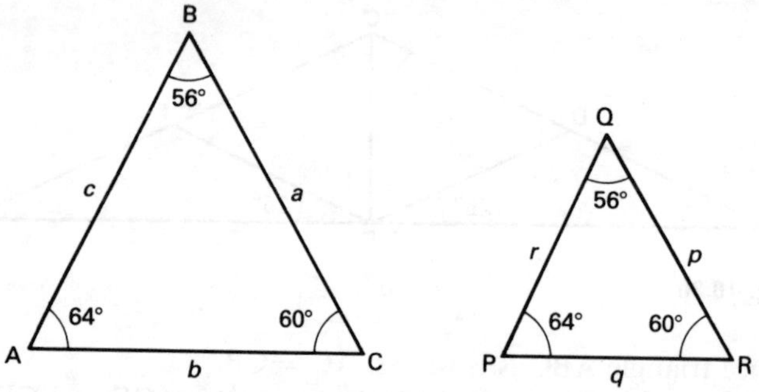

Fig. 10.31

Triangles ABC and PQR of Fig. 10.31 are similar triangles. However, unlike congruent triangles, the corresponding sides of the triangles are not the same, but are in proportion to each other. For example, if AB = 6 cm and PQ = 3 cm then by proportion the lengths of all the sides in triangle PQR will each be one half of the sides of triangle ABC.

Since the sides of the similar triangles are in proportion:

$$\frac{p}{a} = \frac{q}{b} = \frac{r}{c}$$

Worked problems on similar triangles

Problem 1. In Fig. 10.32 find the length of side *b*.

In triangle ABC, $72° + 45° + \angle C = 180°$ (angles in a triangle add up to 180°)

Therefore $\angle C = 180° - 72° - 45°$

$\angle C = 63°$

Similarly, in triangle DEF, $\angle D = 180° - 63° - 45°$

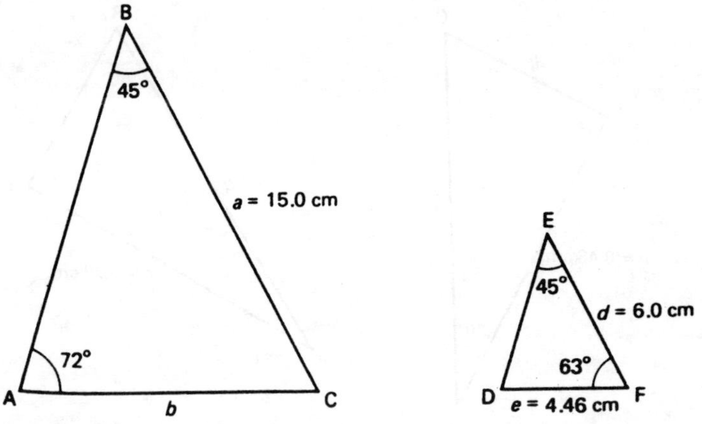

Fig. 10.32

Therefore $\angle D = 72°$

Therefore triangle ABC and triangle DEF are similar, as their angles are the same. Hence by proportion:

$$\frac{a}{d} = \frac{b}{e}$$

$$\text{Therefore } \frac{15.0}{6.00} = \frac{b}{4.46}$$

$$b = \frac{15.0}{6.00}(4.46)$$

$$b = 11.15$$

Hence the length of side b = **11.15 cm**

Problem 2. In Fig. 10.33 find the dimensions a, b and q.

In triangle ABC, $\angle ACB = 180° - 90° - 30° = 60°$

In triangle QPR, $\angle QPR = 180° - 90° - 60° = 30°$

Hence triangle ABC and triangle QPR are similar since three angles in one equal three angles in the other.

Redrawing the triangles gives Fig. 10.34.

Since the two triangles are similar their sides are in proportion:

$$\text{Therefore } \frac{q}{a} = \frac{p}{b} = \frac{r}{c}$$

$$\frac{q}{a} = \frac{5.65}{b} = \frac{9.79}{8.48}$$

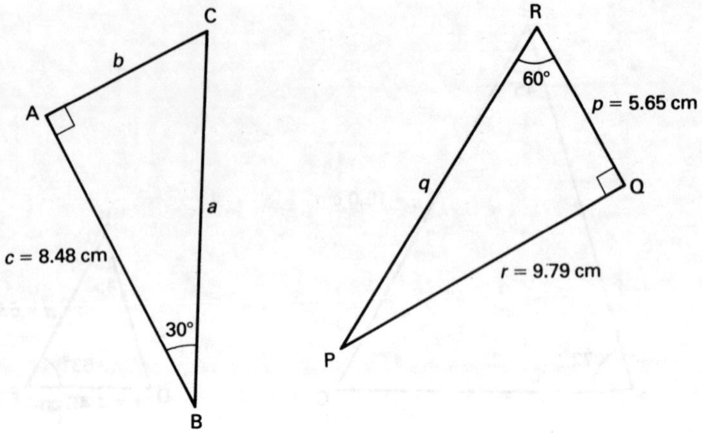

Fig. 10.33

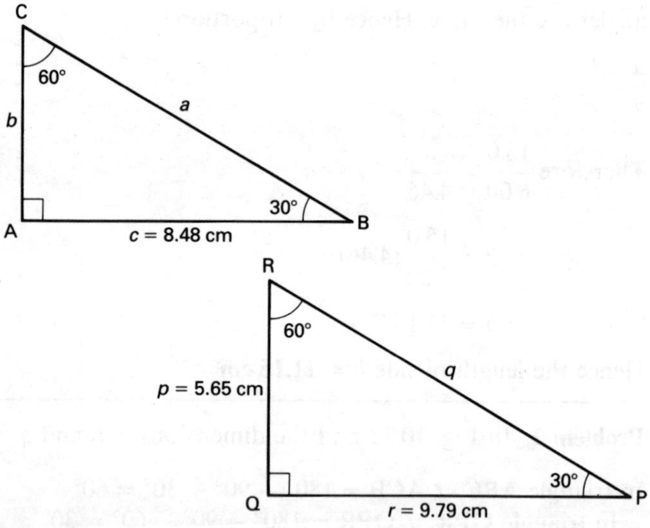

Fig. 10.34

$$b = 8.48\left[\frac{5.65}{9.79}\right]$$

$$b = \mathbf{4.894\ cm}$$

In triangle QPR, q can be found using the theorem of Pythagoras.

That is $q^2 = p^2 + r^2$

$= (5.65)^2 + (9.79)^2$

$= 31.92 + 95.84$

$= 127.76 = 127.8$ (to 4 significant figures)

Therefore $q = \sqrt{127.8} = \pm 11.30$

$q =$ **11.30 cm** (-11.30 can be neglected).

By proportion: $\frac{a}{q} = \frac{c}{r}$

Therefore $\frac{a}{11.30} = \frac{8.48}{9.79}$

$$a = 11.30\left[\frac{8.48}{9.79}\right]$$

$$\mathbf{a = 9.788\ cm}$$

The value of 'a' may be checked using Pythagoras's theorem on triangle ABC.
Hence $a =$ **9.788 cm**, $b =$ **4.894 cm** and $q =$ **11.30 cm**

Problem 3. In Fig. 10.35, PR = 16 cm, PS = 13 cm, RS = 12 cm, RQ = 7 cm and QT is parallel to RS. Show that triangles PQT and PRS are similar and hence find the length of QT and PT.

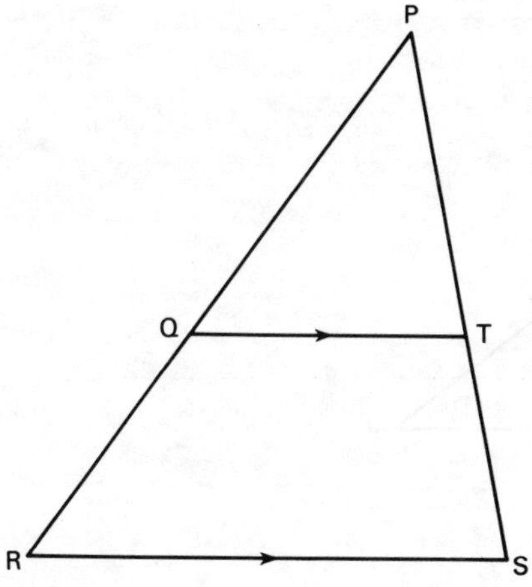

Fig. 10.35

Since QT is parallel to RS then:

∠PQT = ∠PRS (corresponding angles between parallel lines QT and RS) and ∠PTQ = ∠PSR (corresponding angles between parallel lines QT and RS). Also, ∠P is common to triangle PQT and triangle PRS.

Since the three angles in triangle PQT are equal to the three angles in triangle PRS then the triangles are similar.

Hence the sides of triangles PQT and PRS are in proportion.

Therefore $\frac{PQ}{PR} = \frac{PT}{PS} = \frac{QT}{RS}$

$$\frac{16-7}{16} = \frac{PT}{13}$$

$$PT = 13\left(\frac{9}{16}\right)$$

$$\mathbf{PT = 7\tfrac{5}{16}\ cm\ or\ 7.31\ cm}$$

Also $\frac{9}{16} = \frac{QT}{12}$

Therefore $QT = 12\left(\frac{9}{16}\right)$

$$\mathbf{QT = 6\tfrac{3}{4}\ cm\ or\ 6.75\ cm}$$

Problem 4. In Fig. 10.36, prove that triangle QST is similar to triangle QRS. If RT = 10 cm, RS = 6 cm and ST = 8 cm find the length of SQ.

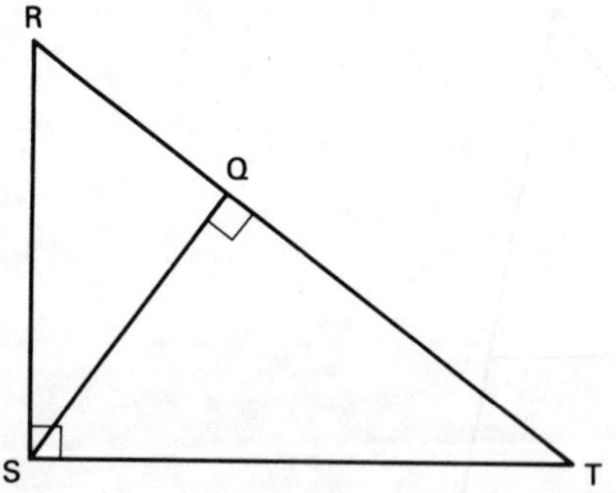

Fig. 10.36

Now, we are given

∠TSQ + ∠QTS = 90° (1)

Also

∠SRT + ∠QTS = 90° (The two other angles in a right-angled triangle add up to 90°) (2)

Therefore ∠SRT = ∠TSQ

Hence triangles QST and SRT each contain a right angle and another pair of corresponding angles. Therefore they are similar triangles. Also, ∠QRS and ∠SRT are equal.
Therefore, triangles QRS and STQ are similar.
Redrawing two of the triangles, RST and SQT separately gives Fig. 10.37.

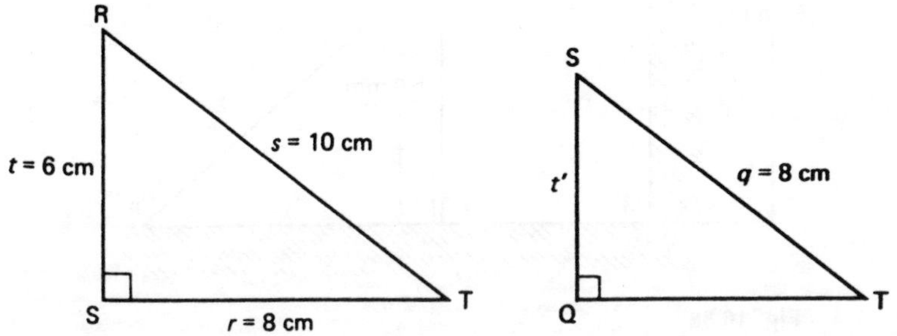

Fig. 10.37

Hence, by proportion:

$$\frac{q}{s} = \frac{t'}{t}$$

Therefore $\frac{8}{10} = \frac{t'}{6}$

$$t' = 6\left(\frac{8}{10}\right)$$

$$= 4.8$$

Hence the length SQ = **4.8 cm**

Problem 5. A rectangular water tank 4.0 m wide and 5.0 m high stands against a perpendicular building of height 8.5 m. A ladder is used to gain access to the roof of the building. Find the minimum length of ladder required, and, using this ladder the distance between the bottom of the ladder and the building.

A side view of the arrangement is shown in Fig. 10.38. The minimum distance between the bottom of the ladder and the building is given by AE.

BD and AE are parallel since the building and the water tank are perpendicular.
∠CDB = ∠DEF (corresponding angles between parallel lines BD and AE). Similarly, ∠BCD = ∠FDE.

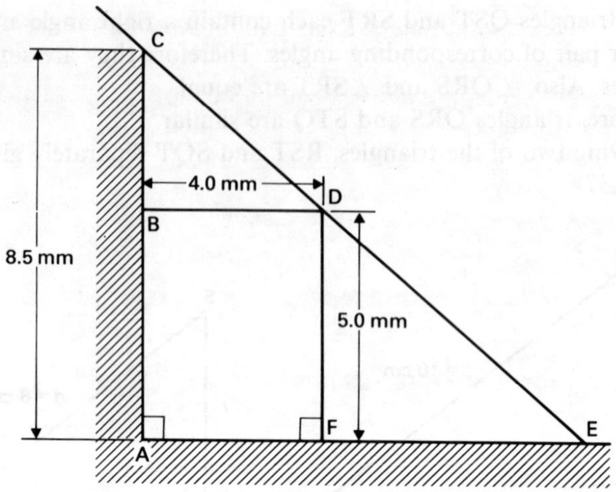

Fig. 10.38

Therefore, triangle BCD is similar to triangle FDE since three angles in one equal three angles in the other.

If AC = 8.5 m and AB (=DF) = 5.0 m

then BC = (8.5 − 5.0) m = 3.5 m

By proportion: $\frac{BC}{FD} = \frac{BD}{FE}$

Therefore $\frac{3.5}{5.0} = \frac{4.0}{FE}$

$$FE = 4.0\left[\frac{5.0}{3.5}\right]$$

$$= 5.71 \text{ m}$$

Therefore, **the distance of the bottom of the ladder from the building is 5.71 m + 4.0 m = 9.71 m**

If AC = 8.5 m and AE = 9.71 m then the length CE (i.e. the minimum length of the ladder) may be found using Pythagoras's theorem.

Hence $(CE)^2 = (AC)^2 + (AE)^2$

$= (8.5)^2 + (9.71)^2$

$= 72.25 + 94.28$

$= 166.53 = 167$ (to 3 significant figures)

Therefore $CE = \sqrt{167} = \pm 12.9$

$= 12.9$ m (-12.9 can be neglected)

Hence **a ladder of minimum length 12.9 m is required.**

The U.K. Health and Safety Executive recommends that the foot of a ladder should be about 1 m out from the building for every 4 m in height. Thus the above position of the ladder is hardly a safe one!

Further problems on similar triangles may be found in Section 10.6 (Problems 24–33), page 251.

10.5 Construction of triangles

To construct any triangle the following drawing instruments are needed:

(*i*) ruler and/or straight edge;
(*ii*) compass;
(*iii*) protractor;
(*iv*) pencil; and
(*v*) set square (possibly)

The method of constructing triangles depends on the available information. It was mentioned in Section 10.3 that six facts determine a triangle completely – the three sides and the three angles. When constructing triangles only three particular facts are required and the only four possible combinations are shown, by example, in cases 1–4 below.

Case 1. Given the lengths of the three sides

Example: To construct a triangle whose sides are 7 cm, 5 cm and 4 cm.

(*i*) Let one side, say that 7 cm long be the base. Draw a straight line of any length, and, with a pair of compasses, mark out a 7 cm length and label it AB as shown in Fig. 10.39(*i*)

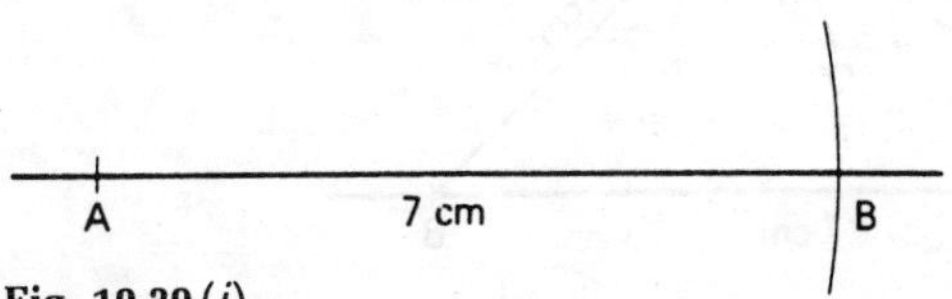

Fig. 10.39 (*i*)

(*ii*) Using a compass set to 5 cm and, with centre at A describe the arc MN as shown in Fig. 10.39(*ii*)

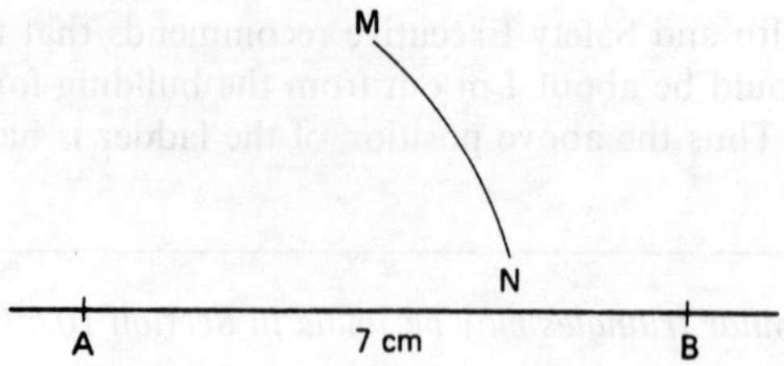

Fig. 10.39 (*ii*)

(*iii*) Set the compass to 4 cm and, with centre at B, describe the arc OP as shown in Fig. 10.39(*iii*)

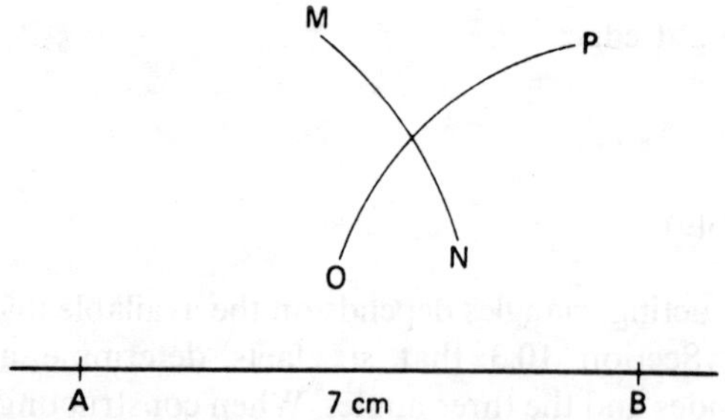

Fig. 10.39 (*iii*)

(*iv*) The intersection of the two arcs (produced if necessary) at C is a vertex of the required triangle. Thus join AC and BC by straight lines as shown in Fig. 10.39(*iv*)

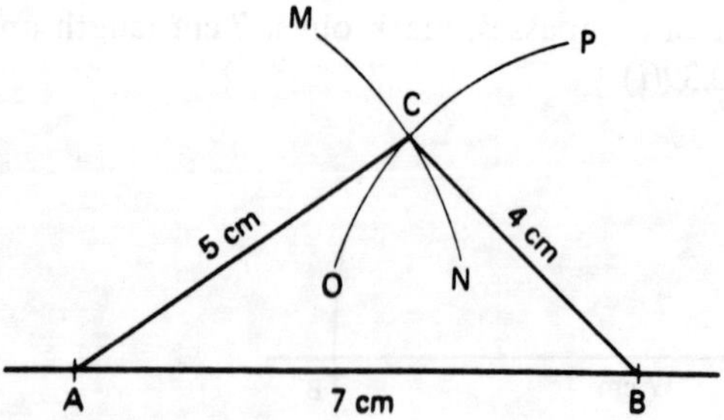

Fig. 10.39 (*iv*)

Thus triangle ABC with sides 7 cm, 5 cm and 4 cm has been constructed.

It was stated in Section 10.1 that the ratio of the angles of a triangle is not equal to the ratio of the sides. This can be proved by referring to Fig. 10.39(*iv*).

BC = 4 cm. As a proportion of the perimeter of the triangle ABC

$$BC = \frac{4}{4+5+7} = \frac{4}{16} = \frac{1}{4}$$

Similarly, AC = $\frac{5}{16}$ of the perimeter and

AB = $\frac{7}{16}$ of the perimeter.

If it is assumed that the ratio of the angles is equal to the ratio of the sides then:

$\angle$A (the angle opposite side BC) should be $\frac{1}{4}$ of 180° that is, 45°;
$\angle$B (the angle opposite AC) should be $\frac{5}{16}$ of 180°, that is, 56°; and
$\angle$C (the angle opposite AB) should be $\frac{7}{16}$ of 180°, that is, 79°.

Using a protractor, the angles of triangle ABC can be measured, with the following values resulting:

$$\angle A = 34°; \ \angle B = 44\tfrac{1}{2}°; \ \angle C = 101\tfrac{1}{2}°$$

It can therefore be clearly seen that **the ratio of the angles of a triangle is not equal to the ratio of the sides.**

Case 2. Given the lengths of two sides and the included angle

Example: To construct a triangle given two sides, of 6 cm and 3 cm, and the angle between these two sides as 50°.

(*i*) Let one side, say that 6 cm long, be the base. Draw a line 6 cm long and label it PQ as shown in Fig. 10.40(*i*)

Fig. 10.40 (*i*)

(*ii*) Using a protractor centred at P make an angle of 50° to PQ as shown in Fig. 10.40(*ii*)

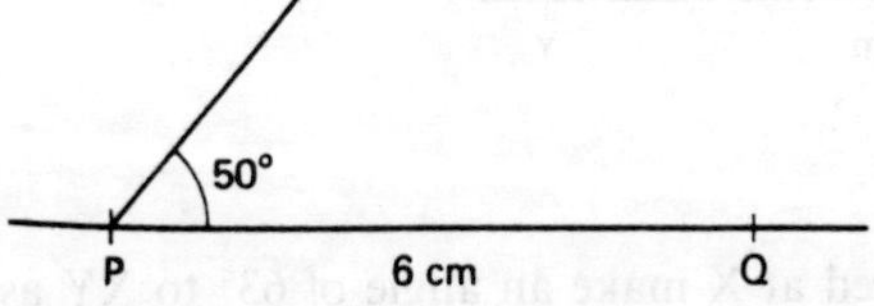

Fig. 10.40 (*ii*)

(*iii*) From P, measure a length of 3 cm and label the length PR, as shown in Fig. 10.40(*iii*)

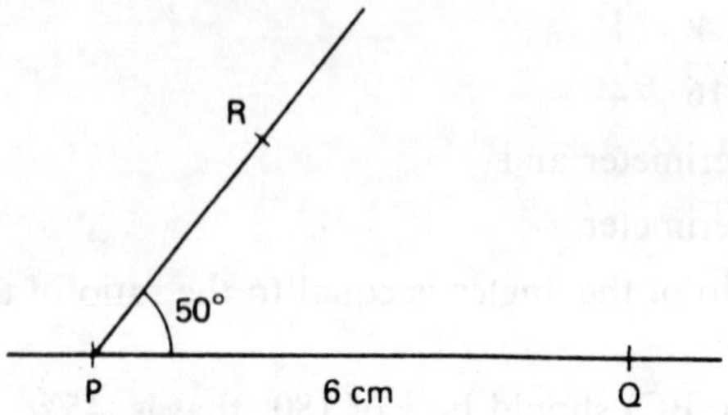

Fig. 10.40 (*iii*)

(*iv*) Join the point Q to R by a straight line as shown in Fig. 10.40(*iv*)

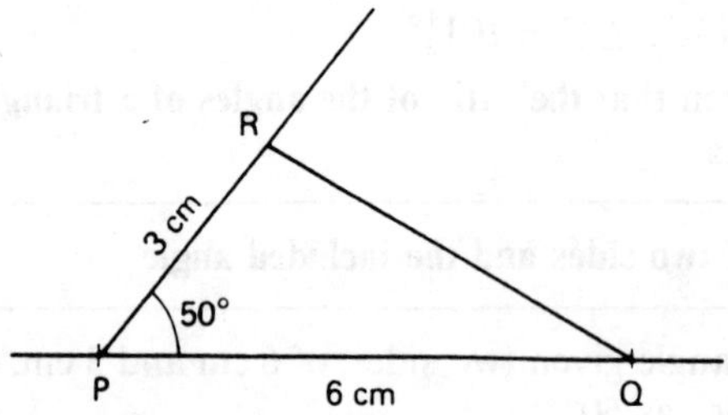

Fig. 10.40 (*iv*)

Thus, given two sides, of 6 cm and 3 cm, and an included angle of 50°, a triangle PQR has been constructed.

Case 3. Given the length of one side and the value of two angles

---

Example (a): To construct a triangle XYZ given that XY = 8 cm, $\angle X = 63°$ and $\angle Y = 42°$

(*i*) Let the 8 cm side be the base. Draw a line 8 cm long and label it XY as shown in Fig. 10.41(*i*)

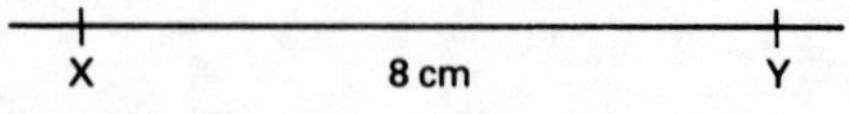

Fig. 10.41 (*i*)

(*ii*) Using a protractor centred at X make an angle of 63° to XY as shown in Fig. 10.41(*ii*)

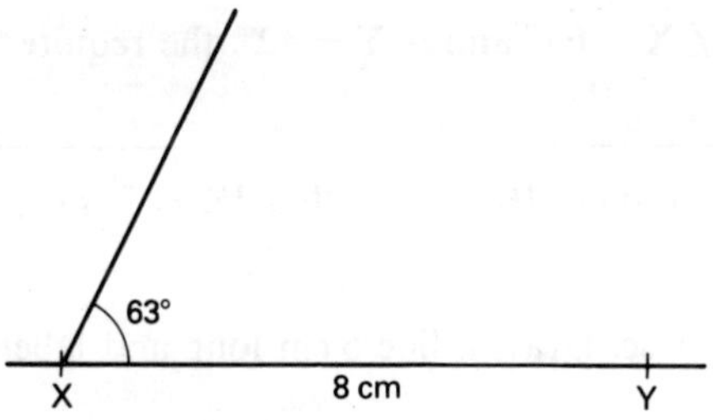

Fig. 10.41 (*ii*)

(*iii*) Using a protractor centred at Y make an angle of 42° to XY as shown in Fig. 10.41(*iii*)

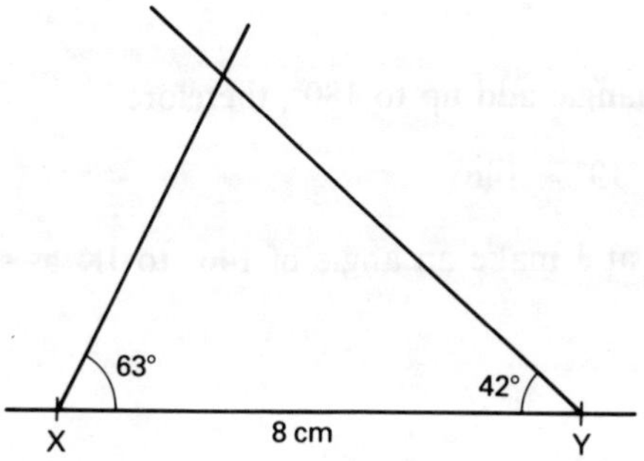

Fig. 10.41 (*iii*)

(*iv*) The intersection of the arms (produced if necessary) forms the vertex, Z, of the triangle as shown in Fig. 10.41(*iv*)

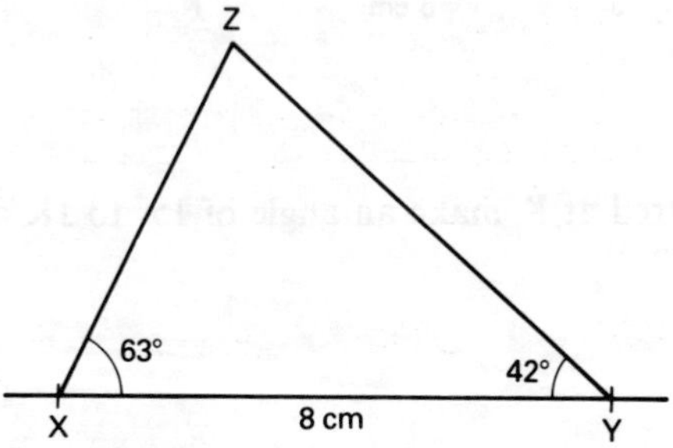

Fig. 10.41 (*iv*)

Wherever two angles of a triangle are given the third angle can immediately be calculated. In this example the third angle $\angle Z$ is given by:

$$\angle Z = 180° - 63° - 42° = 75°$$

Using a protractor centred at Z this angle can readily be checked.

Thus, given XY = 8 cm, ∠X = 63° and ∠Y = 42°, the required triangle XYZ has been constructed.

Example (b): To construct a triangle JKL given that JK = 5 cm, ∠K = 15° and ∠L = 19°

(*i*) Let the 5 cm side be the base. Draw a line 5 cm long and label it JK as shown in Fig. 10.42(*i*)

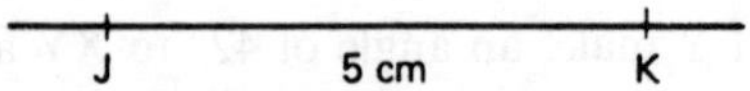

Fig. 10.42 (*i*)

(*ii*) The three angles of a triangle add up to 180°, therefore

$$\angle J = 180° - 15° - 19° = 146°$$

Using a protractor centred at J make an angle of 146° to JK as shown in Fig. 10.42(*ii*)

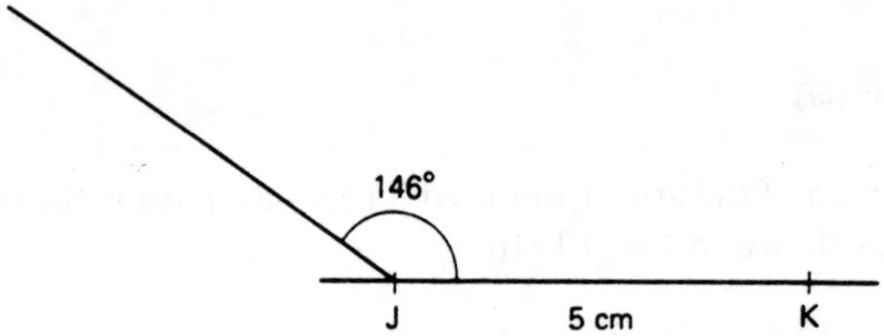

Fig. 10.42 (*ii*)

(*iii*) Using a protractor centred at K make an angle of 15° to JK as shown in Fig. 10.42(*iii*)

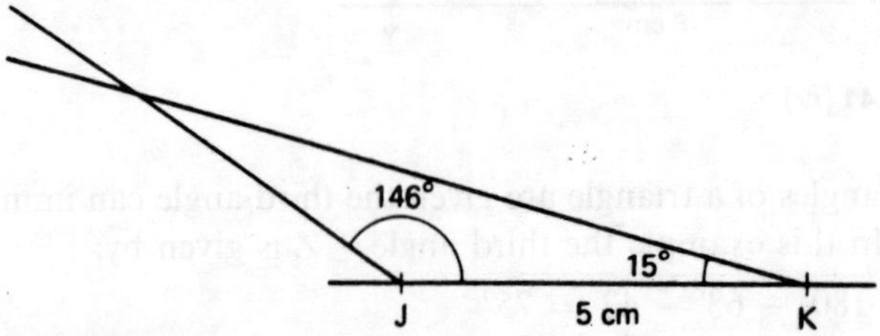

Fig. 10.42 (*iii*)

(*iv*) The intersection of the arms (produced if necessary) forms the vertex, *L*, of the triangle as shown in Fig. 10.42(*iv*). (The angle L must be equal to 19° and this should be checked using a protractor centred at L.)

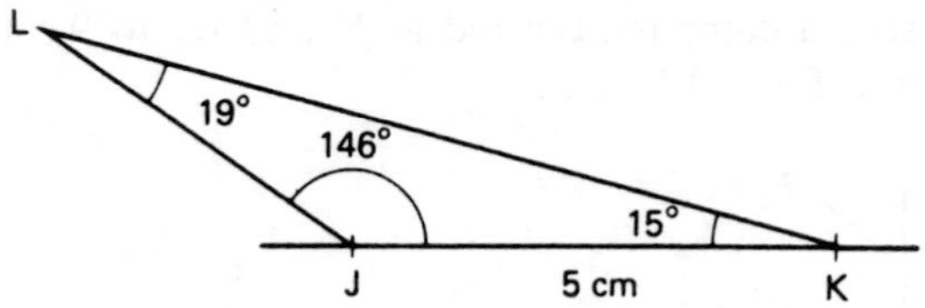

Fig. 10.42 (*iv*)

Thus, given JK = 5 cm, ∠K = 15° and ∠L = 19°, the required triangle JKL has been constructed.

Case 4. To construct a triangle given one side, the hypotenuse and a right angle

Example: To construct a triangle MNO given that MN = 6 cm, the hypotenuse, NO = 9 cm, and ∠M = 90°

(*i*) Let the 6 cm side be the base. Draw a line 6 cm long and label it MN as shown in Fig. 10.43(*i*)

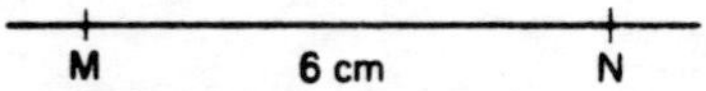

Fig. 10.43 (*i*)

(*ii*) Produce the line MN any distance to P. With compass centred at M make an arc at A and A′. (The length AM is arbitrary.) With compass centred at A, draw an arc BC greater in length than AM. With the same compass setting, and centred at A′, draw the arc DE. This is shown in Fig. 10.43(*ii*).

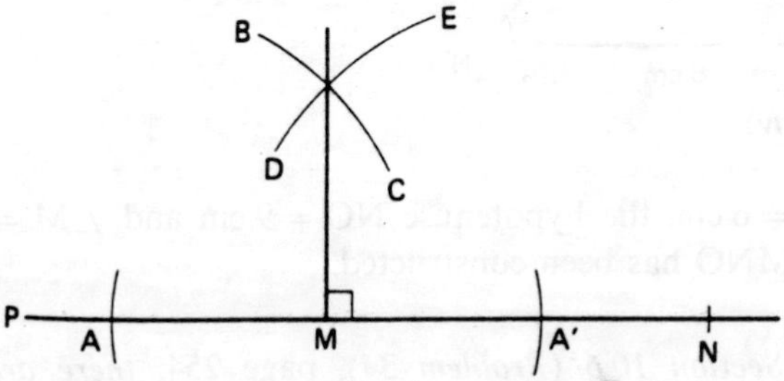

Fig. 10.43 (*ii*)

Join the intersection of the arcs to M and a right angle to MN is produced at M, as shown in Fig. 10.43(*iii*). (Alternatively, a protractor or set square can be used to construct a 90° angle.)
(*iii*) The hypotenuse is always opposite the right angle. Thus the side NO is opposite ∠M. Using a compass, centred at N and set to 9 cm describe the arc EF as shown in Fig. 10.43(*iii*)

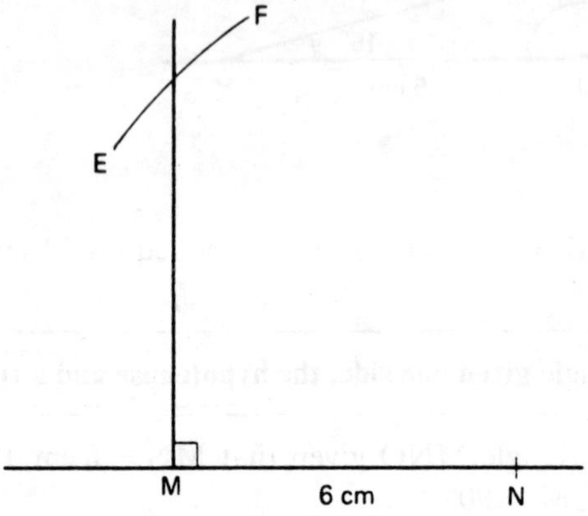

Fig. 10.43 (*iii*)

(*iv*) The intersection of the arms (produced if necessary) forms the vertex, O, of the required triangle. Join ON by a straight line as shown in Fig. 10.43(*iv*)

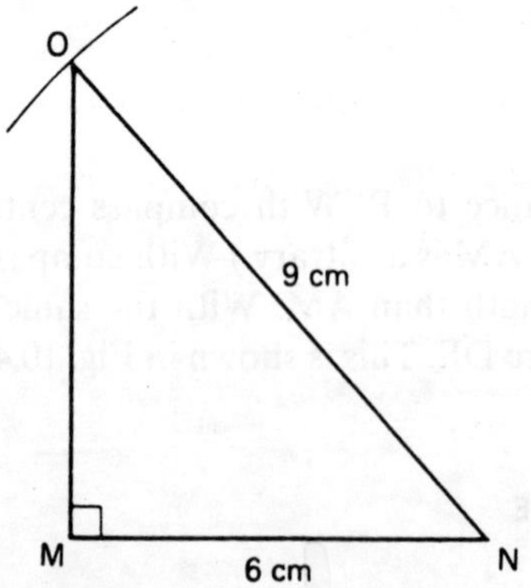

Fig. 10.43 (*iv*)

Thus, given MN = 6 cm, the hypotenuse NO = 9 cm and ∠M = 90°, the required triangle MNO has been constructed.

In the following Section 10.6 (*Problem 34*), page 254, *there are further examples of triangles for construction.*

10.6 Further problems

Properties of triangles

1. Find angles a to f in Fig. 10.44

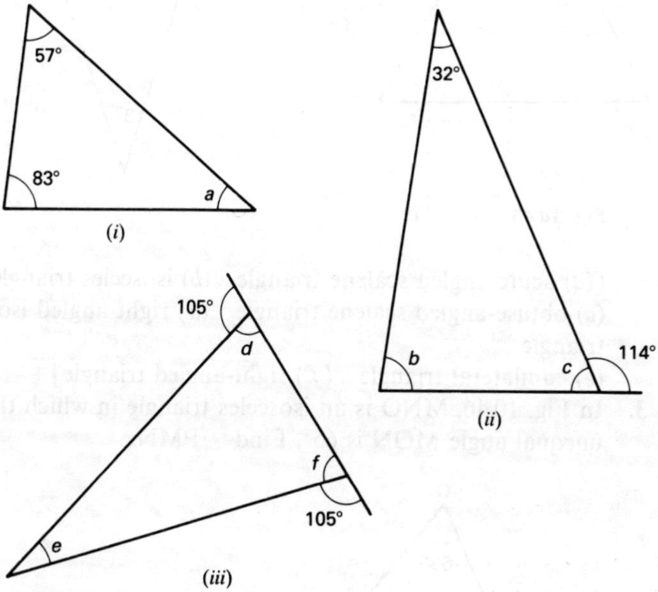

Fig. 10.44

[$a = 40°$ $b = 82°$ $c = 66°$ $d = 75°$ $e = 30°$ $f = 75°$]

2. Name the types of triangles shown in Fig. 10.45

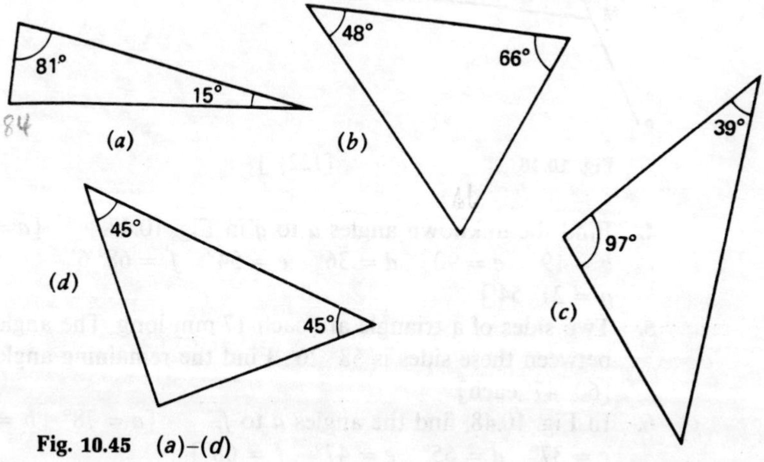

Fig. 10.45 (*a*)–(*d*)

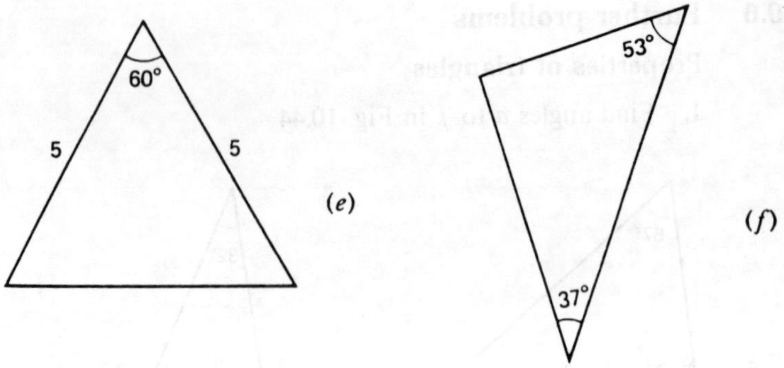

Fig. 10.45 (e)–(f)

[(a) acute-angled scalene triangle (b) isosceles triangle
(c) obtuse-angled scalene triangle (d) right-angled isosceles triangle
(e) equilateral triangle (f) right-angled triangle]

3. In Fig. 10.46, MNO is an isosceles triangle in which the unequal angle MON is 65°. Find ∠PMN.

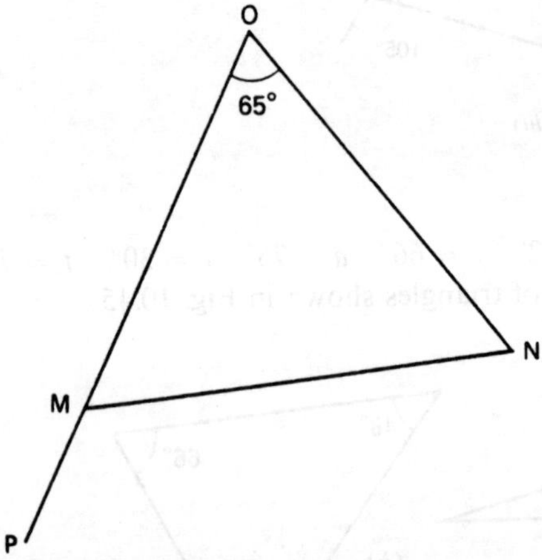

Fig. 10.46 [$122\frac{1}{2}$°]

4. Find the unknown angles a to g in Fig. 10.47. [$a = 71°$ $b = 19°$ $c = 90°$ $d = 36°$ $e = 54°$ $f = 68° 6'$ $g = 21° 54'$]
5. Two sides of a triangle are each 17 mm long. The angle between these sides is 58° 26′. Find the remaining angles. [60° 47′ each]
6. In Fig. 10.48, find the angles a to f. [$a = 78°$ $b = 65°$ $c = 37°$ $d = 65°$ $e = 47°$ $f = 68°$]

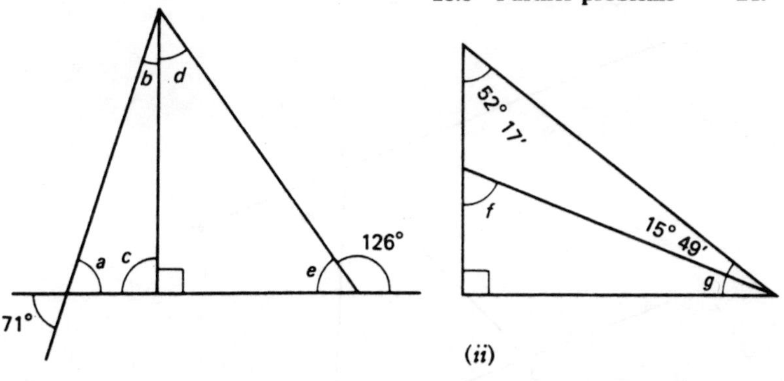

Fig. 10.47

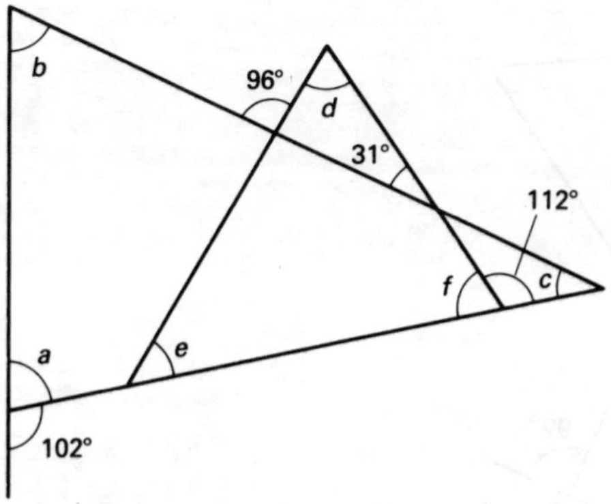

Fig. 10.48

7. In a triangle PQR a line SR is drawn perpendicular to PQ such that ∠PRS is 27°. Calculate the value of ∠QPR. [63°]
8. Triangle ABC, shown in Fig. 10.49, is isosceles with ∠BAC the unequal angle. Find the three interior angles of the triangle [A = 48° B = 66° C = 66°]
9. Triangle JKL has a right angle at K and ∠KJL is 28°. KL is produced to M. If the bisectors of ∠JKL and ∠JLM meet at N find ∠KNL. [14°]
10. If, in Fig. 10.50, triangle ABC is equilateral, find the interior angles of triangle DCE. [C = 60° D = 32° E = 88°]

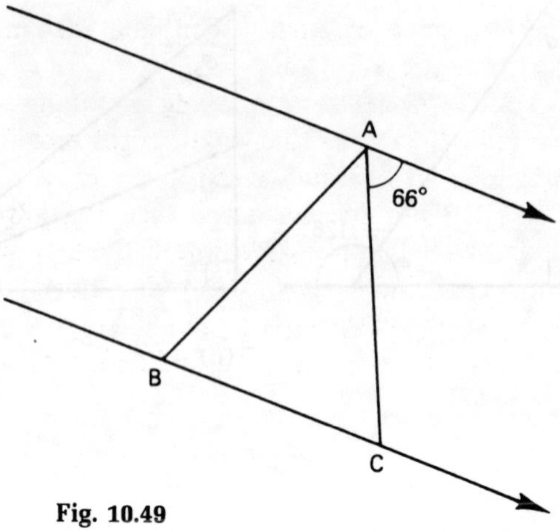

Fig. 10.49

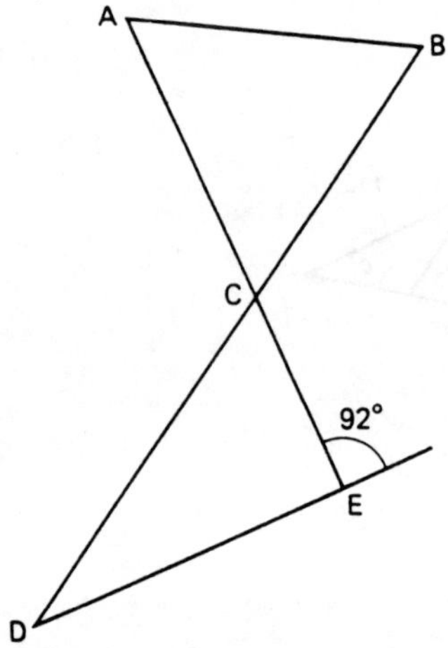

Fig. 10.50

Theorem of Pythagoras

11. In a triangle ABC, ∠A is a right angle, $b = 9.20$ mm and $c = 7.10$ mm. Find side a [11.62 mm]

12. In a triangle PQR, ∠Q = 90°, $r = 17.5$ mm and $q = 29.6$ mm. Find side p [23.87 mm]

13. In a triangle MNQ, ∠N = 90°, $m = 5.26$ cm and $n = 9.58$ cm. Find side q [8.007 cm]

14. A tent peg is 4.0 m away from a 6.0 m high tent. What length of rope runs from the top of the tent to the peg? [7.21 m]

15. Show that if a triangle has sides 7, 24 and 25 cm respectively, it is right-angled.

16. A triangle has sides of 9.6 m, 18 m and 20.4 m. Is it a right-angled triangle? [Yes]

17. Triangle ABC is isosceles, ∠A being right-angled. If the hypotenuse is 11.46 mm find the length of the remaining sides of the triangle. [8.103 mm each]

18. A man walks 15.0 km due north and then 19.0 km due east. Another man, starting at the same time as the first man, walks 25.0 km due east and then 9.00 km due north. Find the distance between the two men. [8.485 km]

19. Figure 10.51 shows a plan view of a kite design. Calculate the dimensions *d* and *l*

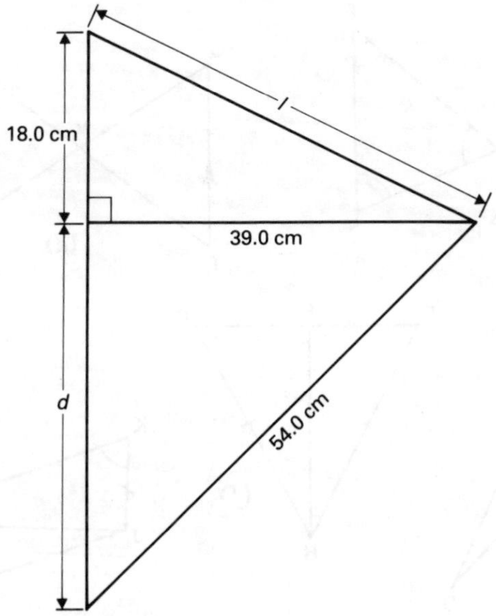

Fig. 10.51

[d = 37.35 cm l = 42.95 cm]

20. A ladder 4.0 m long is placed against a wall with its foot 1.5 m from the wall. How far up the wall does the ladder reach? If the foot of the ladder is now pulled 1.0 m further away from the wall, calculate how far the top of the ladder falls.
[3.71 m, 59 cm]

21. Two aircraft leave an airfield at the same time. One travels due south at an average speed of 350 km/h and the other travels due west at an average speed of 250 km/h. Find their distance apart after 2 hours 30 minutes. [1 075 km]

Congruent triangles

22. State which of the following pairs of triangles in Fig. 10.52 are congruent and name them in correct sequence.

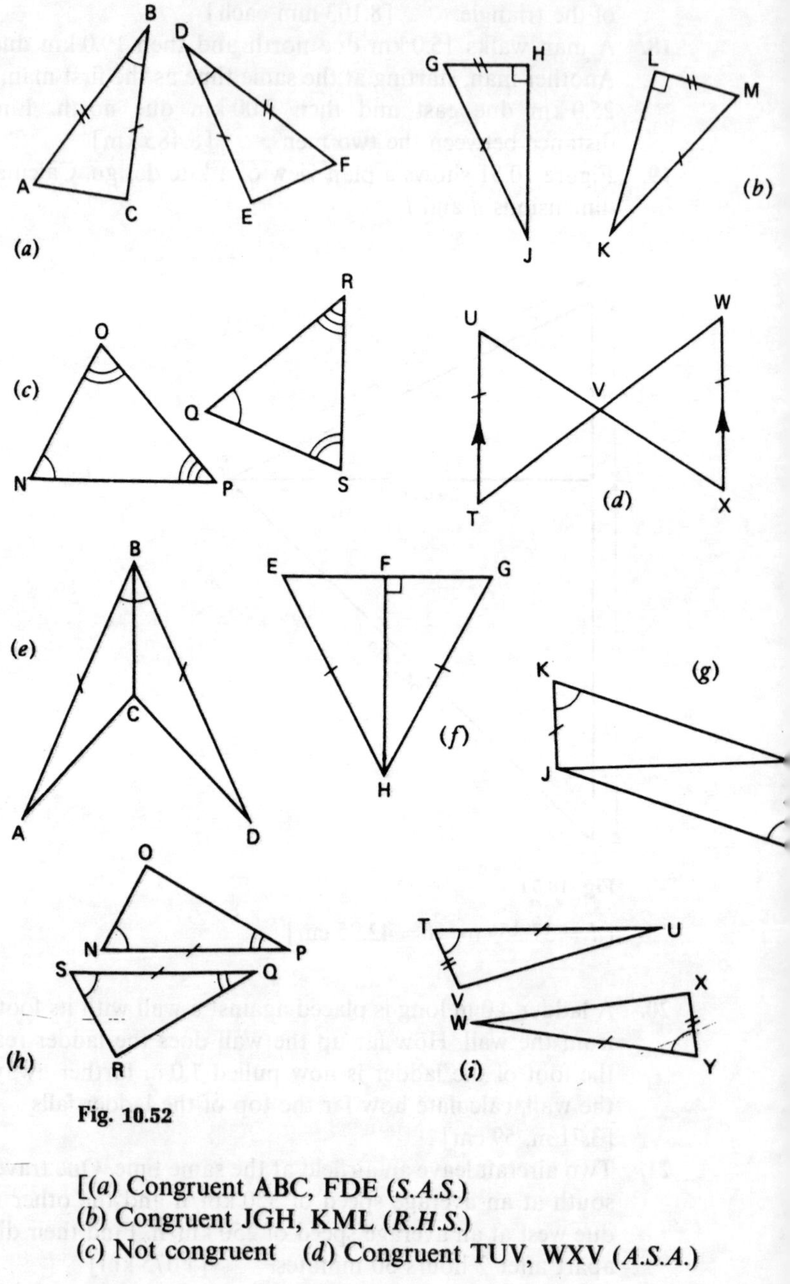

Fig. 10.52

[(*a*) Congruent ABC, FDE (*S.A.S.*)
(*b*) Congruent JGH, KML (*R.H.S.*)
(*c*) Not congruent (*d*) Congruent TUV, WXV (*A.S.A.*)

(*e*) Congruent ABC, DBC (*S.A.S.*)
(*f*) Congruent EFH, GFH (*R.H.S.*)
(*g*) Congruent LMJ, KJM (*A.S.A.*)
(*h*) Congruent NOP, SRQ (*A.S.A.*)
(*i*) Congruent TUV, YWX (*S.A.S.*)]

23. In triangle MNO, MN = NO and P and Q are points on MN and NO respectively such that MP = OQ. Show that triangles MQN and OPN are congruent.

Similar triangles

24. In Fig. 10.53, determine the length of *a*.

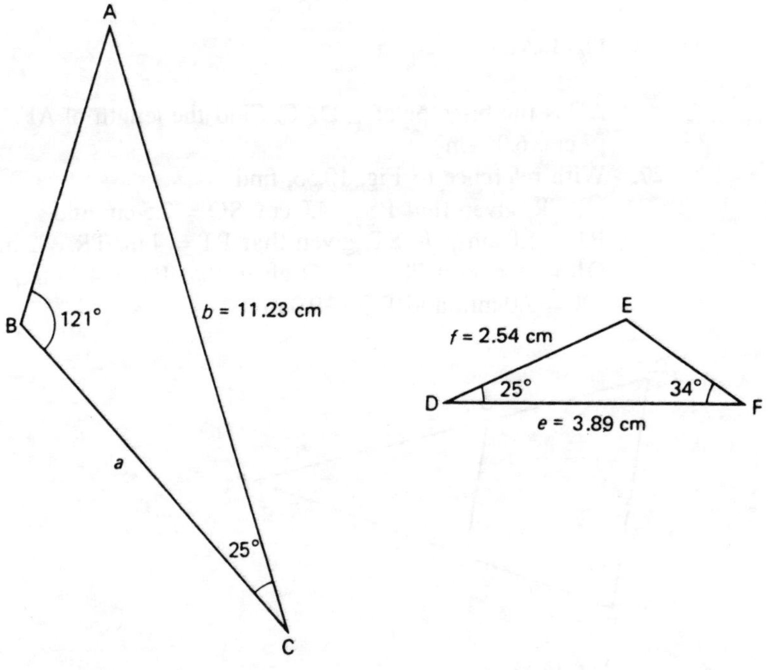

Fig. 10.53

[7.333 cm]

25. In Fig. 10.54, find the lengths of *a*, *c* and *d*.
[$a = 15.38$ cm $c = 9.468$ cm $d = 5.581$ cm]

26. ABC is an equilateral triangle of side 3.0 cm. When AB and AC are produced to D and E respectively, DE is found to be parallel with BC. If AD is 7.0 cm, find the length of DE. F is a point on the line DE, between D and E, such that the line

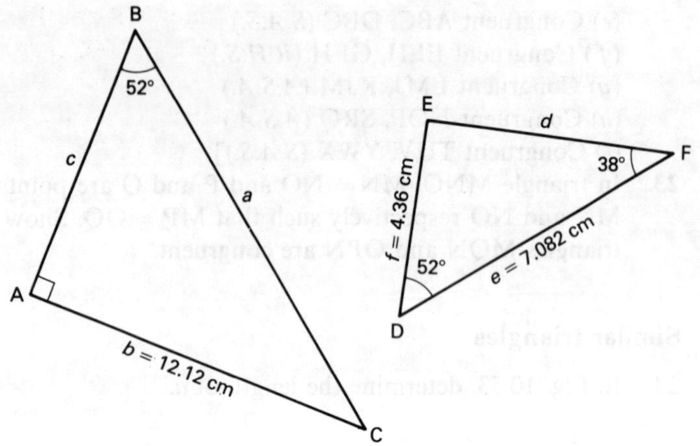

Fig. 10.54

AF is the bisector of ∠DAE. Find the length of AF.
[7 cm, 6.06 cm]

27. With reference to Fig. 10.55, find
(*a*) TR, given that PS = 4.0 cm, SQ = 2.5 cm and PT = 5.0 cm (*b*) ST, given that PT = 4 m, TR = 2 m and QR = 9 m (*c*) PS and SQ given that PT = 4.0 mm, TR = 7.0 mm and PQ = 9.0 mm

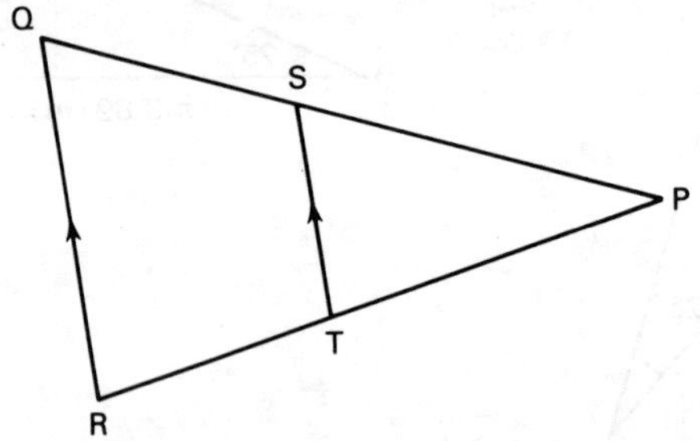

Fig. 10.55

(*a*) [3.125 cm] (*b*) [6 m] (*c*) [SP = 3.27 m, SQ = 5.73 m]

28. In a triangle ABC, AB = 3.5 cm, BC = 2.4 cm and AC = 3.2 cm. DE is drawn parallel to BC so that AD = 2.8 cm. (D lies on AB and E lies on AC.) Find the lengths of the remaining sides of triangle ADE.
[DE = 1.92 m, AE = 2.56 cm]

29. The triangle MNO is right-angled at N. If MN = 3.40 cm and NO = 1.80 cm find NP, the length of the perpendicular

to the hypotenuse and also the length of MP and PO.
[NP = 1.591 cm PO = 0.842 cm MP = 3.005 cm]

30. In Fig. 10.56, find:
(*a*) the length of OQ when MN = 8.0 m, PQ = 10.0 m and MO = 3.0 m (*b*) the length of PQ when NO = 1.8 cm, PO = 5.4 cm, and MN = 10.5 cm

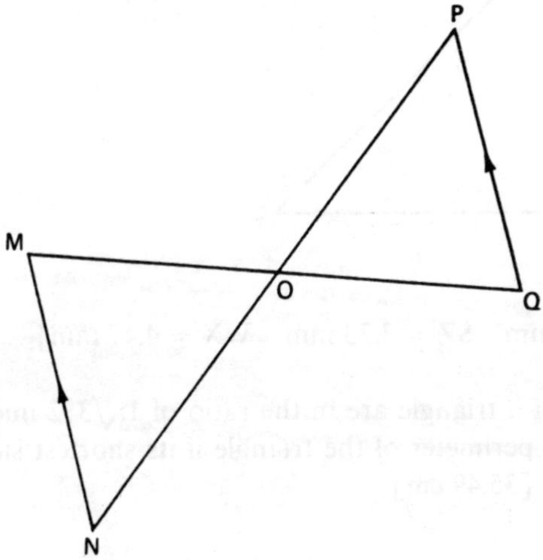

Fig. 10.56

(*a*) [3.75 m] (*b*) [31.5 cm]

31. In Fig. 10.57, OQ = 9 cm, NO = 6 cm and MN = 4 cm. Find the length of NR. [3.6 cm]

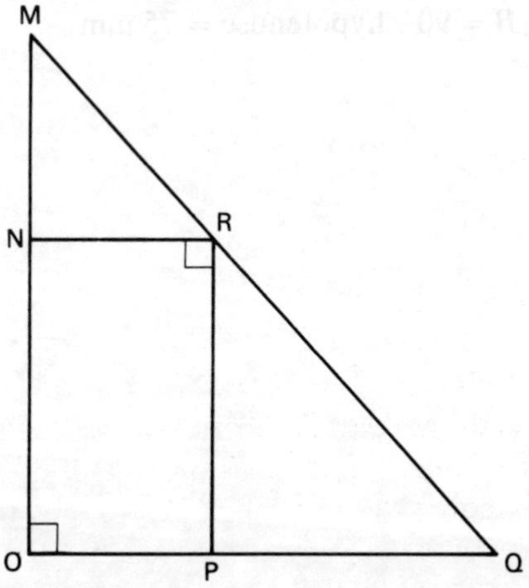

Fig. 10.57

32. If, in Fig. 10.58, WR = 6.0 mm, RX = 4.0 mm and SR = 5.0 mm, find the length of RZ, SZ and WX.

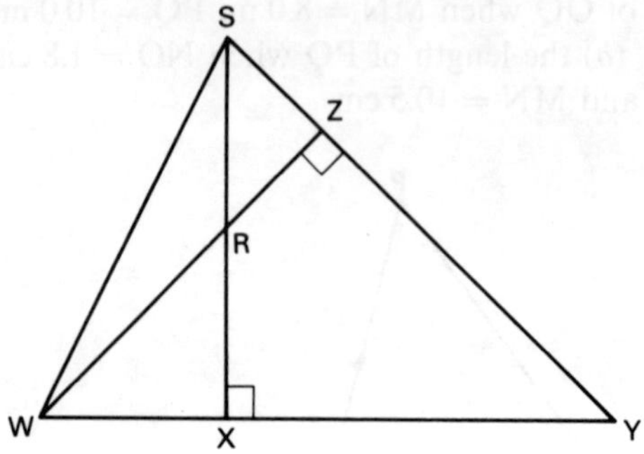

Fig. 10.58

[RZ = 3.33 mm SZ = 3.73 mm WX = 4.47 mm]

33. If the sides of a triangle are in the ratio of $1{:}\sqrt{3}{:}2$ find the length of the perimeter of the triangle if its shortest side is 7.50 cm. [35.49 cm]

Construction of triangles

34. Construct the following triangles:
(*a*) $a = 9$ cm, $b = 7$ cm and $c = 5$ cm
(*b*) $a = 4.5$ cm, $b = 6.25$ cm and $\angle C = 62°$
(*c*) $a = 53$ mm, $\angle C = 49°$ and $\angle B = 81°$
(*d*) $c = 3.8$ cm, $\angle A = 125°$ and $\angle C = 20°$
(*e*) $a = 48$ mm, $\angle B = 90°$, hypotenuse = 75 mm

Chapter 11

The circle

11.1 Terminology and properties of the circle

1. A circle is a plane figure enclosed by a curved line, every point on which is equidistant from a point within, called the **centre**. In Fig. 11.1, O is the centre of the circle.
2. The distance from the centre to the curve, OA, is called the **radius,** r, of the circle.
3. Any straight line passing through the centre and touching the circumference at each end as BC does in Fig. 11.1 is called the **diameter**, d. The diameter is twice the length of the radius. Thus $d = 2r$.
4. The boundary of a circle, that is, the perimeter, is called the **circumference**, c.
   It is found that if the length of the circumference, c, of any circle is divided by the length of the diameter, d, the answer is always the same, that is $\frac{c}{d}$ is a constant.

   The constant is denoted by a Greek letter π (pronounced as 'pie'), where $\pi = 3.141\,59$ to 5 decimal places.

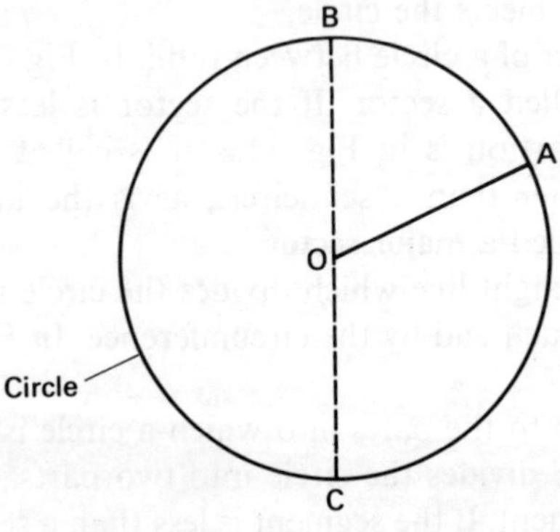

Fig. 11.1

Thus $\frac{c}{d} = \pi$

or $c = \pi d$

But $d = 2r$

Therefore $c = 2\pi r$

5. A **semicircle** is one half of the whole circle.
6. A **quadrant** is one quarter of a whole circle.
7. A **tangent** to a circle is a straight line which meets the circle in one point only and does not cut the circle when produced. In Fig. 11.2, FG is a tangent to the circle, since it touches the circle at the point H only. If radius OH is drawn, as in Fig. 11.3, then ∠FHO is a right angle, that is, 90°

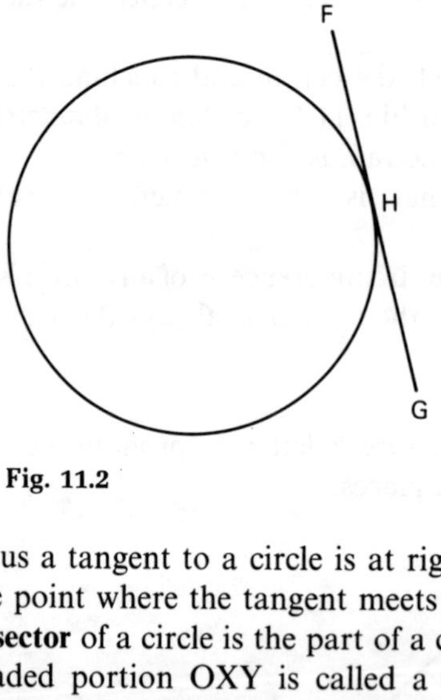

Fig. 11.2

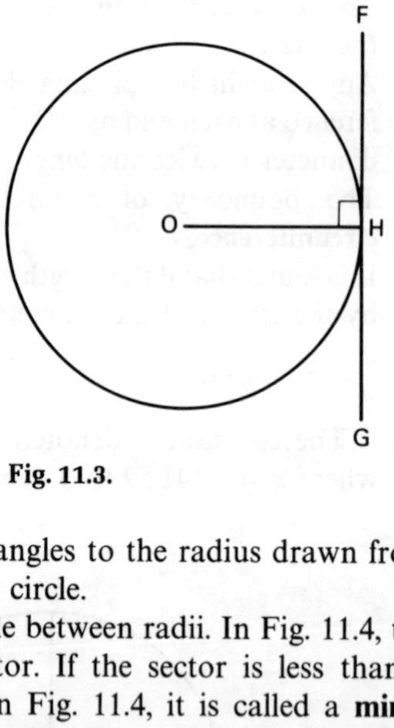

Fig. 11.3.

Thus a tangent to a circle is at right angles to the radius drawn from the point where the tangent meets the circle.

8. A **sector** of a circle is the part of a circle between radii. In Fig. 11.4, the shaded portion OXY is called a sector. If the sector is less than a semicircle, as the shaded portion is in Fig. 11.4, it is called a **minor sector**. If the sector is greater than a semicircle, as is the unshaded portion of Fig. 11.4, it is called a **major sector**.
9. A **chord** of a circle is any straight line which divides the circle into two parts, and is terminated at each end by the circumference. In Fig. 11.5, the line DE is a chord.
10. A **segment** is the name given to the parts into which a circle is divided by a chord. In Fig. 11.5, DE divides the circle into two parts. Each of the two parts is called a segment. If the segment is less than a semicircle it is called a **minor segment**. The shaded area of Fig. 11.5 is therefore a

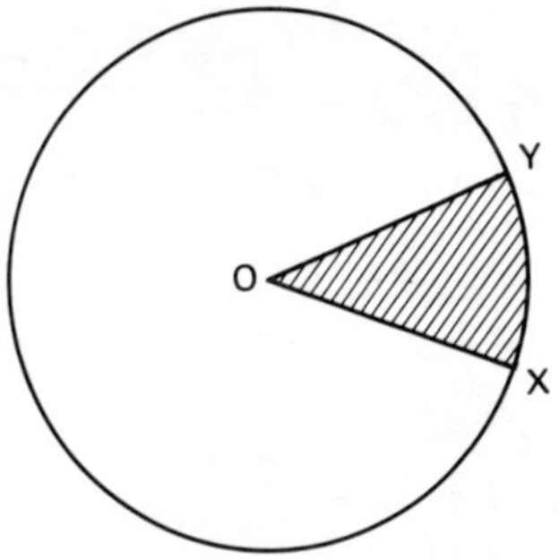

Fig. 11.4

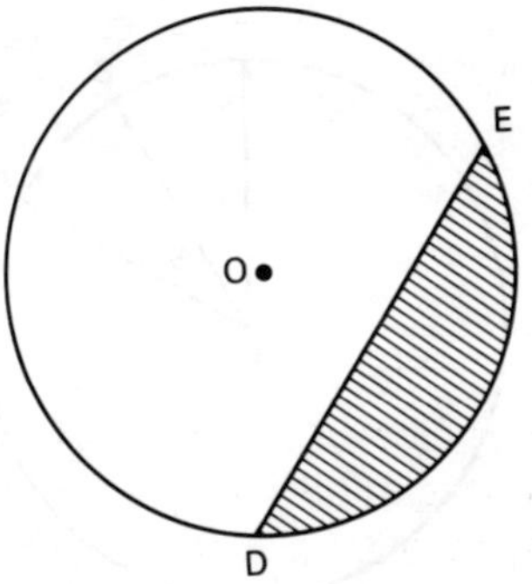

Fig. 11.5

minor segment. If the segment is greater than a semicircle it is called a **major segment**. The unshaded area of Fig. 11.5 is therefore called a major segment.

11. An **arc** is a portion of the circumference of a circle. In Fig. 11.6, PR is a chord which divides the circle into a minor segment PQR and a major segment PSR.

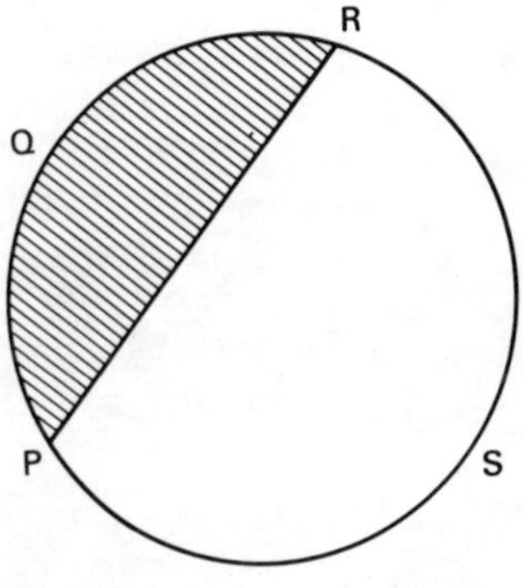

Fig. 11.6

The shorter distance from P to R along the circumference, that is, through Q, is called the **minor arc**. The longer distance from P to R along the circumference, that is, through S, is called the **major arc**.

There is a direct relationship between the length of an arc and the angle subtended at the centre of a circle. In Fig. 11.7 let $\angle AOB = 30°$ and $\angle BOC = 30°$.

As the angles subtended at the centre of the circle are the same for each of the sectors AOB and BOC, the length of the arc AB is equal to the length of the arc BC, that is, **the arc length is proportional to the angle which it subtends at the centre**. For example, the length of arc

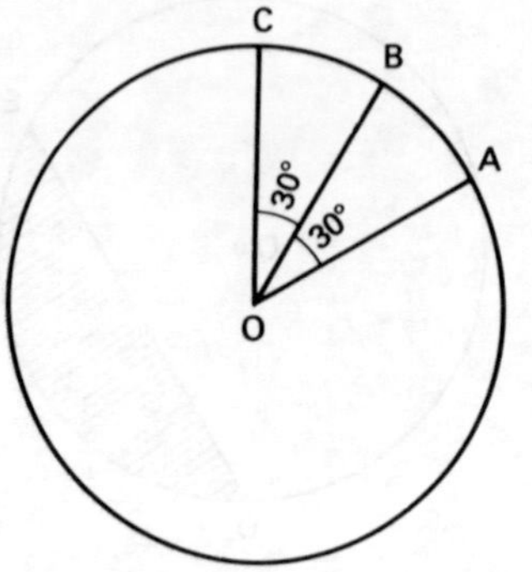

Fig. 11.7

for a 60° sector is twice that of a 30° sector. If the angle subtended at the centre of a sector is 60°, then, since the circumference subtends an angle of 360°, the length of the arc would, by proportion, be $\frac{60}{360}$, that is $\frac{1}{6}$, of the total circumference.

12. The angle at the centre of a circle, subtended by an arc, is double the angle at the circumference subtended by the same arc, i.e. from Fig. 11.8, **∠AOB = 2 × ∠ACB**

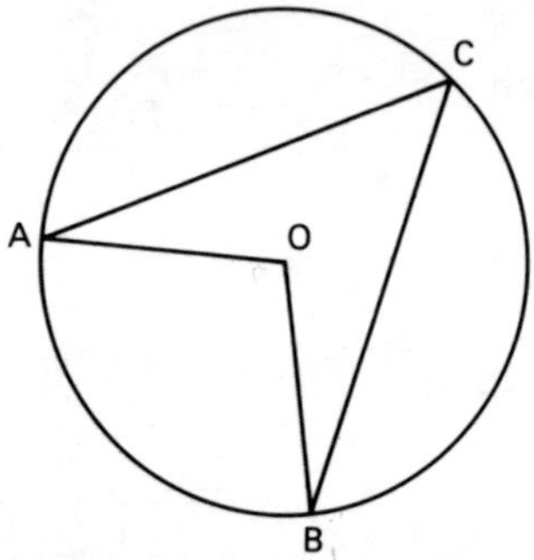

Fig. 11.8

13. An angle may be measured in degrees or **radians.**
A radian is defined as the angle subtended at the centre of a circle by an arc equal in length to the radius.
Thus, in Fig. 11.9, when $r = l$ then angle θ is 1 radian. This angle is the unit of measurement in so-called circular measure. It is of constant size whatever the length of the radius.
The length of arc, l, for 1 radian is equal to r
The length of arc, l, for 2 radians is equal to $2r$
The length of arc, l, for θ radians is equal to $r\theta$

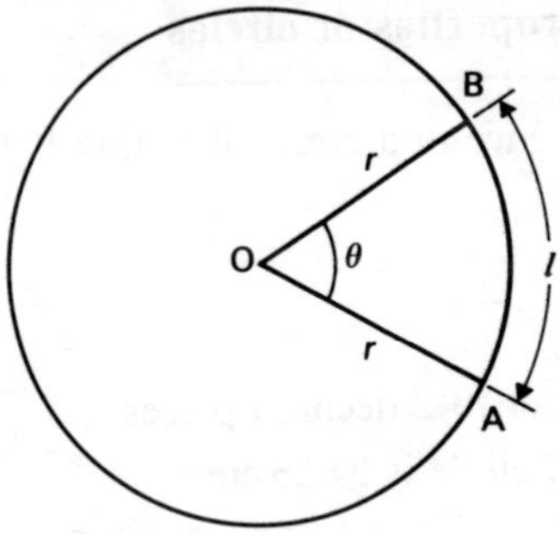

Fig. 11.9

Therefore generally, $l = r\theta^c$ or $\theta^c = \frac{l}{r}$

'radians' may be written as 'θ^c', the 'c' standing for circular measure.

The number of radians subtended by the whole circumference of a circle is given by the number of times the radius is contained in the circumference.

Now the circumference $= 2\pi r$. Hence the number of radians in one complete revolution is given by:

$$\frac{2\pi r}{r} = 2\pi = 6.283 \text{ (correct to 3 decimal places)}$$

Hence **2π radians = 360°**

$$\textbf{1 radian} = \left(\frac{360}{2\pi}\right)^\circ$$

$$= \mathbf{57.30^\circ} \text{ (correct to 2 decimal places)}$$

There are therefore just over 6 radians contained within a complete revolution.

Whenever conversion from radians to degrees or vice versa is necessary, either of the above relationships may be used. Some calculators have the facility to convert from radians to degrees and vice versa at the press of a button.

The following relationships are often used and can be derived from $2\pi^c = 360°$:

$\pi^c = 180°$; $\quad \frac{\pi^c}{2} = 90°$; $\quad \frac{\pi^c}{3} = 60°$;

$\frac{\pi^c}{4} = 45°$; $\quad \frac{\pi^c}{6} = 30°$; $\quad \frac{\pi^c}{12} = 15°$;

and so on.

Worked problems on the properties of circles

Problem 1. Find the circumference of a circle of radius 15.0 cm

Circumference, $c = 2\pi(\text{radius})$

That is $c = 2\pi r$

$= (2)(\pi)(15.0)$

$= 94.25$, correct to 2 decimal places.

Hence the circumference of the circle is **94.25 cm**

Problem 2. Find the circumference of a circle of diameter 12.0 mm

The circumference $c = \pi d$

$= (\pi)(12.0)$

$= 37.70$

Hence the circumference of the circle is **37.70 mm**

Problem 3. Find the radius of a circle if its perimeter is 54.0 cm

Perimeter = circumference, $c = 54.0$ cm

$c = 2\pi r$

Therefore $r = \dfrac{c}{2\pi}$

$= \dfrac{54.0}{(2)(\pi)}$

$r = 8.594$

Hence the radius of the circle is **8.594 cm**

Problem 4. Figure 11.10 shows a tangent XY touching a circle at Y. If the radius of the circle is 50 mm and the distance XY is 152 mm, find the distance OX in centimetres.

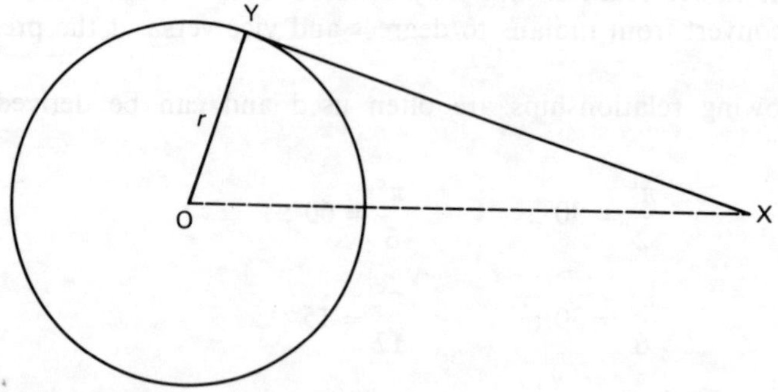

Fig. 11.10

A tangent to a circle is at right angles to a radius drawn from the point of contact, in this case Y. Hence $\angle XYO = 90°$.

Therefore, as XYO is a right-angled triangle, Pythagoras's theorem may be used.

Thus, $(OX)^2 = (XY)^2 + (OY)^2$

$= (152)^2 + (50)^2$

$= 23\,104 + 2\,500$

$= 25\,604$

Therefore $OX = \sqrt{25\,604}$

$= 160.0$ mm

Hence the distance OX is **16.0 cm**

Problem 5. Convert:
(*a*) 119° to radians (*b*) 73° 33′ to radians

(*c*) 2.681 radians to degrees (*d*) $\frac{3\pi}{7}$ radians to degrees.

(*a*) If $180° = \pi$ rad

then $1° = \frac{\pi}{180}$ rad

Therefore $119° = 119\left[\frac{\pi}{180}\right]$ rad

$= 2.077$ rad

Therefore 119 degrees = **2.077 radians**

(*b*) 73° 33′ is a mixture of degrees and minutes and it is first necessary to convert to degrees and decimals of a degree.

$33' = \frac{33°}{60} = 0.55°$

Therefore $73°\ 33' = 73.55°$

If $180° = \pi$ rad

then $73.55° = \frac{73.55(\pi)}{180}$

$= 1.284$ rad

Therefore 73° 33′ = **1.284 radians**

(*c*) π rad = 180°

Therefore 1 rad = $\dfrac{180°}{\pi}$

$$2.681 \text{ rad} = 2.681\left[\frac{180}{\pi}\right]^\circ$$
$$= 153.61^\circ$$
$$0.61^\circ = (0.61 \times 60)'$$
$$= 37', \text{ correct to the nearest minute}$$
$$153.61^\circ = 153^\circ\ 37'$$

Therefore 2.681 radians = **153° 37′**

(*d*) π rad = 180°

Therefore $1 \text{ rad} = \left[\dfrac{180}{\pi}\right]^\circ$

$$\frac{3\pi}{7}\text{ rad} = \left[\frac{3\pi}{7}\right]\left[\frac{180}{\pi}\right]^\circ$$
$$= \frac{3}{7}(180)^\circ$$
$$= 77.143^\circ$$

Now $0.143^\circ = (0.143 \times 60)'$

$$= 8.58'$$

and $0.58' = (0.58 \times 60)''$

$$= 35''$$

So $77.143^\circ = 77^\circ\ 8'\ 35''$

Therefore $\dfrac{3\pi}{7}$ radians = **77.14° or 77° 8′ 35″**

Problem 6. Express the following angles in radians in terms of π: (*a*) 150° (*b*) 270° (*c*) 37.5° (*d*) 383° 17′ 23″

(*a*) 180° = π rad

Therefore $150^\circ = \dfrac{150}{180}\pi$ rad

$$= \frac{5\pi}{6}\text{ rad}$$

Hence 150° = $\mathbf{\dfrac{5\pi}{6}}$ **radians (or 0.833 3π radians)** (*b*) $180° = \pi$ rad

Therefore $270° = \frac{270}{180}\pi$ rad

$= \frac{3\pi}{2}$ rad

Hence $270° = \mathbf{\frac{3\pi}{2}}$ **radians (or 1.5π radians)**

(*c*) $180° = \pi$ rad

Therefore $37.5° = \left(\frac{37.5}{180}\right)\pi$ rad

$= \left(\frac{75}{360}\right)\pi$ rad

$= \left(\frac{5\pi}{24}\right)$ rad

Hence $37.5° = \mathbf{\frac{5\pi}{24}}$ **radians (or 0.208 3π radians)**

(*d*) $23'' = \frac{23'}{60} = 0.38'$

$17.38' = \frac{17.38°}{60} = 0.289\,7° = 0.29°$

Therefore $383°\ 17'\ 23'' = 383.29° = 383.3°$ (to 4 significant figures)
Now $180° = \pi$ rad

Therefore $383.3° = \frac{383.3}{180}\pi$ rad

$= 2.129\pi$ rad

Hence $383°\ 17'\ 23'' =$ **2.129π radians**

Further problems on the properties of circles may be found in the following Section 11.2 (Problems 1–13).

11.2 Further problems

1. Calculate the length of the circumference of a circle of radius 6.50 cm. [40.84 cm]
2. If the diameter of a circle is 65.86 mm calculate the length of the circumference. [206.9 mm]

3. Find the length of the radius of a circle whose circumference is 11.48 cm. [1.827 cm]
4. Calculate the length of the diameter of a circle whose perimeter is 72.40 cm. [23.05 cm]
5. How many degrees are there in:
(*a*) 1 revolution (*b*) $3\frac{1}{2}$ revolutions (*c*) a semicircle
(*d*) a quadrant
(*a*) [360°] (*b*) [1 260°] (*c*) [180°] (*d*) [90°]
6. From a point A, a tangent to a circle is drawn, touching the circle at the point B (see Fig. 11.11). If the diameter of the circle is 6.0 cm and the length of AB is 9.0 cm find the distance from A to the centre of the circle, O.

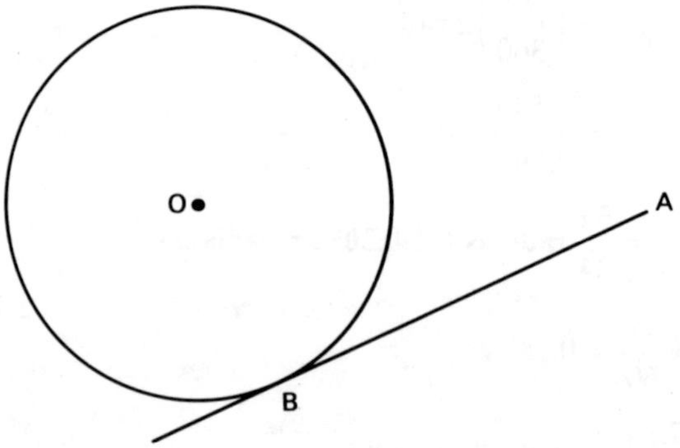

Fig. 11.11

[9.49 cm]

7. A tangent XY meets a circle at Y as shown in Fig. 11.12. The line YZ divides the angle XYO such that OZ = XZ.

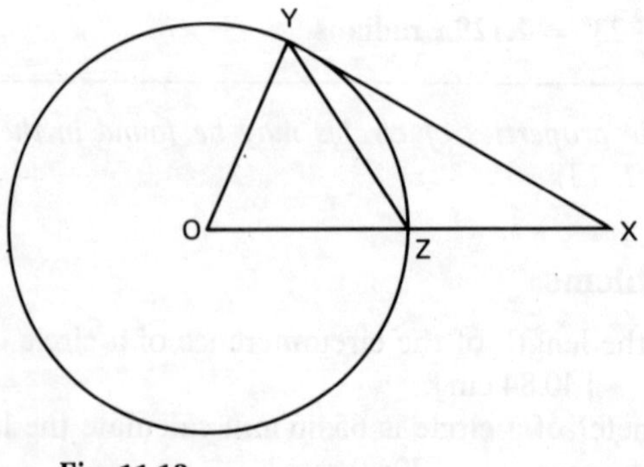

Fig. 11.12

If the diameter of the circle is 15.0 cm find the length of the tangent XY. [12.99 cm]

8. In Fig. 11.13, find the length of x. [20 cm]

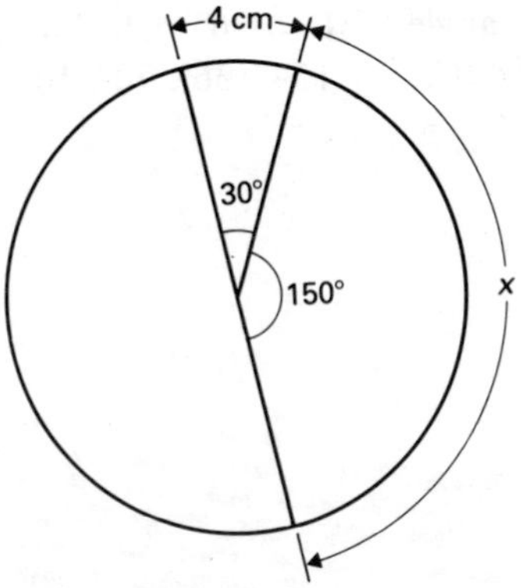

Fig. 11.13

9. Convert the following angles from degrees to radians in terms of π:

(*a*) 90° (*b*) 225° (*c*) 75° (*d*) 60° (*e*) 420° (*f*) 1 110° (*g*) 263° 15′ 25″

(*a*) $\left[\frac{\pi}{2} \text{ or } 0.5\pi\right]$ (*b*) $\left[\frac{5\pi}{4} \text{ or } 1.25\pi\right]$ (*c*) $\left[\frac{5\pi}{12} \text{ or } 0.416\,7\pi\right]$

(*d*) $\left[\frac{\pi}{3} \text{ or } 0.333\,3\pi\right]$ (*e*) $\left[\frac{7\pi}{3} \text{ or } 2.333\pi\right]$ (*f*) $\left[\frac{37\pi}{6} \text{ or } 6.167\pi\right]$

(*g*) [1.463π]

10. Convert the following angles from degrees to radians, correct to 4 significant figures:

(*a*) 65° (*b*) 34° 17′ (*c*) 111° 5′ (*d*) 251° 52′ (*e*) 192° 15′ (*f*) 323° 47′ 15″

(*a*) [1.134] (*b*) [0.598 4] (*c*) [1.939] (d) [4.396] (*e*) [3.355] (*f*) [5.651]

11. Convert the following angles from radians to degrees:

(*a*) 2π (*b*) $\frac{\pi}{2}$ (*c*) $\frac{5\pi}{6}$ (*d*) $\frac{5\pi}{3}$ (*e*) $\frac{9\pi}{8}$ *f*) $\frac{5\pi}{9}$

(*a*) [360°] (*b*) [90°] (*c*) [150°] (*d*) [300°] (*e*) [$202\frac{1}{2}$°] (*f*) [100°]

12. Convert the following angles from radians to degrees, minutes and seconds:

(*a*) 2.48 (*b*) 3.912 (*c*) 0.013 2 (*d*) 17.82 (*e*) 5.75 (*f*) 7.325
(*a*) [142° 5′ 37″] (*b*) [224° 8′ 28″] (*c*) [0° 45′ 23″]
(*d*) [1 021° 0′ 39″] (*e*) [329° 27′ 3″] (*f*) [419° 41′ 30″]

13. The lines AOC and BOD are diameters of the circle ABCD, centre O. Given that the angle ABD = 36°, calculate angles OBC, OAB, AOD and DOC. [54°; 36°, 72°; 108°]

Chapter 12

Mensuration

12.1 Definition

Mensuration is a branch of mathematics concerned with the determination of lengths, areas and volumes.

12.2 Properties of quadrilaterals

A **polygon** is a closed plane figure bounded by straight lines. A polygon which has four sides is called a **quadrilateral**. Figure 12.1 shows some typical quadrilaterals.

If the opposite corners of a quadrilateral are joined by a straight line, two triangles are produced. Figure 12.2 shows a quadrilateral ABCD. If A is joined to C by a straight line two triangles ABC and ADC, are produced.

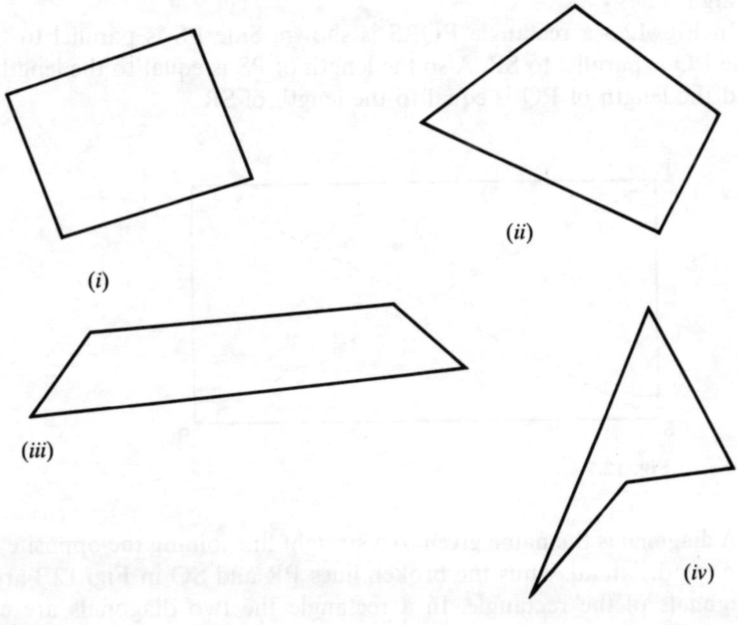

Fig. 12.1

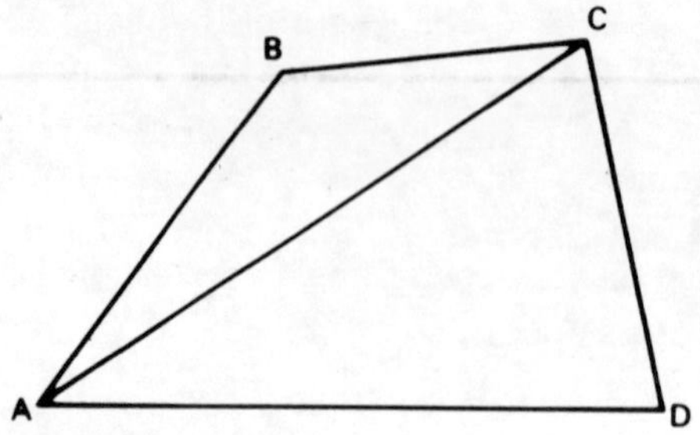

Fig. 12.2

There are five types of quadrilaterals, which are given special names:

1. Rectangle
2. Square
3. Parallelogram
4. Rhombus
5. Trapezium

1. A rectangle

A **rectangle** is a plane rectilinear four-sided figure with all four of its internal angles equal to 90°. (The word **rectilinear** means it is composed entirely of straight lines.)

In Fig. 12.3 a rectangle PQRS is shown. Side PS is parallel to QR and side PQ is parallel to SR. Also the length of PS is equal to the length of QR and the length of PQ is equal to the length of SR.

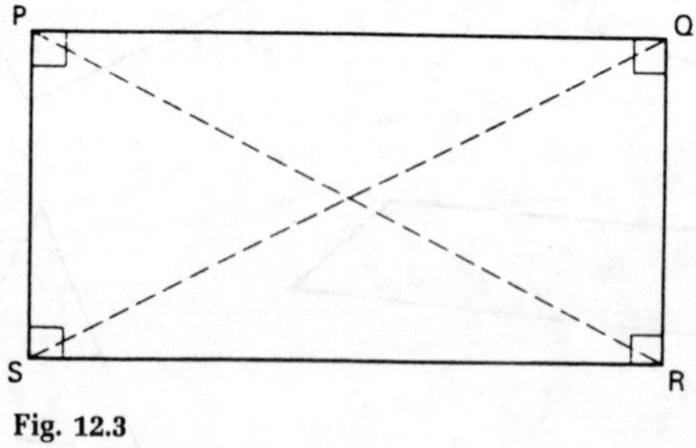

Fig. 12.3

A **diagonal** is the name given to a straight line joining the opposite corners of a quadrilateral. Thus the broken lines PR and SQ in Fig. 12.3 are called diagonals of the rectangle. In a rectangle the two diagonals are equal in length. Thus, PR = SQ in Fig. 12.3.

If the sides of a rectangle are known, the length of the diagonal may be calculated using the theorem of Pythagoras (see Section 10.2, page 221).

In Fig. 12.3, the diagonal PR divides the rectangle into two congruent triangles PSR and PQR. Similarly, the diagonal SQ divides the rectangle into two congruent triangles SPQ and QRS.

As each of the four internal angles of a rectangle are 90°, the sum of the interior angles is 360°.

2. A square

A **square** is a rectangle in which all four sides are of equal length.

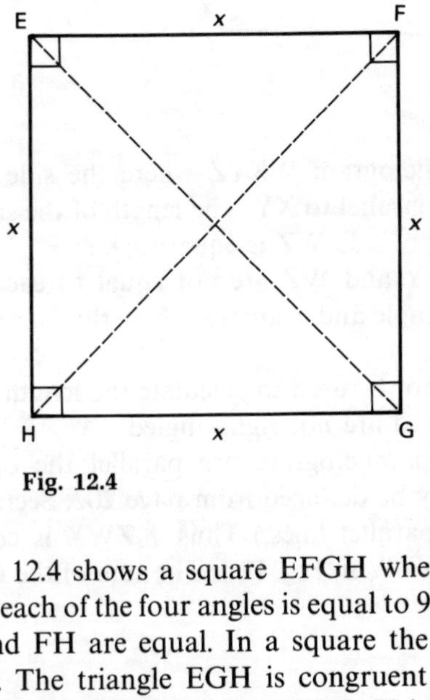

Fig. 12.4

Figure 12.4 shows a square EFGH where the sides are all equal to length x, and each of the four angles is equal to 90°. The lengths of the two diagonals EG and FH are equal. In a square the diagonals bisect each of the right angles. The triangle EGH is congruent to triangle GEF and the triangle EFH is congruent to triangle GHF. Using Pythagoras's theorem on triangle EGH gives:

$$(\text{EG})^2 = (\text{EH})^2 + (\text{HG})^2$$

Therefore $(\text{EG})^2 = x^2 + x^2$

$$= 2x^2$$

$$\text{EG} = \sqrt{(2x^2)}$$

$$\mathbf{EG} = (\sqrt{2})\boldsymbol{x}$$

Similarly the diagonal $\mathbf{FH} = (\sqrt{2})\boldsymbol{x}$

Thus the length of a diagonal of a square is always equal to $\sqrt{2}$ (i.e. 1.414 to 3 decimal places) times the length of a side.

The sum of the interior angles of a square is again 360°.

3. A parallelogram

A **parallelogram** is a plane rectilinear four-sided figure with its opposite sides parallel. From this definition it is evident that the rectangle and the square are both particular kinds of parallelogram with additional properties.

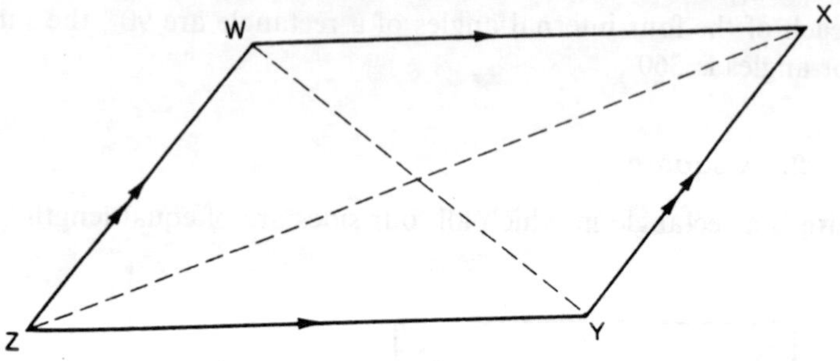

Fig. 12.5

Figure 12.5 shows a typical parallelogram WXYZ where the side WX is parallel to ZY and the side WZ is parallel to XY. The length of the side WX is equal to side ZY and the length of side WZ is equal to XY.

The lengths of the diagonals WY and WZ are not equal to each other except in the special cases of a rectangle and a square, where the four interior angles are each equal to 90°.

The theorem of Pythagoras cannot be used to calculate the lengths of the diagonals since the triangles involved are not right-angled.

Since the opposite sides of a parallelogram are parallel the opposite interior angles are equal. (This may be deduced from page 202, Section 9.4, item 5, regarding transversals of parallel lines.) Thus ∠ZWX is equal to ∠XYZ and ∠WXY is equal to ∠YZW in Fig. 12.5. The sum of the interior angles of a parallelogram is 360°.

4. A rhombus

A **rhombus** is a plane rectilinear four-sided figure in which all four sides are of equal length and the opposite sides are parallel. A rhombus is thus a parallelogram having additional properties.

Figure 12.6 shows a typical rhombus where each of the four sides is equal in length to x. The side AB is parallel to DC and the side AD is parallel to BC. The diagonals, AC and BD, are not equal in length and may not be calculated using the theorem of Pythagoras since the triangles involved are not right-angled.

The opposite interior angles are equal. Thus ∠DAB is equal to ∠BCD and ∠ABC is equal to ∠CDA in Fig. 12.6. The diagonals of a rhombus bisect the interior angles. They also bisect each other at right angles.

The sum of the interior angles of a rhombus is 360°.

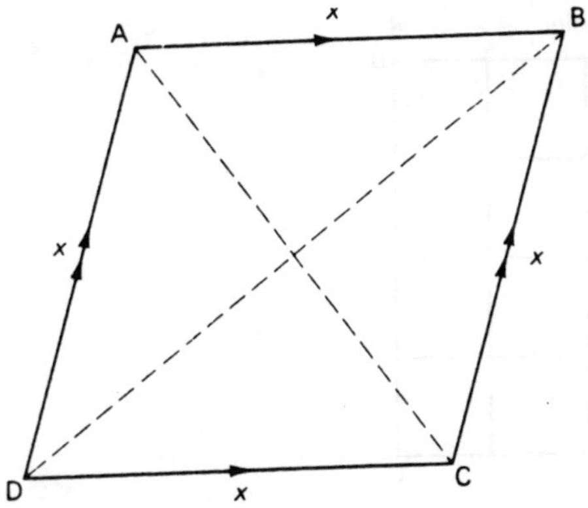

Fig. 12.6

5. A trapezium

A **trapezium** is a plane rectilinear four-sided figure which has a pair of parallel lines which are not of equal length.

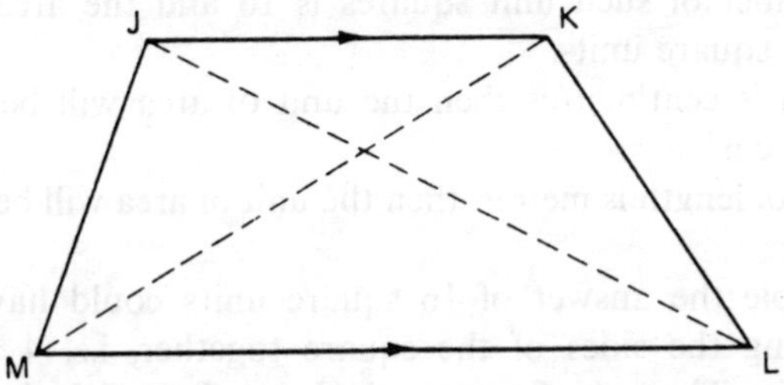

Fig. 12.7

Figure 12.7 shows a typical trapezium JKLM, where JK is parallel to ML. The lengths of the diagonals JL and KM are not equal, except in the special case when $\angle$LMJ is equal to $\angle$MLK and $\angle$MJK is equal to $\angle$LKJ.

The lengths of the diagonals may not be calculated by the theorem of Pythagoras since the triangles involved are not right-angled.

The sum of the interior angles of a trapezium is 360°.

12.3 Areas of plane figures

Area is a measure of the size or extent of a plane surface.

1. Square

Figure 12.8 shows a square ABCD of side 4 units.

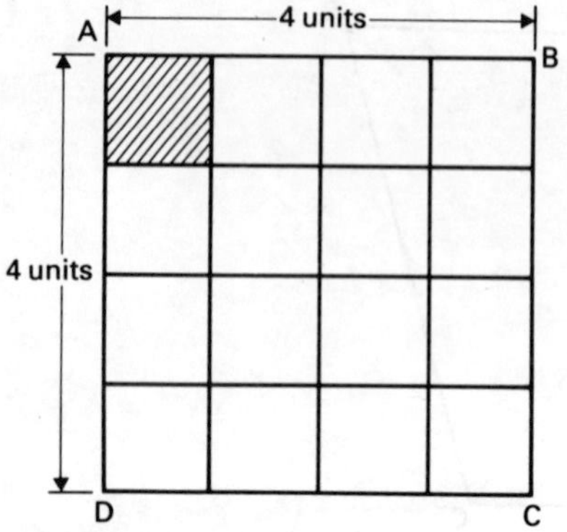

Fig. 12.8

[The SI unit of length is the metre (m). However, as we have already seen, decimal multiples such as millimetres (mm), centimetres (cm), or kilometres (km), are acceptable and much used.]
The area of the square is given by the number of unit squares it contains. In Fig. 12.8 a one unit square, where each side is of one unit length, is shown shaded. Thus the number of such unit squares is 16 and the area of the square is said to be 16 square units.

If the unit of length is centimetres then the unit of area will be square centimetres, written as cm^2.

Similarly, if the unit of length is metres, then the unit of area will be square metres, written as m^2.

In the above example the answer of 16 square units could have been obtained by multiplying the sides of the square together, i.e. 4 units × 4 units = 16 square units. **The area of a square is the product of the lengths of two of its sides**. This method of obtaining the area is true for any size square.

If any square EFGH has each of its sides x units in length as shown in Fig. 12.9, then the area $A = x^2$ **square units.**

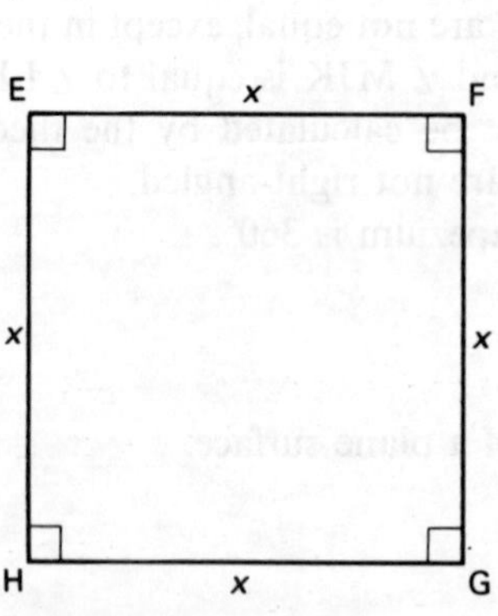

Fig. 12.9

2. A rectangle

Figure 12.10 shows a rectangle JKLM, whose sides are 5 units and 3 units long.

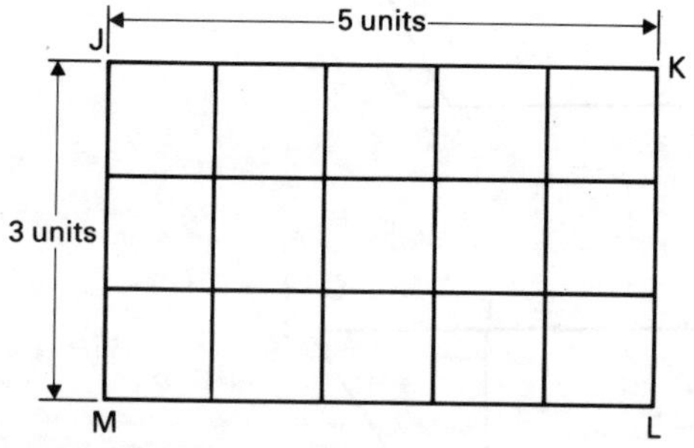

Fig. 12.10

The area of the rectangle is the number of unit squares it contains. Then the area of JKLM is 15 square units. The area of a rectangle may be calculated by multiplying together the lengths of two perpendicular sides, that is, $5 \times 3 = 15$ square units. If any rectangle NOPQ has sides b and l units in length as shown in Fig. 12.11 then area A is given by:

$A = bl$ **square units.**

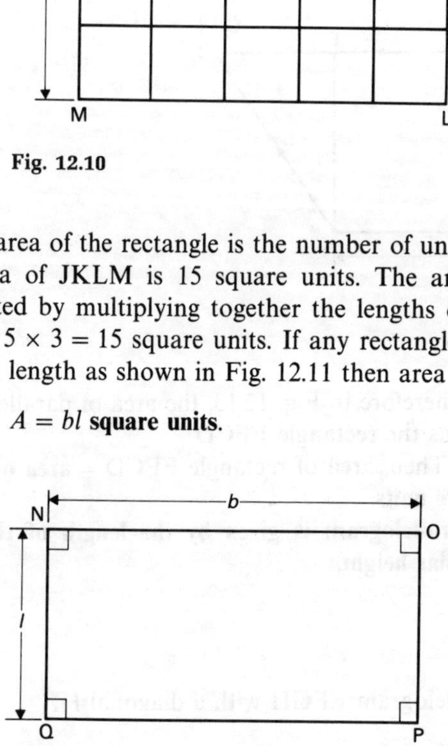

Fig. 12.11

3. A parallelogram

Figure 12.12 shows a parallelogram ABCD those sides are a and b.

DF and CE are constructed perpendicular to AB and AB (produced), as shown in Fig. 12.13.

Now AD = BC and FD = EC and $\angle$AFD = $\angle$BEC = 90°. Hence triangles AFD and BEC are congruent (right angle, hypotenuse and one other side, that is, *R.H.S.*, see Section 10.3). Thus the area of triangle FAD is equal to

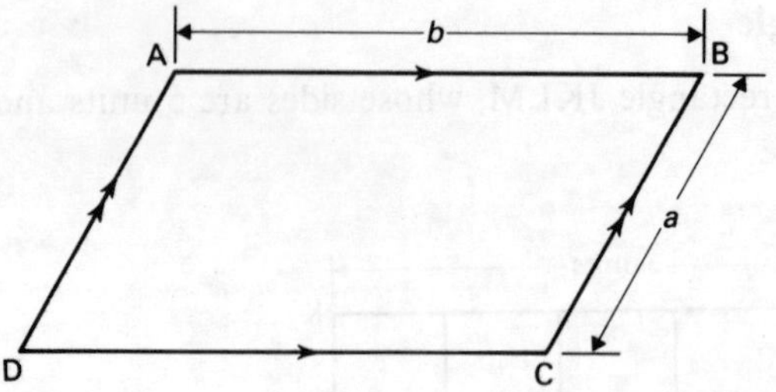

Fig. 12.12

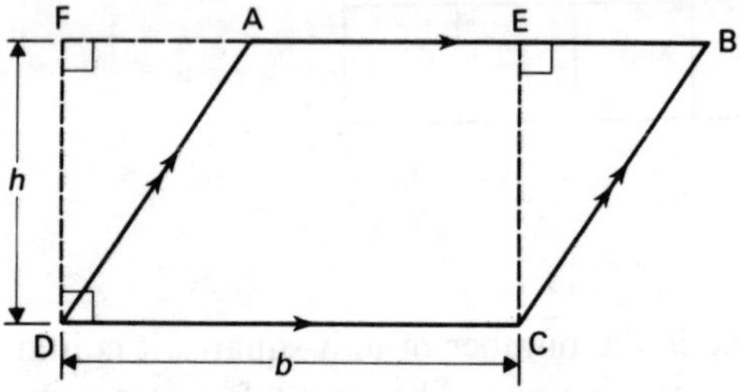

Fig. 12.13

the area of triangle EBC. Therefore in Fig. 12.13, the area of parallelogram ABCD is exactly the same as the rectangle FECD.

Let FD = EC = h units. Then, area of rectangle FECD = **area of parallelogram ABCD = bh square units.**

Hence **the area of a parallelogram is given by the length of the base multiplied by the perpendicular height.**

4. A triangle

Figure 12.14 shows a parallelogram EFGH with a diagonal HF.

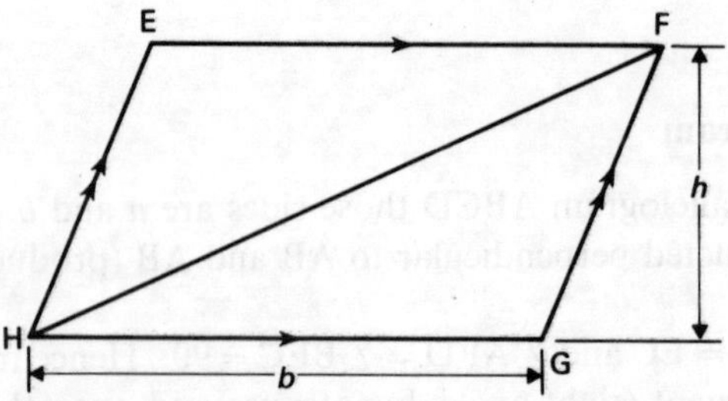

Fig. 12.14

Since triangles EFH and GHF are congruent, their areas are the same.

Therefore, area of triangle EFH = area of triangle GHF

$= \frac{1}{2}$ area of parallelogram EFGH

$= \frac{1}{2}bh$ square units.

Similarly, Fig. 12.15 shows a rectangle ABCD with a diagonal AC.

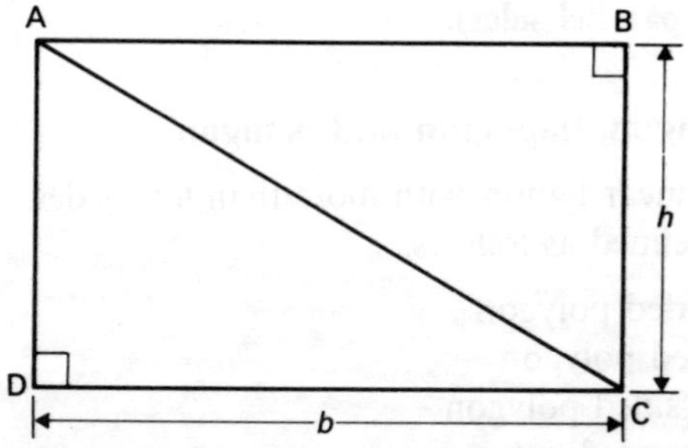

Fig. 12.15

Since triangles ADC and CBA are congruent, their areas are the same.

Therefore, area of triangle ADC = area of triangle CBA

$= \frac{1}{2}$ area of rectangle

$= \frac{1}{2}bh$ square units.

A general expression for the area (A) of any triangle is given by:
$A = \frac{1}{2}$ **× base × perpendicular height.**

5. A trapezium

Figure 12.16 shows a trapezium MNOP with a diagonal NP.

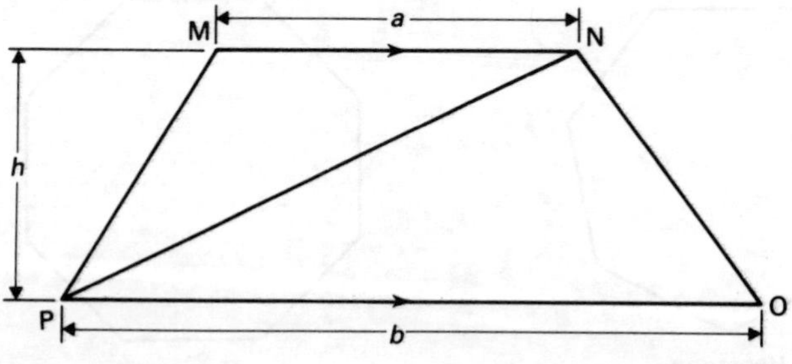

Fig. 12.16

Area of trapezium MNOP = area of triangle MNP + area of triangle NOP.
The area of any triangle = $\frac{1}{2}$ base × perpendicular height.

Hence the area of triangle MNP $= \frac{1}{2}ah$
and the area of triangle NOP $= \frac{1}{2}bh$
Therefore the area of trapezium MNOP $= \frac{1}{2}ah + \frac{1}{2}bh$
$= \frac{1}{2}(a + b)h$ square units

Hence **the area of any trapezium = $\frac{1}{2}$ (sum of the parallel sides) × (perpendicular distance between the parallel sides).**

6. Pentagon, hexagon, heptagon and octagon

There are many plane rectilinear figures with more than four sides. The first four of these **polygons** are defined as follows:

(*i*) A **pentagon** is a five-sided polygon
(*ii*) A **hexagon** is a six-sided polygon
(*iii*) A **heptagon** is a seven-sided polygon
(*iv*) An **octagon** is an eight-sided polygon

If the term 'regular' is applied to any of the above figures it means that all the sides in the polygon are equal. For example, a 'regular hexagon' means that all six sides are of equal length. Figure 12.17 shows typical examples of regular polygons.

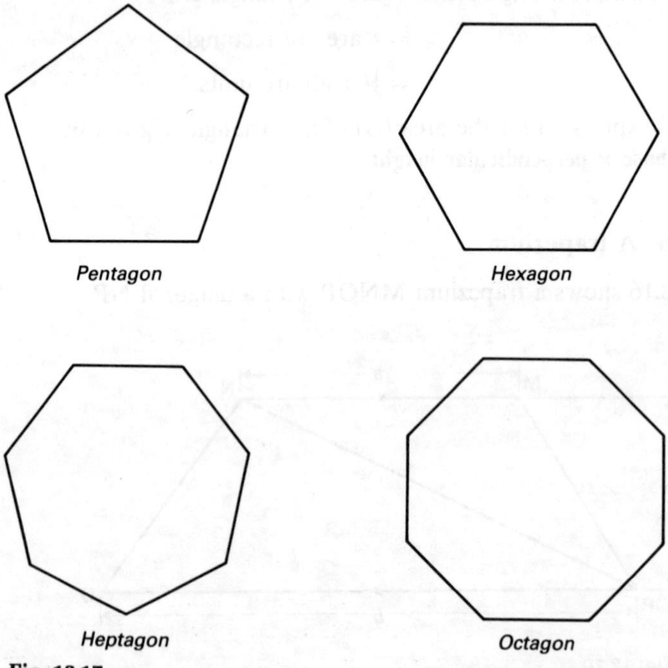

Fig. 12.17

To find the area of any of the regular polygons shown in Fig. 12.17 the procedure adopted is always the same. The bisectors of any two interior angles are drawn to meet at the centre. (The bisectors of each of the interior angles of a regular polygon will meet at the centre.) A radius is drawn from the centre of the figure to each of the vertices. This produces triangles the areas of which are equal and may be calculated. For example, in the hexagon shown in Fig. 12.18 six equal triangles have been produced. Thus the area of the hexagon is equal to six times the area of one of the triangles. (The formula, area $= \frac{1}{2}$ base $\times$ perpendicular height, is used to calculate the area of a triangle.)

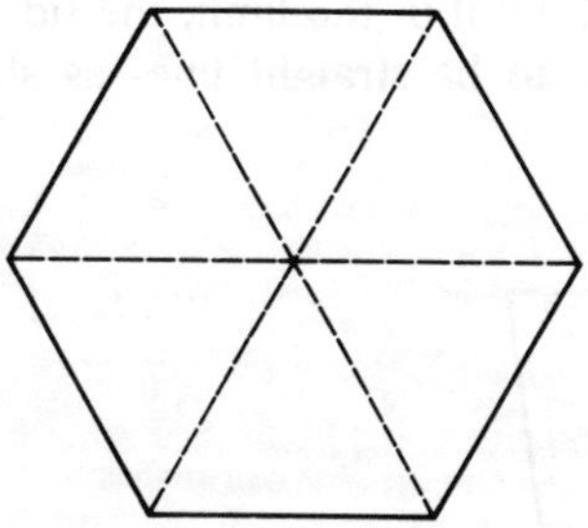

Fig. 12.18

7. A circle

Figure 12.19 shows a circle of radius r divided into eight equal sectors.

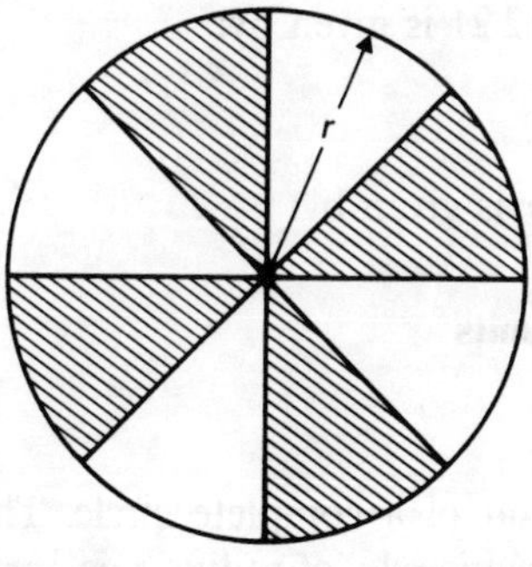

Fig. 12.19

Let the sectors be cut out and arranged as in Fig. 12.20.

The area approximates to a parallelogram of length πr (since the circumference of a circle is $2\pi r$) and perpendicular height r. If instead of eight sectors a very large number of sectors were taken the arc curvature of

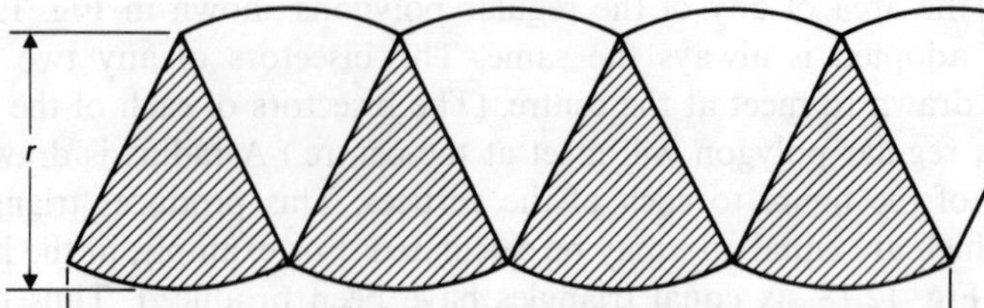

Fig. 12.20

each sector would become less prominent, until in the limit, the sides of the shape shown in Fig. 12.20 would appear to be straight lines as shown in Fig. 12.21

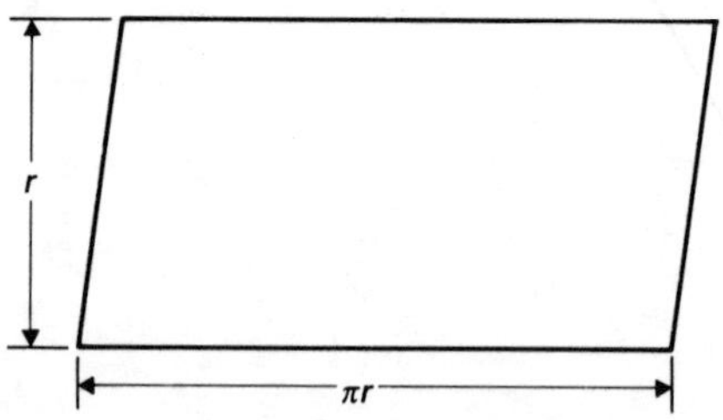

Fig. 12.21

The area A of the parallelogram of Fig. 12.21 is given by:

$$A = (\pi r)r = \pi r^2$$

Hence the area A of any circle of radius r is given by:

$$A = \pi r^2 = \pi\left(\frac{d}{2}\right)^2 = \frac{\pi d^2}{4} \textbf{ square units}$$

where d is the diameter of the circle.

The area of a **semicircle** is one half that of a complete circle. The word 'semi' means 'half'. Hence the area of a semicircle, of radius r, is $\frac{1}{2}\pi r^2$

Worked problems on areas of plane figures

Problem 1. State the types of quadrilateral shown in Fig. 12.22 and find the angles a to l

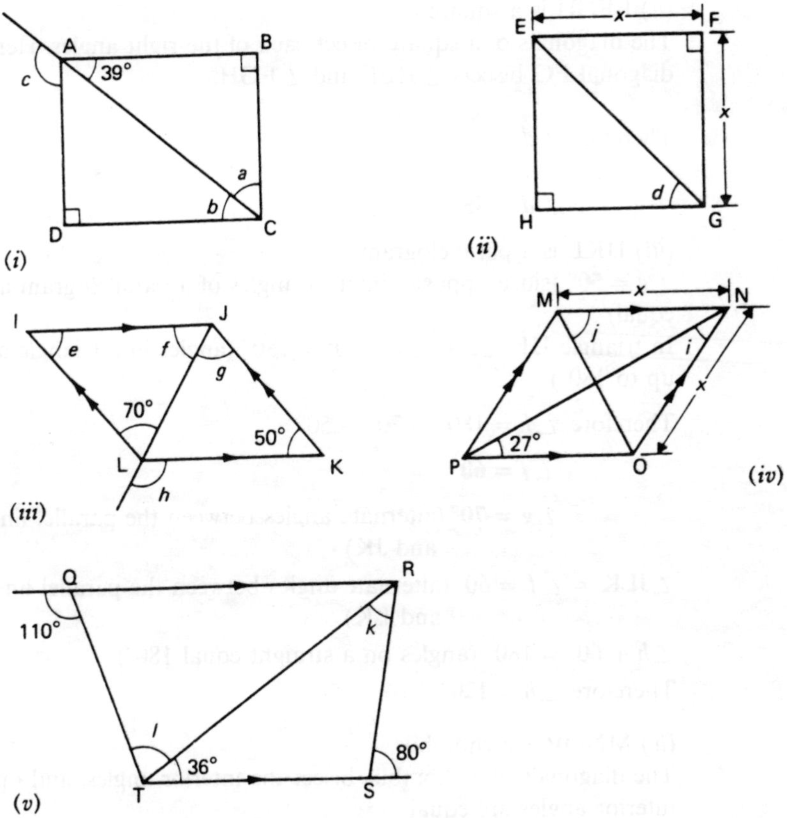

Fig. 12.22

(*i*) ABCD is a rectangle.
In triangle ABC, $39^\circ + 90^\circ + \angle a = 180^\circ$
(angles in a triangle add up to 180°)

Therefore $\angle a = 180^\circ - 90^\circ - 39^\circ$

$\angle a = \mathbf{51^\circ}$

$\angle b = \angle \text{BAC}$ (alternate angles between parallels AB and DC)

$\angle \boldsymbol{b} = \mathbf{39^\circ}$

$\angle c = \angle \text{ADC} + \angle \text{DCA}$ (external angle of a triangle equals the sum of the interior opposite angles)

$\angle c = 90^\circ + 39^\circ$

$\angle \boldsymbol{c} = \mathbf{129^\circ}$

(*ii*) EFGH is a square.
The diagonals of a square bisect each of the right angles. Hence the diagonal EG bisects $\angle$HEF and $\angle$FGH.

$$\text{Therefore } \angle d = \frac{90°}{2}$$

$$\angle \boldsymbol{d} = \mathbf{45°}$$

(*iii*) IJKL is a parallelogram.
$\angle e$ = **50°** (since opposite interior angles of a parallelogram are equal)
In triangle IJL $\angle e + \angle f + 70° = 180°$ (angles in a triangle add up to 180°)

Therefore $\angle f = 180° - 70° - 50°$

$\angle \boldsymbol{f} = \mathbf{60°}$

$\angle \boldsymbol{g} = \mathbf{70°}$ (alternate angles between the parallel lines IL and JK)

$\angle$JLK $= \angle f = 60°$ (alternate angles between the parallel lines IJ and LK)

$\angle h + 60° = 180°$ (angles on a straight equal 180°)
Therefore $\angle h$ = **120°**

(*iv*) MNOP is a rhombus.
The diagonals of a rhombus bisect the interior angles, and opposite interior angles are equal.
Thus $\angle$OPN = $\angle$MPN = $\angle$ONP = $\angle$MNP = 27°
Therefore $\angle \boldsymbol{i}$ = **27°**
If $\angle$MPN = $\angle$MNP = 27° then $\angle$PMN is given by
27° + 27° + $\angle$PMN = 180° (angles in a triangle add up to 180°)
Therefore $\angle$PMN = 180° − 27° − 27°
= 126°
As the diagonal MO bisects $\angle$PMN, then

$$\angle j = \frac{126°}{2} = \mathbf{63°}$$

(*v*) QRST is a trapezium.
$\angle k + 36° = 80°$ (external angle of a triangle is equal to the sum of the two interior opposite angles)
Therefore $\angle k = 80° - 36°$

$\angle \boldsymbol{k} = \mathbf{44°}$

$\angle$QRT = 36° (alternate angles between parallel lines QR and TS)

Therefore $\angle l + 36° = 110°$ (external angle of a triangle is equal to the sum of the two interior opposite angles)

$\angle l = 110° - 36°$

$\boldsymbol{\angle l = 74°}$

Problem 2. Find the areas of the figures shown in Fig. 12.23.

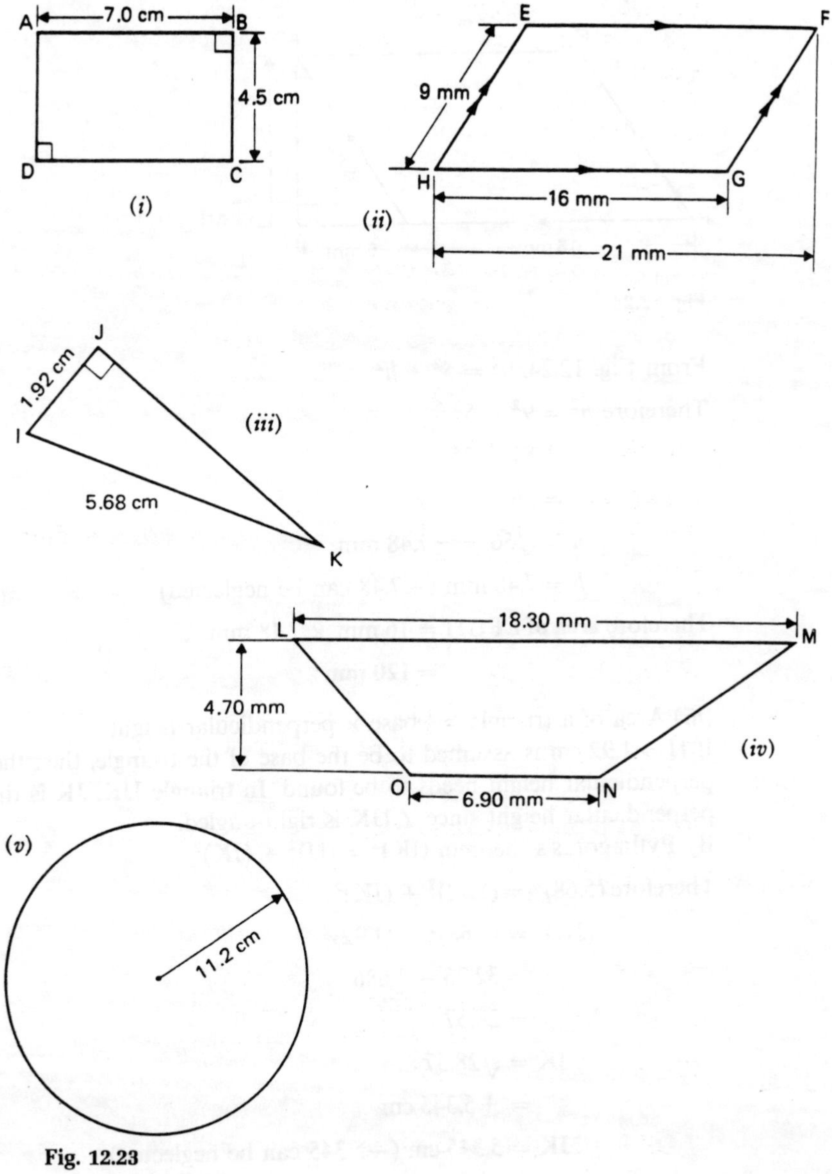

Fig. 12.23

(*i*) Area of a rectangle = length × breadth

Therefore **area of ABCD** = 7.0 cm × 4.5 cm

$= \mathbf{31.5\ cm}$

(*ii*) Area of a parallelogram = base × perpendicular height.
The perpendicular height, *h*, may be found using Pythagoras's theorem.

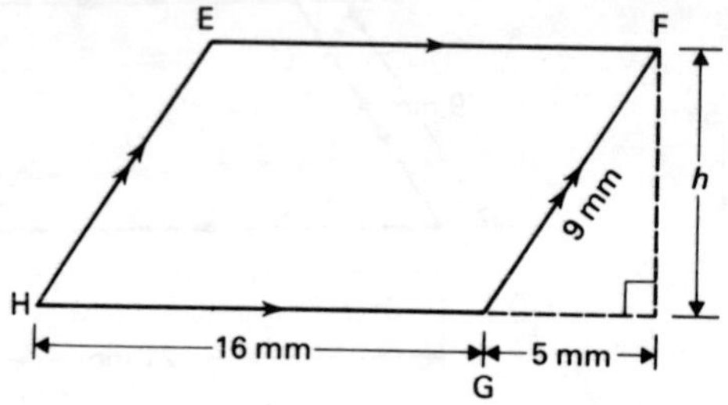

Fig. 12.24

From Fig. 12.24, $9^2 = 5^2 + h^2$

Therefore $h^2 = 9^2 - 5^2$

$= 81 - 25$

$= 56$

$h = \sqrt{56} = \pm 7.48$ mm

$h = 7.48$ mm (-7.48 can be neglected)

Therefore **area of EFGH** = 16 mm × 7.48 mm

$= \mathbf{120\ mm^2}$

(*iii*) Area of a triangle = $\frac{1}{2}$ base × perpendicular height.
If IJ = 1.92 cm is assumed to be the base of the triangle, then the perpendicular height needs to be found. In triangle IJK, JK is the perpendicular height since ∠IJK is right-angled.
By Pythagoras's theorem $(IK)^2 = (IJ)^2 + (JK)^2$

Therefore $(5.68)^2 = (1.92)^2 + (JK)^2$

$(JK)^2 = (5.68)^2 - (1.92)^2$

$= 32.26 - 3.686$

$= 28.57$

$JK = \sqrt{28.57}$

$= \pm 5.345$ cm

JK = 5.345 cm (-5.345 can be neglected)

Therefore **area of IJK** $= \frac{1}{2}(1.92)(5.345)$

$= \mathbf{5.131\ cm^2}$

(*iv*) Area of a trapezium $= \frac{1}{2} \times$ (sum of the parallel sides) $\times$ (perpendicular distance between parallel sides).

Therefore **area of LMNO** $= \frac{1}{2}(18.30 + 6.90)(4.70)$

$= \frac{1}{2}(25.20)(4.70)$

$= \mathbf{59.22\ mm^2}$

(*v*) Area of a circle $= \pi r^2$

Therefore **area of circle with radius 11.2 cm** $= (\pi)(11.2)^2$

$= \mathbf{394.1\ cm^2}$

Problem 3. A hollow shaft has an outside diameter of 6.42 cm and an inside diameter of 2.15 cm. Calculate the cross-sectional area of the shaft.

The cross-sectional area of the shaft is shown by the shaded portion of Fig. 12.25. This area is often called an **annulus.**

If the area of the smaller circle is subtracted from the area of the larger circle the remainder will be the shaded area.

Now the area of a circle $= \pi r^2 = \dfrac{\pi d^2}{4}$

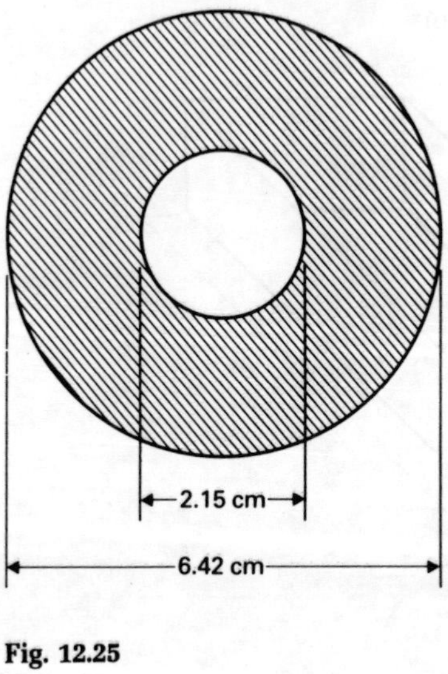

Fig. 12.25

Therefore **cross-sectional area of shaft** $= \frac{\pi}{4}(6.42)^2 - \frac{\pi}{4}(2.15)^2$

$$= \frac{\pi}{4}[(6.42)^2 - (2.15)^2]$$

$$= \frac{\pi}{4}[41.216 - 4.623]$$

$$= \frac{\pi}{4}(36.593)$$

$$= \mathbf{28.74\ cm^2}$$

Problem 4. An open water tank has a rectangular base and sides and is made from steel. The tank is 5 m long by 78.0 cm wide and 37.5 cm high. Find the area of the steel plate required to make the tank in (*a*) cm^2 and (*b*) m^2.

Before the areas can be calculated all measurements should be in the same units. Thus for (*a*) the measurements of the tank are 500 cm by 78.0 cm by 37.5 cm as shown in Fig. 12.26.

Area of base of tank $= 78.0 \times 500 = 39\,000\ cm^2$

Area of 2 ends of tank $= 2(78.0 \times 37.5) = 2(2\,925)\ cm^2$

Area of 2 sides of tank $= 2(37.5 \times 500) = 2(18\,750)\ cm^2$

Therefore **area of metal required** $= 39\,000 + 2(2\,925) + 2(18\,750)\ cm^2$

$$= \mathbf{82\,350\ cm^2}$$

Now 100 cm = 1 m
and 100 cm × 100 cm = 1 m^2

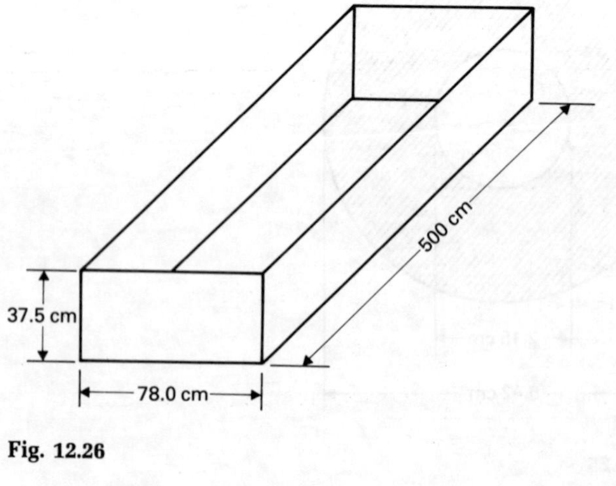

Fig. 12.26

Therefore $10\,000\ \text{cm}^2 = 1\ \text{m}^2$

$$\mathbf{82\,350\ cm^2} = \frac{82\,350}{10\,000}\ \text{m}^2 = \mathbf{8.235\ m^2}$$

Thus the area of steel required is **$82\,350\ \text{cm}^2$** or **$8.235\ \text{m}^2$**

Problem 5. A swimming pool is 55.0 m long and 10.0 m wide. The perpendicular depth at the deep end is 4.20 m and at the shallow end 140.0 cm, the slope being uniform. The pool needs two coats of a protective paint inside. Find how many litres of paint will be required if 1 litre covers $12\ \text{m}^2$.

The dimensions of the swimming pool are shown in Fig. 12.27 in metre units.

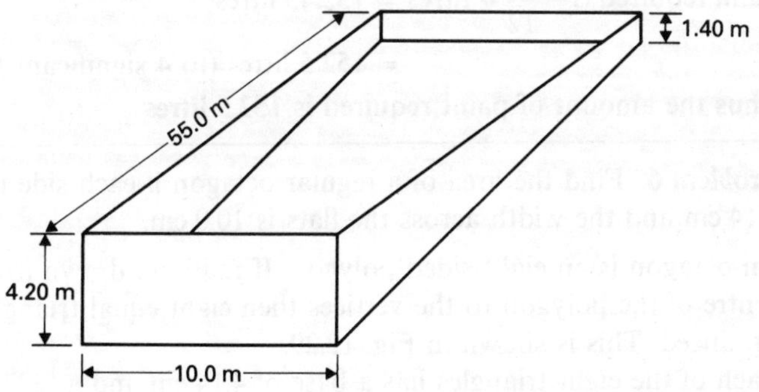

Fig. 12.27

The side view of the swimming pool is shown in Fig. 12.28.

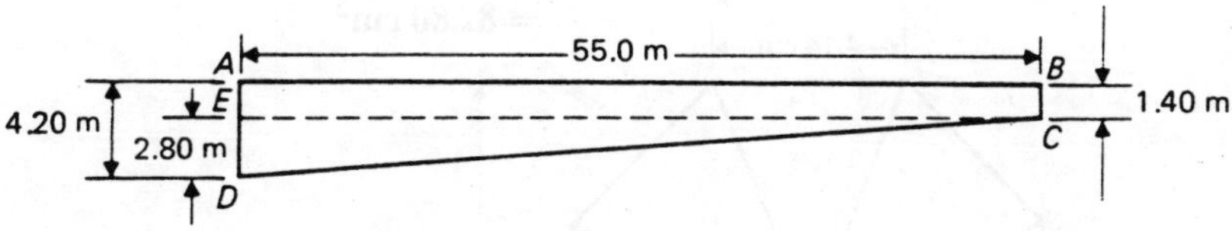

Fig. 12.28

The area ABCD is a trapezium.
Therefore area of ABCD $= \frac{1}{2}(4.20 + 1.40)55.0$

$$= 154\ \text{m}^2$$

To find the length CD, Pythagoras's theorem is used:

$(\text{CD})^2 = (\text{ED})^2 + (\text{CE})^2$

$$\text{Therefore } (CD)^2 = (2.80)^2 + (55.0)^2$$
$$= 7.84 + 3\,025$$
$$= 3\,032.84 = 3\,033 \text{ (correct to 4 significant figures)}$$
$$CD = \sqrt{3\,033} = \pm 55.07$$
$$= 55.07 \text{ m } (-55.07 \text{ can be neglected})$$

The area of the bottom of the pool = (55.07)(10.0) = 550.7 m^2
The area of the two sides of the pool = 2(154) = 308 m^2
The area of the deep end face of the pool = 4.20(10.0) = 42 m^2
The area of the shallow end face of the pool = 1.40(10.0) = 14 m^2
The total area of the pool = (550.7 + 308 + 42 + 14) m^2
= **914.7 m^2**

Two coats of paint would therefore cover 2 × 914.7, that is, 1 829.4 m^2. If 1 litre of paint covers 12 m^2 then the amount of paint required is $\frac{1\,829.4}{12}$ litres = 152.45 litres

= 152.5 litres (to 4 significant figures)

Thus the amount of paint required is **152.5 litres**

Problem 6. Find the area of a regular octagon if each side is 4.14 cm and the width across the flats is 10.0 cm.

An octagon is an eight-sided polygon. If radii are drawn from the centre of the polygon to the vertices then eight equal triangles are produced. This is shown in Fig. 12.29.
Each of the eight triangles has a base of 4.14 cm and a perpendicular height of 5.0 cm.
Hence the area of one triangle = $\frac{1}{2}$(4.14)(5.0)
= 10.35 cm^2
Therefore **area of the octagon** = 8(10.35)
= **82.80 cm^2**

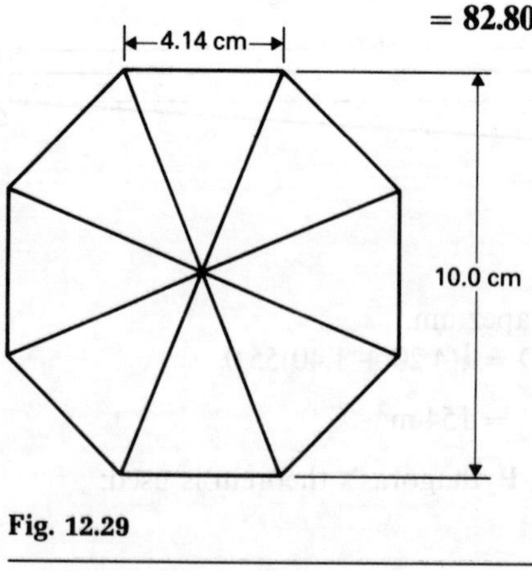

Fig. 12.29

Problem 7. Find the area of a regular hexagon which has sides 6.0 cm long.

A hexagon is a six-sided polygon. If radii are drawn from the centre of the polygon to the vertices then six equal triangles are produced. This is shown in Fig. 12.30.

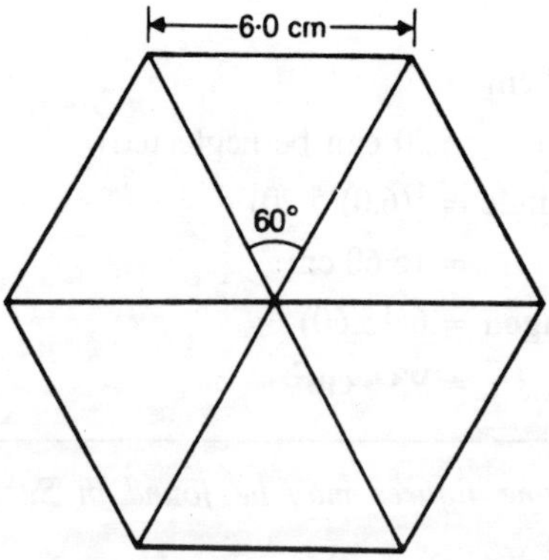

Fig. 12.30

The angle subtended at the centre by each triangle is $\dfrac{360°}{6} = 60°$.

The two other angles in the triangle add up to 120° and are equal to each other. Therefore each triangle has three 60° angles, that is, each of the triangles is equilateral.

To find the area of one of the equilateral triangles, a perpendicular is drawn from one vertex to bisect the opposite side as shown in Fig. 12.31

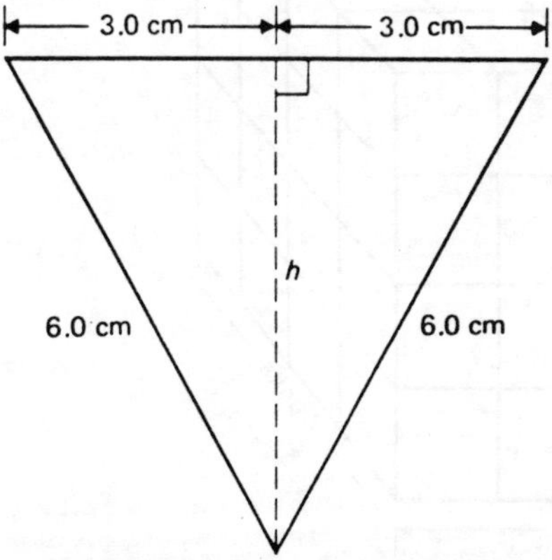

Fig. 12.31

Using Pythagoras's theorem:

$$(6.0)^2 = (3.0)^2 + h^2$$

Therefore
$$h^2 = (6.0)^2 - (3.0)^2 = 36 - 9 = 27$$

$$h = \sqrt{27} = \pm 5.20 \text{ cm}$$

$$h = 5.20 \text{ cm} \quad (-5.20 \text{ can be neglected})$$

Hence the area of one triangle $= \frac{1}{2}(6.0)(5.20) = 15.60 \text{ cm}^2$

Hence **the area of the hexagon** $= 6(15.60) = \mathbf{93.6\ cm^2}$

Further problems on areas of plane figures may be found in Section 12.6 (Problems 1–23), page 303.

12.4 Volumes of regular solids

The **volume** of any solid is a measure of the space occupied by the solid. If a cube measures 4 units by 4 units by 4 units, as shown in Fig. 12.32, then the volume is found by counting the number of unit cubes contained within the space occupied by the whole cube.

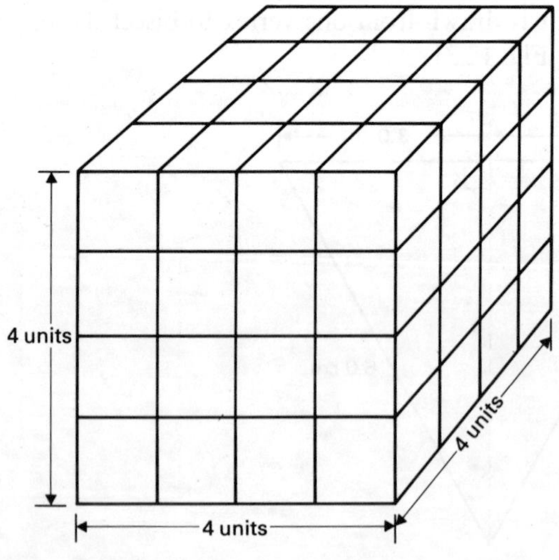

Fig. 12.32

Thus the volume of the cube shown is 64 cubic units. This volume could be obtained by multiplying the three perpendicular sides of the cube together, that is, 4 units × 4 units × 4 units = 64 cubic units. In general, if each side of a cube is of length x units then the volume of the cube is x^3 cubic units.

If the unit of length is the centimetre, then the unit of volume will be cm × cm × cm, that is, cubic centimetres, written as cm^3.

Similarly, if the unit of length is the metre then the unit of volume will be m × m × m, that is, cubic metres, written as m^3.

1. Prisms

A prism is a solid with a constant cross-section, and with the two ends parallel.

The shape of the end is used to describe the prism. Figure 12.33 shows four typical prisms.

The volume V of a prism of cross-sectional area A and perpendicular height h is given by

$V = Ah$ **cubic units**

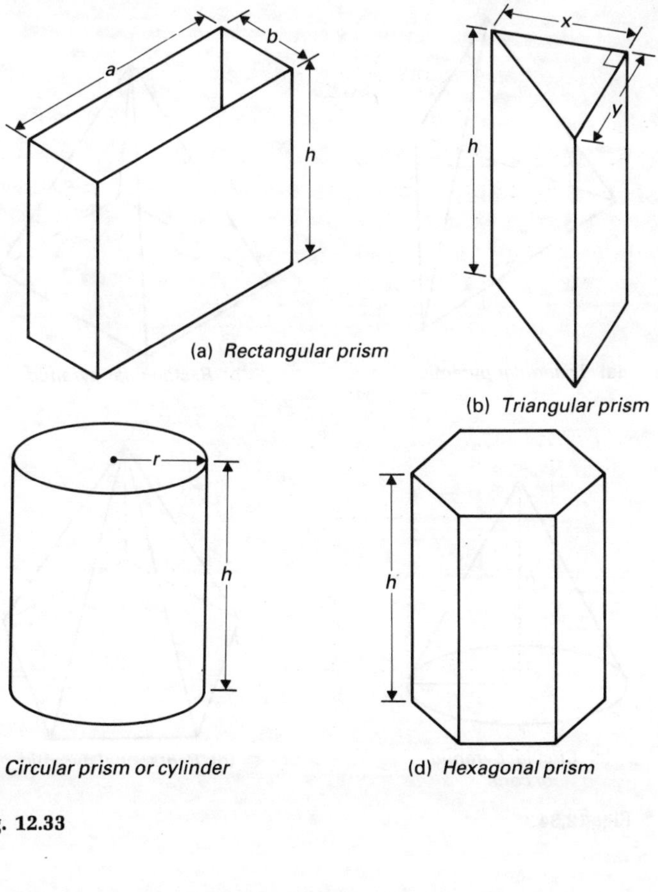

(a) *Rectangular prism*

(b) *Triangular prism*

(c) *Circular prism or cylinder*

(d) *Hexagonal prism*

Fig. 12.33

This formula for volume is true for any prism irrespective of the shape of the base. So, referring to Fig. 12.33:

(*a*) **Volume of rectangular prism** $= (ab)h$

$= abh$ **cubic units**

(*b*) **Volume of triangular prism** $= (\frac{1}{2}xy)h$

$= \frac{1}{2}xyh$ **cubic units**

(*c*) **Volume of cylinder** $= (\pi r^2)h$

$= \pi r^2 h$ **cubic units**

(*d*) **Volume of hexagonal prism** $=$ **(Area of hexagon)**h **cubic units.**

2. Pyramids

A solid with a plane end and straight sides meeting in a point is called a pyramid. As in the case of prisms, the shape of the plane end is used to describe the pyramid. Figure 12.34 shows four typical pyramids.

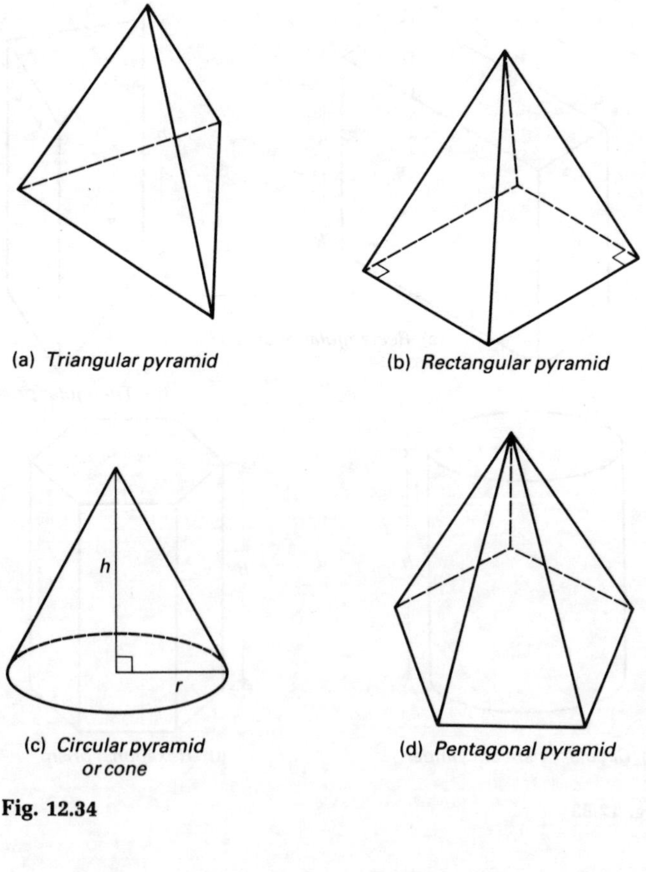

(a) *Triangular pyramid*

(b) *Rectangular pyramid*

(c) *Circular pyramid or cone*

(d) *Pentagonal pyramid*

Fig. 12.34

The volume V of any pyramid of base area A and perpendicular height h is given by

$$V = \frac{1}{3}Ah \text{ cubic units}$$

A circular pyramid is given the special name of a **cone**. The volume V of a cone is given by

$$V = \frac{1}{3}(\pi r^2)h \text{ cubic units}$$

since the base is a circle of area πr^2.

3. Spheres

The volume V of a sphere of radius r is given by

$$V = \frac{4}{3}\pi r^3 \text{ cubic units.}$$

Worked problems on volumes of regular solids

Problem 1. Find the volume of each of the solids shown in Fig. 12.35

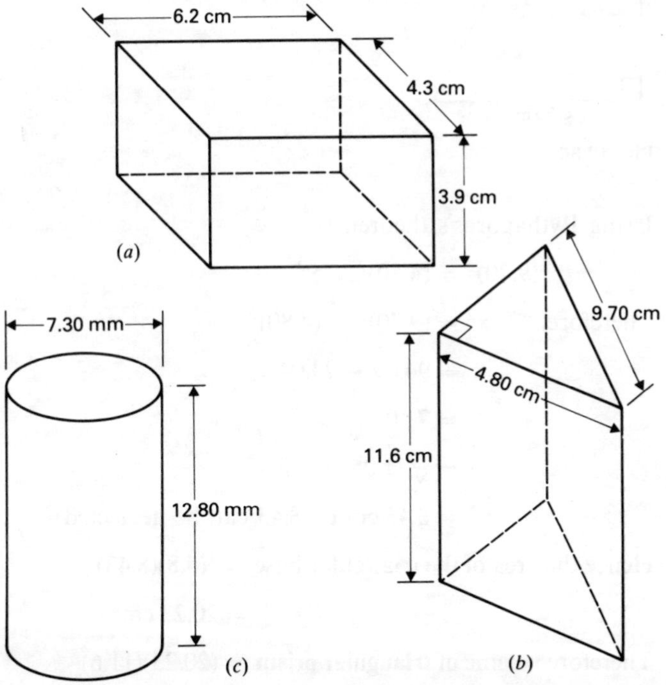

Fig. 12.35

(*a*) The solid is a **rectangular prism**.
Volume = area of base × height (any of the faces can be termed the base).
Therefore **volume** = (6.2)(4.3)(3.9)

= **104 cm³**

(*b*) The solid is a **triangular prism**.
Volume = area of base × perpendicular height
Area of a triangle = $\frac{1}{2}$ base × perpendicular height
If the base of the triangle is taken as 4.80 cm then the perpendicular height x is required (see Fig. 12.36).

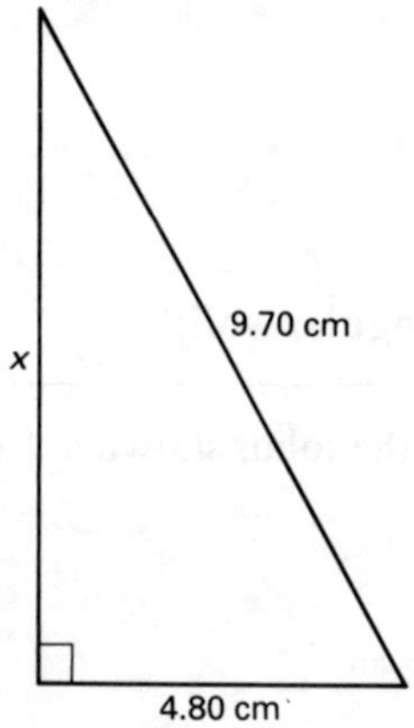

Fig. 12.36

Using Pythagoras's theorem

$$(9.70)^2 = (4.80)^2 + x^2$$

Therefore

$$x^2 = (9.70)^2 - (4.80)^2$$
$$= 94.09 - 23.04$$
$$= 71.05$$
$$x = \sqrt{71.05} = \pm 8.43$$
$$= 8.43 \text{ cm } (-8.43 \text{ can be neglected})$$

Hence the area of the triangular base = $\frac{1}{2}(4.8)(8.43)$

= 20.23 cm²

Therefore **volume of triangular prism** = (20.23)(11.6)

= **234.7 cm³**

(*c*) The solid is a **circular prism** or **cylinder**.
Volume of cylinder $= \pi r^2 h$

$$= \frac{\pi d^2}{4}$$

$$= \frac{(\pi)(7.30)^2(12.80)}{4}$$

$$= \mathbf{535.7\ mm^3}$$

Problem 2. Calculate the volume of the solid prism shown in Fig. 12.37

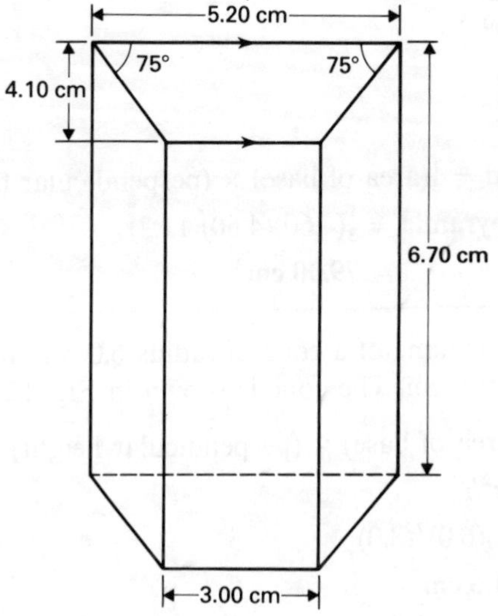

Fig. 12.37

The prism has a trapezoidal base.
Volume of a prism = area of base × perpendicular height.
Area of trapezium $= \frac{1}{2}$ (sum of parallel sides) × (perpendicular distance between them)

$$= \tfrac{1}{2}(5.20 + 3.00)(4.10)$$

$$= \tfrac{1}{2}(8.20)(4.10)$$

$$= 16.81\ \text{cm}^2$$

Therefore **volume of prism** $= (16.81)(6.70)$

$$= \mathbf{112.6\ cm^3}$$

Problem 3. Find the volume of the square pyramid shown in Fig. 12.38

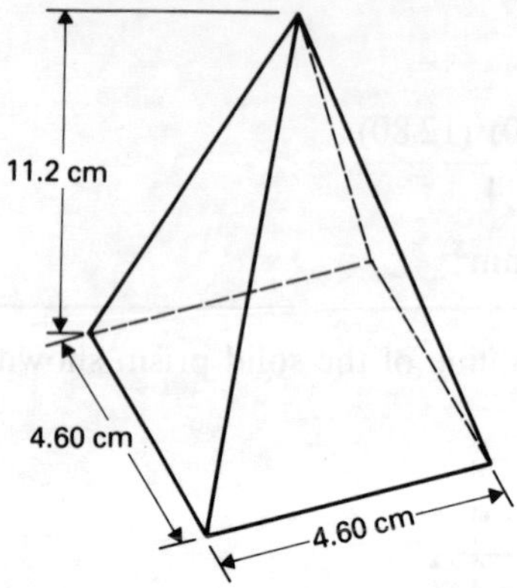

Fig. 12.38

Volume of a pyramid = $\frac{1}{3}$(area of base) × (perpendicular height)

Therefore **volume of pyramid** = $\frac{1}{3}(4.60)(4.60)(11.2)$

= **79.00 cm³**

Problem 4. Find the volume of a cone of radius 6.0 cm and perpendicular height 8.0 cm. The cone is shown in Fig. 12.39

Volume of cone = $\frac{1}{3}$(area of base) × (perpendicular height)

$= \frac{1}{3}\pi r^2 h$

$= \frac{1}{3}(\pi)(6.0)^2(8.0)$

= **301.6 cm³**

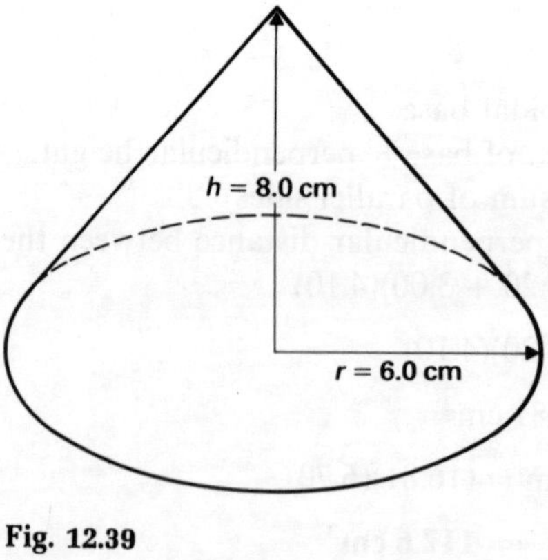

Fig. 12.39

Problem 5. A pyramid stands on a rectangular base of sides of 5.0 cm and 4.0 cm. If the length of each sloping edge is 11.0 cm, calculate the volume of the pyramid.

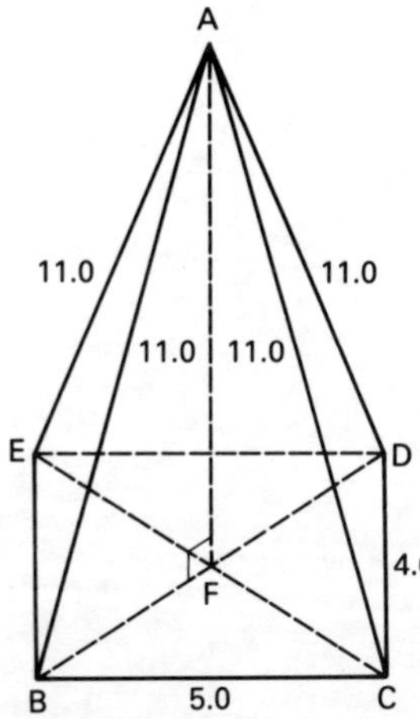

Fig. 12.40

Figure 12.40 shows the rectangular pyramid with equal sloping edges, the dimensions being in centimetres.
Volume of a pyramid $= \frac{1}{3}$ (area of base) $\times$ (perpendicular height)

To find the volume of a pyramid it is therefore necessary to find the perpendicular height. This is shown as AF in Fig. 12.40. AF is one side of triangle AFB shown in Fig. 12.41
FB is half the length of diagonal BD. Using Pythagoras's theorem on triangle BCD of Fig. 12.40 gives

$$BD^2 = (5.0)^2 + (4.0)^2$$
$$= 25 + 16$$
$$= 41$$

Therefore $BD = \sqrt{41}$
$$= \pm 6.40 \text{ cm}$$
$$= 6.40 \text{ cm } (-6.40 \text{ can be neglected})$$

Therefore $FB = \dfrac{6.40}{2} = 3.20$ cm

Now, using Pythagoras's theorem on triangle AFB of Fig. 12.41 gives

$$(11.0)^2 = (3.20)^2 + (AF)^2$$

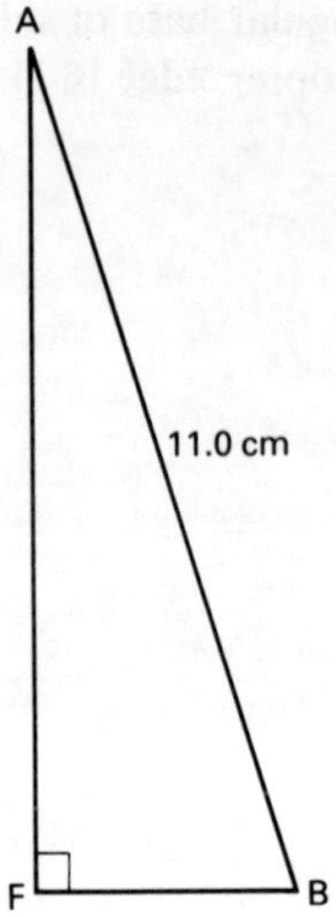

Fig. 12.41

Therefore $AF^2 = (11.0)^2 - (3.20)^2$

$= 121 - 10.24$

$= 110.8$

Therefore $AF = \sqrt{110.8} = \pm 10.53$ cm

$= 10.53$ cm (-10.53 can be neglected)

Hence the perpendicular height of the pyramid is 10.53 cm.

Therefore **volume of pyramid** $= \frac{1}{3}(5.0)(4.0)(10.53)$

$= \mathbf{70.20\ cm^3}$

Problem 6. If a sphere has a diameter of 54.0 mm, find its volume.

If the diameter is 54.0 mm then the radius $= \dfrac{54.0}{2}$

$= 27.0$ mm

Volume of a sphere $= \frac{4}{3}\pi r^3$

$= \frac{4}{3}(\pi)(27.0)^3 = \mathbf{82\,450\ mm^3}$

As 1 cm = 10 mm

$1\ cm^3 = 1\,000\ mm^3$

Therefore **volume of sphere** $= \dfrac{82\,450}{1\,000} = \mathbf{82.45\ cm^3}$

Problem 7. Figure 12.42 shows a 2 m section of metal rod. Find its volume in cm^3 and m^3 each correct to 4 significant figures.

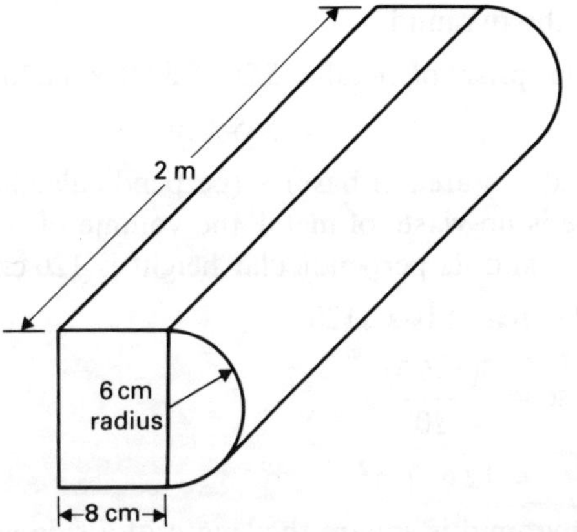

Fig. 12.42

The section of metal is a prism where the end comprises a rectangle and a semicircle. Since the radius of the semicircle is 6 cm, the dimensions of the rectangle are 8 cm by 12 cm.

Thus, area of rectangle $= 8 \times 12 = 96\ \text{cm}^2$

area of semicircle $= \frac{1}{2}\pi r^2 = \frac{1}{2}(\pi)(6)^2$

$= 56.55\ \text{cm}^2$

Hence area of the end of the prism $= 96 + 56.55$

$= 152.55\ \text{cm}^2$

$= 152.6\ \text{cm}^2$

(to 4 significant figures)

Therefore volume of metal section $= (152.6)(200)$

$= 30\,520\ \text{cm}^3$

Now 100 cm = 1 m

Therefore $100\ \text{cm} \times 100\ \text{cm} = 1\ \text{m} \times 1\ \text{m} = 1\ \text{m}^2$

and $100\ \text{cm} \times 100\ \text{cm} \times 100\ \text{cm} = 1\ \text{m} \times 1\ \text{m} \times 1\ \text{m} = 1\ \text{m}^3$

That is, $1\,000\,000\ \text{cm}^3 = 1\ \text{m}^3$

Therefore, $30\,520\ \text{cm}^3 = \dfrac{30\,520}{1\,000\,000}\ \text{m}^3$

$= 0.030\,52\ \text{m}^3$

Thus the volume of the metal section is **30 520 cm^3** or **0.030 52 m^3**

Problem 8. A rectangular piece of metal with dimensions 5.20 cm by 8.60 cm by 11.30 cm is melted down and recast into a square pyramid of perpendicular height 1.20 m. Find the length of the sides of the base of the pyramid.

Volume of rectangular prism of metal $= 5.20 \times 8.60 \times 11.30$

$= 505.3\ \text{cm}^3$

Volume of a pyramid $= \frac{1}{3}$(area of base) $\times$ (perpendicular height)
Assuming that there is no waste of metal, the volume of the pyramid is 505.3 cm^3 and its perpendicular height is 120 cm.
Therefore $505.3 = \frac{1}{3}$ (area of base) 120

$$\text{Therefore area of base} = \frac{3(505.3)}{120}$$

$$= 12.63\ \text{cm}^2$$

As the base of the pyramid is square the length of a side will be $\sqrt{12.63} = 3.554$ cm.
Hence **the length of side of the base of the pyramid is 3.554 cm.**

Problem 9. A block of copper weighing 30 kg is drawn out to make 400 m of wire of uniform circular cross-section. Given that 1 cm^3 of copper weighs 8.91 g calculate:
(*a*) the volume of the copper in cm^3;
(*b*) the area of the cross-section of the wire in cm^2;
(*c*) the radius of the cross-section of the wire in mm.

(*a*) If 8.91 g of copper has a volume of 1 cm^3

then 1 g of copper has a volume of $\frac{1}{8.91}\ \text{cm}^3$

hence **30 000 g of copper has a volume of** $\frac{30000}{8.91}\ \text{cm}^3$

$= \mathbf{3\,367\ cm^3}$

(*b*) The wire may be regarded as a very long circular prism or cylinder. Hence the volume of wire = (area of cross-section) $\times$ (length of wire)
Therefore 3 367 = (area of cross-section) (40 000)

Therefore **area of cross-section** $= \frac{3\,367}{40\,000}\ \text{cm}^2$

$= \mathbf{0.084\ cm^2}$

(*c*) Area of circle $= \pi r^2$.
Therefore $0.084\ \text{cm}^2 = 8.4\ \text{mm}^2 = \pi r^2$

Therefore $r^2 = \dfrac{8.4}{\pi} = 2.6738$

$$r = \sqrt{2.6738} = \pm 1.635 \text{ mm}$$

$$= 1.635 \text{ mm } (-1.635 \text{ can be neglected})$$

Hence **the radius of cross-section is 1.635 mm.**

Problem 10. A solid metal cylinder of diameter 10 cm and height 25 cm is melted down and recast into the shape shown in Fig. 12.43 with 15% of the metal wasted in the process. The new shape is a hemisphere attached to the circular base of a cone. If the radius of the cone is to be equal to the height of the cone find the overall length l of the new shape.

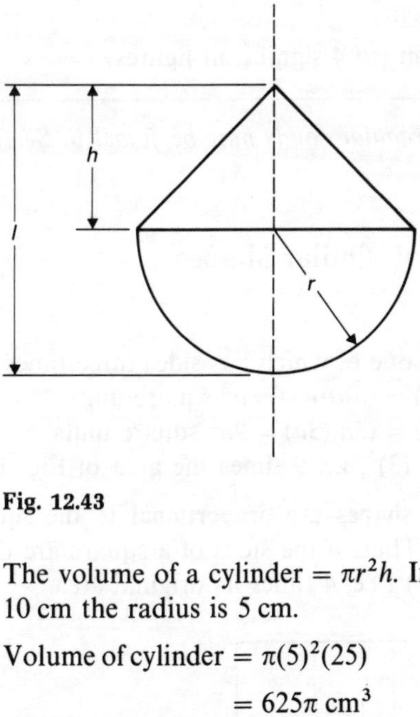

Fig. 12.43

The volume of a cylinder $= \pi r^2 h$. If the diameter of the cylinder is 10 cm the radius is 5 cm.

Volume of cylinder $= \pi(5)^2(25)$

$= 625\pi \text{ cm}^3$

If 15% of the metal is lost then 85% of 625π forms the volume of the new shape. That is, volume of new shape $= (0.85)(625\pi)$ cm^3.
The volume of Fig. 12.43 consists of:

(*a*) a hemisphere (i.e. half a sphere).
Therefore volume of hemisphere $= \frac{1}{2}(\frac{4}{3}\pi r^3)$

$= \frac{2}{3}\pi r^3$

(*b*) a cone.
Volume of cone $= \frac{1}{3}\pi r^2 h$. But $h = r$

Therefore volume of cone $= \frac{1}{3}\pi r^2 r$
$= \frac{1}{3}\pi r^3$

Therefore the total volume of the new shape $= \frac{2}{3}\pi r^3 + \frac{1}{3}\pi r^3$
$= \pi r^3$ cm^3

Hence $(0.85)(625\pi) = \pi r^3$

$r^3 = (0.85)(625)$
$= 531.25$
$= 531.3$ (to 4 significant figures)
$r = \sqrt[3]{531.3}$
$= 8.099$ cm

The overall length l of the new shape is $h + r$
that is, $2r = 2(8.099)$
$= 16.198$ cm
$=$ **16.20 cm** (to 4 significant figures)

Further problems on volumes of regular solids may be found in Section 12.6 (Problems 24–46), page 306.

12.5 Areas and volumes of similar shapes

Areas

Figure 12.44 shows two squares, one of which has sides three times as long as the other. Area of Fig. 12.44(*a*) $= (a)(a) = a^2$ square units.
Area of Fig. 12.44(*b*) $= (3a)(3a) = 9a^2$ square units.
Hence Fig. 12.44(*b*) has an area $(3)^2$, i.e. 9 times the area of Fig. 12.44(*a*).

In general, **the areas of similar shapes are proportional to the squares of corresponding linear dimensions**. Thus, if the sides of a square are doubled, for example, the area becomes $(2)^2$, i.e. 4 times its original area.

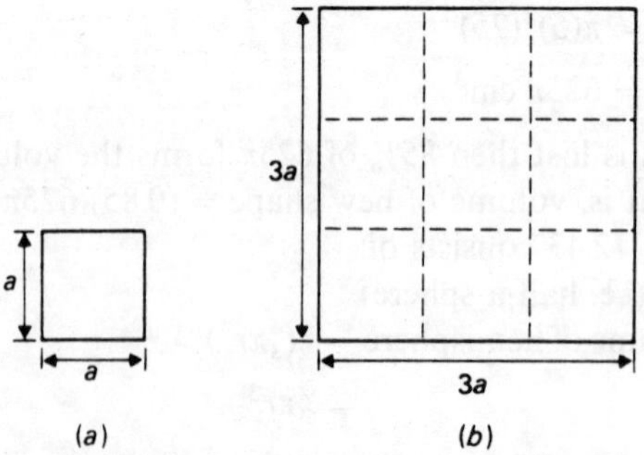

Fig. 12.44

Volumes

Figure 12.45 shows two cubes, one of which has sides three times as long as those of the other. Volume of Fig. 12.45(*a*) = $(a)(a)(a) \quad = a^3$ cubic units.
Volume of Fig. 12.45(*b*) = $(3a)(3a)(3a) = 27a^3$ cubic units.

Hence Fig. 12.45(*b*) has a volume $(3)^3$, i.e. 27 times the volume of Fig. 12.45(*a*).

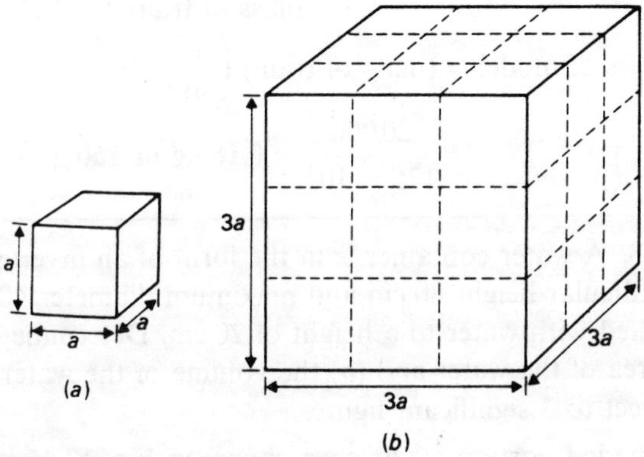

Fig. 12.45

In general, **the volumes of similar bodies are proportional to the cubes of corresponding linear dimensions.** Thus, if the sides of a cube are doubled the volume becomes $(2)^3$, i.e. 8 times its original volume.

Worked problems on areas and volumes of similar shapes

Problem 1. A rectangular building is shown on a building plan having dimensions 10 mm by 15 mm. If the plan is drawn to a scale of 1 to 300, determine the true area of the building, in square metres.

Area of building on the plan = 10 mm × 15 mm = 150 mm^2.

Since the areas of similar shapes are proportional to the square of corresponding dimensions then:

true area of building $= 150 \times (300)^2 = 13.5 \times 10^6$ mm^2

$= \mathbf{13.5\ m^2}$

(since 1 m^2 = 10^6 mm^2)

Problem 2. A train has a mass of 20 000 kg. A model of the train is made to a scale of 1 to 50. Determine the mass of the model if the train and its model are made of the same material.

$\left(\dfrac{\text{Volume of model}}{\text{Volume of train}}\right) = \left(\dfrac{1}{50}\right)^3$, since the volumes of similar bodies are proportional to the cube of corresponding dimensions.

Mass = density × volume, and since both train and model are made of the same material then $\left(\dfrac{\text{mass of model}}{\text{mass of train}}\right) = \left(\dfrac{1}{50}\right)^3$

Hence mass of model = (mass of train) $\left(\dfrac{1}{50}\right)^3$

$$= \frac{20\,000}{125 \times 10^3} = \mathbf{0.16\ kg\ or\ 160\ g}$$

Problem 3. A water container is in the form of an inverted cone of perpendicular height 80 cm and maximum diameter 40 cm. The cone is filled with water to a height of 20 cm. Determine (*a*) the surface area of the water and (*b*) the volume of the water in dm^3, each correct to 3 significant figures.

(*a*) The shaded portion of the cone shown in Fig. 12.46 represents the water and this is similar in shape to the complete cone but the dimensions are $\frac{20}{80}$, i.e. $\frac{1}{4}$ of the size of the whole cone.

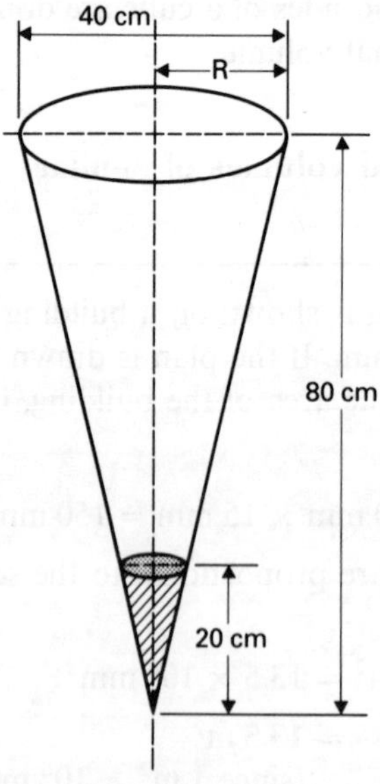

Fig. 12.46

Hence $\left(\dfrac{\text{surface area of water}}{\text{area of end of cone}}\right) = \left(\dfrac{1}{4}\right)^2$, since the areas of similar shapes are proportional to the squares of corresponding dimensions.

Thus surface area of water $= \left(\dfrac{1}{4}\right)^2 (\pi R^2) = \left(\dfrac{1}{4}\right)^2 \pi(20)^2 = 78.54\ \text{cm}^2$

$= \mathbf{78.5\ cm^2}$, correct to 3 significant figures.

(*b*) $\dfrac{\text{Volume of water}}{\text{volume of cone}} = \left(\dfrac{1}{4}\right)^3$, since the volumes of similar bodies are proportional to the cube of corresponding dimensions.

Hence volume of water $= \left(\dfrac{1}{4}\right)^3$ (volume of cone)

$$= \left(\frac{1}{4}\right)^3\left(\frac{1}{3}\pi R^2 h\right)$$

$$= \left(\frac{1}{4}\right)^3\left(\frac{1}{3}\pi(20)^2 80\right)$$

$$= 523.6\ \text{cm}^3$$

$$= \frac{523.6}{1\,000}\ \text{dm}^3 = \mathbf{0.524\ dm^3},$$

correct to 3 significant figures.

Further problems on areas and volumes of similar shapes may be found in the following Section (12.6) (Problems 47–52), page 309.

12.6 Further problems

Areas of plane figures

1. Convert 18.623 m into (*a*) mm (*b*) cm
   (*a*) [18 623 mm] (*b*) [1 862.3 cm]
2. Convert 72 460 mm into (*a*) cm (*b*) m
   (*a*) [7 246 cm] (*b*) [72.46 m]
3. Add 4.23 m, 92 cm, 143 mm and 1.609 m [6.902 m]
4. A metal plate is 91.4 mm long and 37.5 mm wide. Find its area (*a*) in mm^2 (*b*) in cm^2
   (*a*) [3 428 mm^2] (*b*) [34.28 cm^2]
5. A tray is 841.3 mm long and 412.7 mm wide. Find its area (*a*) in cm^2 (*b*) in m^2
   (*a*) [3 472 cm^2] (*b*) [0.347 2 m^2]
6. The area of a metal plate is 426 mm^2. If its length is 47.6 mm find its width in mm (to 3 significant figures). [8.95 mm]

7. A rectangular plate measures 36 mm by 2.4 cm. Find its area in cm^2 and its perimeter in mm. [8.64 cm^2; 120 mm]
8. Calculate the area (correct to 4 significant figures) of a parallelogram whose base is 9.41 cm and whose vertical height is 5.63 cm. [52.98 cm^2]
9. Find the area of each of the angle iron sections (*a*)–(*d*) shown in Fig. 12.47, all the dimensions being in mm.
(*a*) [272 mm^2 or 2.72 cm^2] (*b*) [560 mm^2 or 5.60 cm^2]
(*c*) [1 624 mm^2 or 16.24 cm^2] (*d*) [432 mm^2 or 4.32 cm^2]
10. Find the area of a triangle whose base is 9.80 cm and whose perpendicular height is 7.40 cm. [36.26 cm^2]

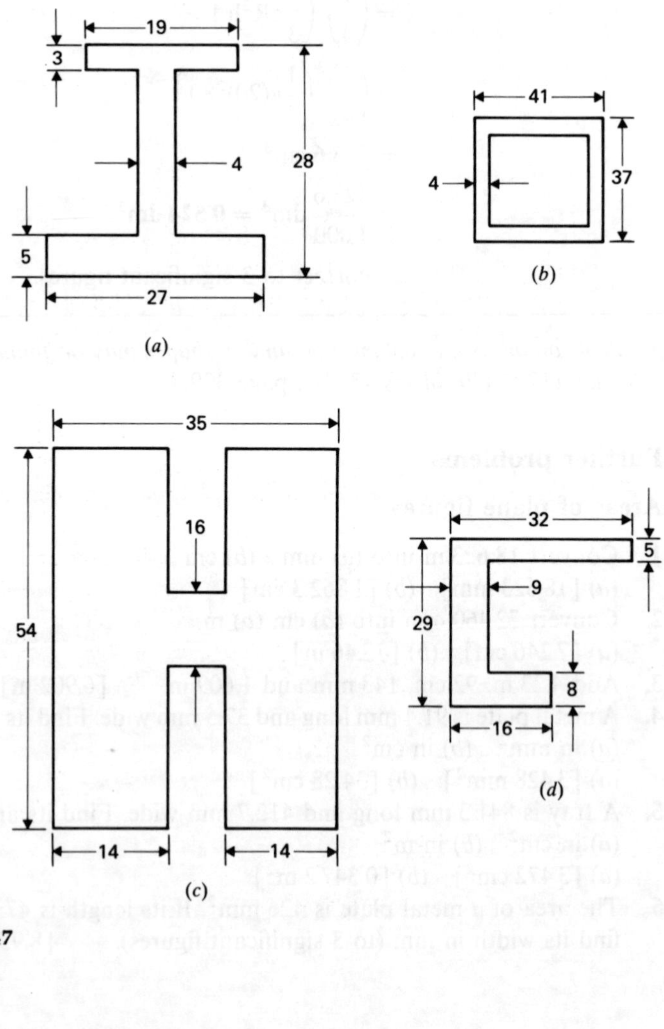

Fig. 12.47

11. Find the area (to 4 significant figures), of a triangle XYZ if
(*a*) $x = 3.10$ cm, $z = 4.30$ cm and $\angle y = 90°$
(*b*) $x = 11.62$ cm, $y = 4.92$ cm and $\angle x = 90°$
(*a*) [6.665 cm^2] (*b*) [25.90 cm^2]
12. A rectangular garden measures 50 m by 20 m. A 1.5 m flower border is made round the two longest sides and one short side. A circular fish pond of diameter 7 m is constructed in the middle of the garden. Find to the nearest square metre the area remaining. [786 m^2]
13. Find the area of an equilateral triangle of side 15.0 cm. [97.43 cm^2]
14. Find the area of a trapezium whose parallel sides measure 4.60 cm and 8.10 cm. The perpendicular distance between these two sides is 5.40 cm. [34.29 cm^2]
15. A square has an area of 79.23 cm^2. Calculate the length of a side (to 2 decimal places). [8.90 cm]
16. Find the length of the side of a square whose area is equal to that of a parallelogram with a 6.0 cm base and a perpendicular height to the base of 4.70 cm. [5.31 cm]
17. The area of a trapezium is 19.4 cm^2 and the perpendicular distance between its parallel sides is 2.80 cm. If the length of one of the parallel sides is 4.23 cm find the length of the other parallel side. [9.63 cm]
18. Find the area of the following circles given that
(*a*) the radius is 15.0 mm in length
(*b*) the diameter is 2.80 cm in length
(*c*) the circumference is 42.4 cm in length.
(*a*) [706.9 mm^2] (*b*) [6.158 cm^2] (*c*) [143.1 cm^2]
19. If the area of a circle is 153 mm^2 find the length of
(*a*) the radius, (*b*) the diameter, (*c*) the circumference.
(*a*) [6.979 mm] (*b*) [13.96 mm] (*c*) [43.85 mm]
20. An annulus has an outside diameter of 49.0 mm and an inside diameter of 15.0 mm. Find its area (correct to 4 significant figures). [1 709 mm^2]
21. Name the types of quadrilateral, shown in Fig. 12.48(*a*)–(*e*), and calculate the area and the length of the perimeter of each one.
(*a*) [Parallelogram, 162 mm^2, 54.97 mm]
(*b*) [Square, 16 cm^2, 16 cm]
(*c*) [Rhombus, 23.85 cm^2, 20 cm]
(*d*) [Rectangle, 3 906 mm^2, 270 mm]
(*e*) [Trapezium, 97.82 cm^2, 44.79 cm]
22. A square has sides of 4.31 cm. Find the length of a diagonal. [6.095 cm]

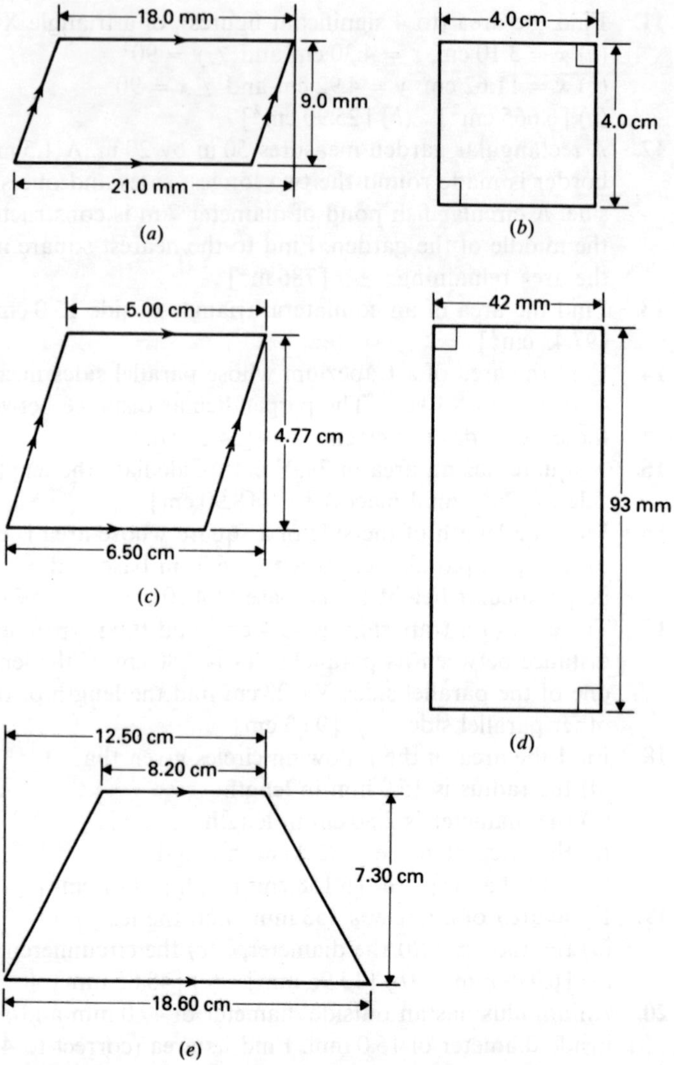

Fig. 12.48

23. Find the angles a to e in Fig 12.49
[$a = 108°$] [$b = 26°$] [$c = 138°$] [$d = 87°$] [$e = 81°$]

Volumes of regular solids

24. A rectangular block of metal has dimensions of 37.6 mm by 25.4 mm by 17.2 mm. Find its volume.
[16 430 mm³ or 16.43 cm³]

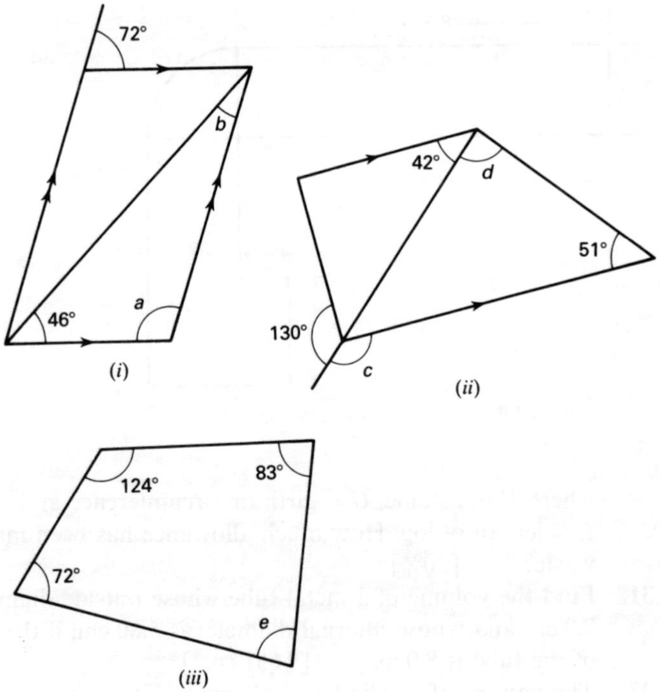

Fig. 12.49

25. A rectangular fish tank measures 42.0 cm by 36.3 cm by 4.20 m. Find its volume in m^3. [0.640 3 m^3]
26. A concrete drive to a garage is to be 10 m long, 2 m wide and $\frac{1}{6}$ m thick. How much concrete is required for its construction? [$3\frac{1}{3}$ m^3]
27. The dimensions of a brick are 65.0 mm × 102.5 mm × 215.0 mm. If 5 mm is allowed all round for mortar how many bricks are required to build a wall 9.6 m × 1.8 m × 0.225 m? [2 048]
28. Determine the area of the path shown in Fig. 12.50. If the path is $\frac{1}{6}$ m thick, what volume of concrete is required for its construction? [34 m^2, 5.67 m^3]
29. (*a*) Determine the volume and surface area of a cube of side length 6 cm. (*b*) If the volume of a cube is 512 mm^3 what is the length of one of its sides?
(*a*) [216 cm^3; 216 cm^2] (*b*) [8 mm]
30. The formula for measuring round logs devised by Hoppus is

$$V = \frac{G^2 L}{5\pi}$$

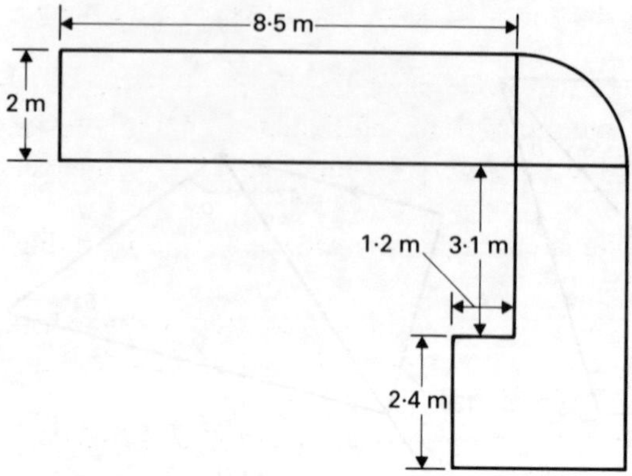

Fig. 12.50

where V = volume, G = girth or circumference and L = length of log. How much allowance has been made for waste? [20%]

31. Find the volume of a metal tube whose outside diameter is 7.0 cm and whose internal diameter is 5.80 cm, if the length of the tube is 8.0 m. [9 651 cm^3]

32. The volume of a cylinder is 348 cm^3. If its radius is 4.0 cm find its height. [6.923 cm]

33. If the height of a cylindrical can is 5.50 cm and its volume is 252 cm^3 find its diameter. [7.638 cm]

34. If a metal cone has a diameter of 92 mm and a perpendicular height of 124 mm find its volume in cm^3 (correct to 4 significant figures). [274.8 cm^3]

35. A cylinder is cast from a piece of alloy 4.60 cm by 7.30 cm by 12.60 cm. If the length of the cylinder is to be 80.0 cm, find its circumference. [8.152 cm]

36. If a cylindrical tin is to have a volume of 520 cm^3 find its height if the height and diameter are equal. [8.716 cm]

37. A square pyramid has a perpendicular height of 24.0 mm. If a side of the base is 3.20 cm, find the volume of the pyramid (to 4 significant figures). [8.192 cm^3]

38. A hemisphere has a diameter of 3.60 cm. Find its volume. [12.21 cm^3]

39. A trench is 12.0 m long. Its width at the top is 4.0 m and at the bottom 2.70 m and it has a depth of 2.0 m. Find the volume of the trench, assuming its sides slant uniformly. [80.4 m^3]

40. Find the weight in kilograms of a hemispherical bowl of copper whose external and internal radii are 15.0 cm and

11.0 cm. Assume that 1 cm^3 of copper weighs 8.9 g. [38.10 kg]

41. A rivet consists of a cylindrical head, of diameter 1 cm and depth 5 mm, and a shaft of diameter 1.5 mm and length 1.5 cm. Find the volume of metal in 1 000 such rivets. [419.2 cm^3]

42. If the volume of a sphere is 476.0 cm^3 find its radius. [4.844 cm]

43. Calculate the volume of metal required to make a pipe of length 11.50 m with external diameter of 24.0 cm and thickness 4.0 mm. [34 105 cm^3]

44. A pyramid stands on a square base of side 15.0 cm. If the length of each sloping edge is 21.0 cm calculate the volume of the pyramid. [1 359 cm^3]

45. The solid shown in Fig. 12.51 comprises a hemisphere and a cone. The radius of the base of the cone is the same as the radius of the hemisphere, and the perpendicular height of the cone is twice its radius. If the volume of the solid is 382.0 cm^3 find the diameter and length l of the solid. [Diameter = 9.002 cm l = 13.50 cm]

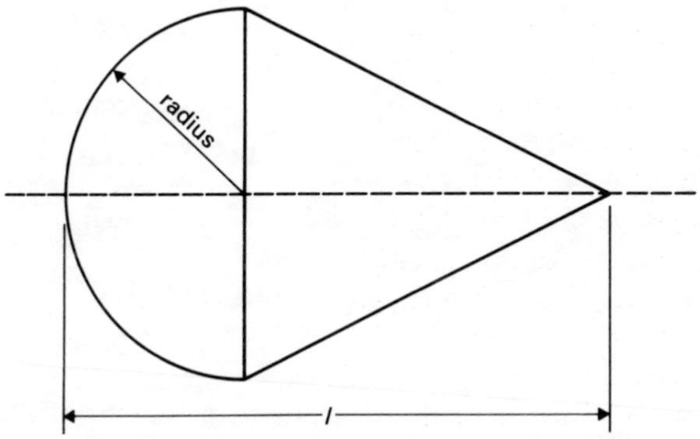

Fig. 12.51

46. A metal washer has an outside diameter of 2.0 cm and an inside diameter of 1.20 cm and is 3.0 mm thick. Find the weight in kilograms of 2 000 washers if 1 cm^3 weighs 9 grams. [10.86 kg]

Areas and volumes of similar shapes

47. A rectangular garage is shown on a building plan having dimensions 15 mm by 24 mm. If the plan is drawn to a scale of 1 to 200, determine the true area of the garage, in square metres. [14.4 m^2]

48. The area of a plot of land on a map is 400 mm^2. If the scale of the map is 1 to 50 000, determine the true area of the land in hectares (1 hectare = 10^4 m^2). [100 ha]

49. A car has a mass of 800 kg. A model of the car is made to a scale of 1 to 20. Determine the mass of the model if the car and its model are made of the same material. [100 g]

50. The diameters of two spherical bearings are in the ratio of 3:5. What is the ratio of their volumes? [27:125]

51. A component has a mass of 600 g. If each of its dimensions is reduced by 25%, determine its new mass. [253.1 g]

52. A container is in the form of an inverted cone of perpendicular height 500 mm and maximum diameter 200 mm. The cone is filled with a liquid to a height of 100 mm. Determine (*a*) the surface area of the liquid, and (*b*) the volume of the liquid, each correct to 4 significant figures. (*a*) [1 257 mm^2](*b*) [41 890 mm^3]

Chapter 13

Trigonometry

13.1 Definition

Trigonometry is the branch of mathematics which deals with the measurement of sides and angles of triangles, and their relationships with each other.

13.2 Trigonometric ratios of acute angles

In Chapter 10 it was shown that the third side of a right-angled triangle could be calculated using the theorem of Pythagoras, provided that the other two sides were known. However, the theorem does not provide a method for finding the angles of a triangle.

Given certain information, a triangle can be constructed by scale drawing and the unknown sides or angles measured. However, in the case of a right-angled triangle, a knowledge of the trigonometric ratios enables both sides and angles to be calculated. Consider a right-angled triangle ABC shown in Fig. 13.1 and let the angle CAB be denoted by θ. The three sides of the triangle are given special names.

The longest side, that is the side opposite the 90° angle, is called the **hypotenuse**. The side opposite the angle θ is called the **opposite side**. The third side is called the **adjacent side**. Hence, referring to angle θ in Fig. 13.1:

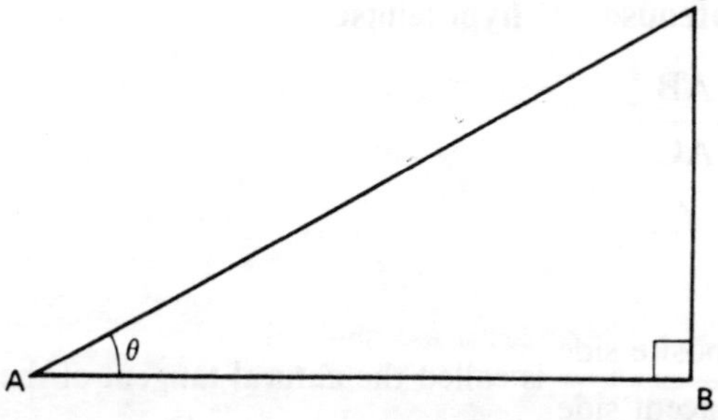

Fig. 13.1

AC = hypotenuse
BC = opposite side
AB = adjacent side

Note that the hypotenuse is always the side opposite the right angle. However, the opposite and adjacent sides in a triangle will depend on which angle is being referred to. If the angle ACB in Fig. 13.1 is referred to, the opposite side would then be AB and the adjacent side, BC.

The following three definitions of trigonometric ratios are true for any right-angled triangle:

1. *Natural sine*

The ratio $\dfrac{\text{length of opposite side}}{\text{length of hypotenuse}}$ is called the **natural sine** of the angle θ.

(The abbreviation 'sin' is usually used for sine.)
Thus,

$$\sin\theta = \frac{\textbf{opposite side}}{\textbf{hypotenuse}} \textbf{ or } \frac{\textbf{opposite}}{\textbf{hypotenuse}}$$

In Fig. 13.1, $\sin\theta = \dfrac{BC}{AC}$

2. *Natural cosine*

The ratio $\dfrac{\text{length of adjacent side}}{\text{length of hypotenuse}}$ is called the **natural cosine** of the angle θ. (The abbreviation 'cos' is usually used for cosine.)
Thus,

$$\cos\theta = \frac{\textbf{adjacent side}}{\textbf{hypotenuse}} \textbf{ or } \frac{\textbf{adjacent}}{\textbf{hypotenuse}}$$

In Fig. 13.1, $\cos\theta = \dfrac{AB}{AC}$

3. *Natural tangent*

The ratio $\dfrac{\text{length of opposite side}}{\text{length of adjacent side}}$ is called the **natural tangent** of the angle θ.

(The abbreviation 'tan' is usually used for tangent.)

Thus,

$$\tan\theta = \frac{\textbf{opposite side}}{\textbf{adjacent side}} \textbf{ or } \frac{\textbf{opposite}}{\textbf{adjacent}}$$

In Fig. 13.1, $\tan\theta = \dfrac{BC}{AB}$

Validity of trigonometric ratios

Sine, cosine and tangent are the three fundamental trigonometric ratios used to determine the size of angles. As they are each a ratio of length divided by length, they have no units.

The definitions of sine, cosine and tangent are true **only** for a right-angled triangle. If a triangle is not right-angled the definitions must not be used.

Worked problems on trigonometric ratios of acute angles

Problem 1. Find sin θ, cos θ and tan θ in the triangle PQR shown in Fig. 13.2.

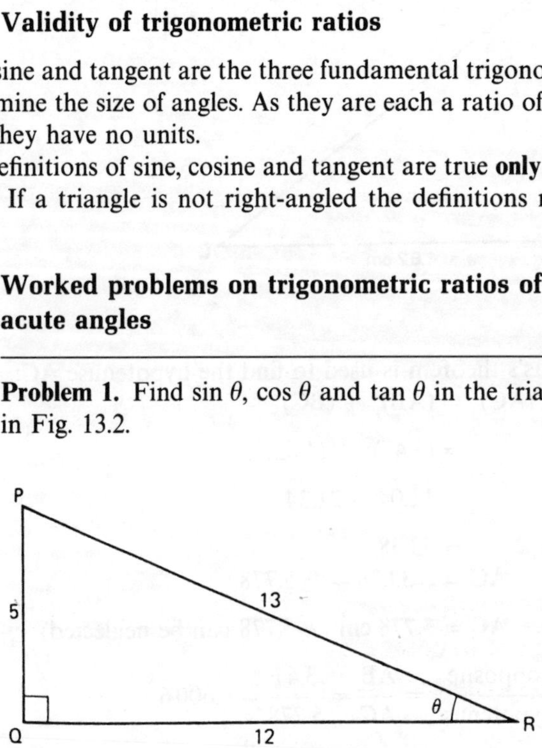

Fig. 13.2

$$\sin\theta = \frac{\text{opposite side}}{\text{hypotenuse}} = \frac{PQ}{PR} = \frac{5}{13} = 0.384\,6$$

$$\cos\theta = \frac{\text{adjacent side}}{\text{hypotenuse}} = \frac{QR}{PR} = \frac{12}{13} = 0.923\,1$$

$$\tan\theta = \frac{\text{opposite side}}{\text{adjacent side}} = \frac{PQ}{QR} = \frac{5}{12} = 0.416\,7$$

Hence, **sin θ = 0.384 6, cos θ = 0.923 1 and tan θ = 0.416 7**

Problem 2. In a triangle ABC, $\angle B = 90°$, $a = 4.62$ cm and $c = 3.47$ cm. Find sin C, cos C and tan C.

The triangle ABC is shown in Fig. 13.3

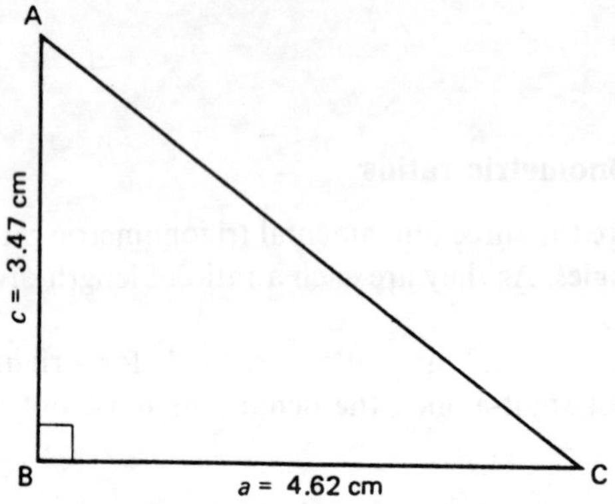

Fig. 13.3

Pythagoras's theorem is used to find the hypotenuse AC:

$$(AC)^2 = (AB)^2 + (BC)^2$$

$$= (3.47)^2 + (4.62)^2$$

$$= 12.04 + 21.34$$

$$= 33.38$$

Therefore $\quad AC = \sqrt{33.38} = \pm 5.778$

AC = 5.778 cm $(-5.778$ can be neglected)

$$\sin C = \frac{\text{opposite}}{\text{hypotenuse}} = \frac{AB}{AC} = \frac{3.47}{5.778} = 0.600\,6$$

$$\cos C = \frac{\text{adjacent}}{\text{hypotenuse}} = \frac{BC}{AC} = \frac{4.62}{5.778} = 0.799\,6$$

$$\tan C = \frac{\text{opposite}}{\text{adjacent}} = \frac{AB}{BC} = \frac{3.47}{4.62} = 0.751\,1$$

Thus, **sin C = 0.600 6, cos C = 0.799 6 and tan C = 0.751 1**

Problem 3.
(*a*) If $\cos A = \frac{11}{61}$, find sin A and tan A
(*b*) If $\tan B = \frac{8}{15}$, find sin B and cos B

(*a*) Figure 13.4 shows a right-angled triangle ABC with $\cos A = \frac{11}{61}$

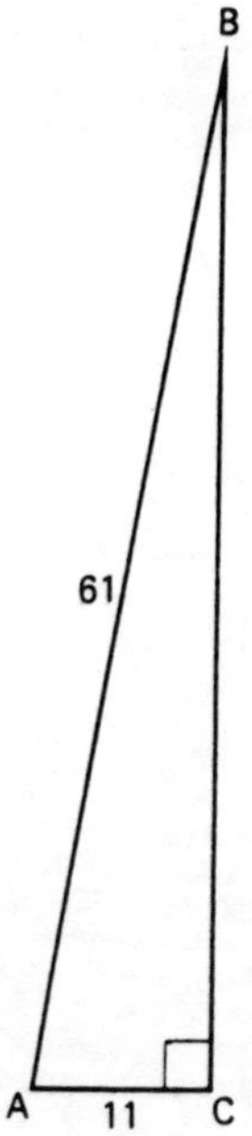

Fig. 13.4

Pythagoras's theorem is used to find the opposite side BC:

$$(AB)^2 = (AC)^2 + (BC)^2$$

Therefore $(61)^2 = (11)^2 + (BC)^2$

$$(BC)^2 = (61)^2 - (11)^2$$
$$= 3\,721 - 121$$
$$= 3600$$

Therefore $BC = \sqrt{3\,600}$

$$= \pm 60$$

$\mathbf{BC = 60}$ $(-60$ can be neglected)

Therefore $\sin A = \dfrac{BC}{AB} = \dfrac{60}{61} = 0.983\,6$

and $\tan A = \dfrac{BC}{AC} = \dfrac{60}{11} = 5\tfrac{5}{11} = 5.455$

Thus, **sin A = 0.983 6, and tan A = 5.455**

(*b*) Figure 13.5 shows a right-angled triangle ABC with $\tan B = \frac{8}{15}$
Pythaboras's theorem is used to find the hypotenuse AB:

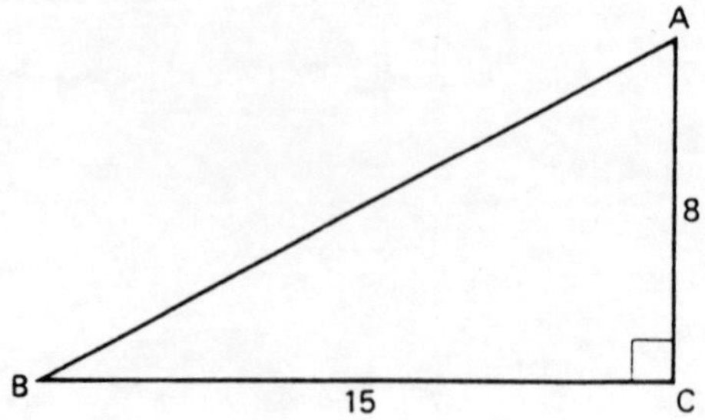

Fig. 13.5

$$
\begin{aligned}
(AB)^2 &= (AC)^2 + (BC)^2 \\
&= (8)^2 + (15)^2 \\
&= 64 + 225 \\
&= 289
\end{aligned}
$$

Therefore $AB = \sqrt{289}$

$= \pm 17$

Therefore **AB = 17** (−17 can be neglected)

Therefore $\sin B = \dfrac{AC}{AB} = \dfrac{8}{17} = 0.4706$

and $\cos B = \dfrac{BC}{AB} = \dfrac{15}{17} = 0.8824$

Hence, **sin B = 0.4706, and cos B = 0.8824**

Problem 4. Find, by drawing suitable triangles, the values of (*a*) sin 40° (*b*) cos 53° (*c*) tan 24°

(*a*) *To find the value of sin 40° by construction*

Since the sine of an angle is the ratio $\dfrac{\text{opposite side}}{\text{hypotenuse}}$ let the hypotenuse be, say, 10 cm long. This will help to make calculations easier. Thus, a triangle with angles of 90°, 40° (and thus another of 50°) and a hypotenuse of 10 cm length has to be constructed using a set square or protractor.

Let any length AB be drawn horizontally. From A a line AC is drawn 10 cm long, the arm AC being drawn at an angle of 40° to AB. From C a perpendicular to AB is drawn to intersect AB at D. The resulting triangle ACD, with hypotenuse 10 cm in length is shown in Fig. 13.6.

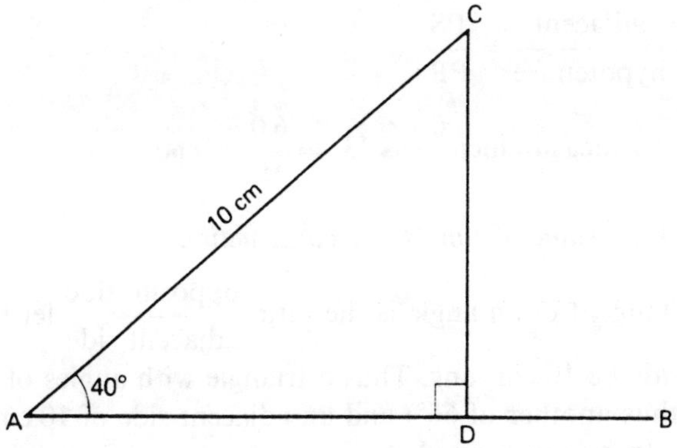

Fig. 13.6

$$\sin 40^\circ = \frac{\text{opposite}}{\text{hypotenuse}} = \frac{CD}{AC}$$

Therefore by measurement, $\mathbf{\sin 40^\circ = \frac{6.43}{10} = 0.643}$

(*b*) *To find the value of cos 53° by construction*

Since the cosine of an angle is the ratio $\frac{\text{adjacent side}}{\text{hypotenuse}}$ let the hypotenuse again be 10 cm long. Thus a triangle with angles of 90°, 53° (and thus another of 37°) and a hypotenuse of 10 cm length has to be constructed. This is done in a similar way to (*a*) above to produce a triangle PRS as shown in Fig. 13.7

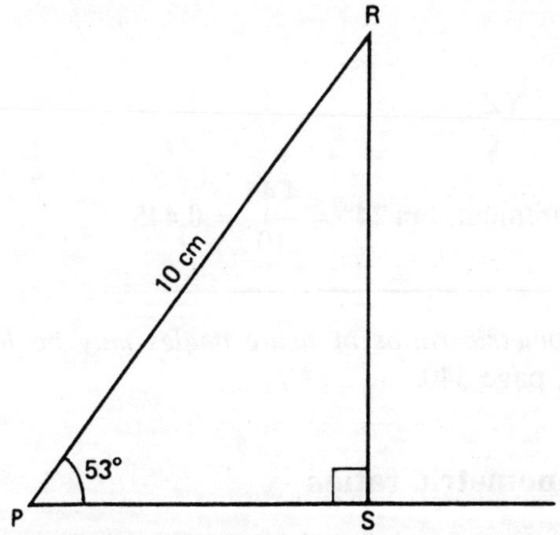

Fig. 13.7

$$\cos 53° = \frac{\text{adjacent}}{\text{hypotenuse}} = \frac{PS}{PR}$$

Therefore by measurement, $\mathbf{\cos 53°} = \frac{6.0}{10} = \mathbf{0.60}$

(c) *To find the value of tan 24° by construction*

Since the tangent of an angle is the ratio $\frac{\text{opposite side}}{\text{adjacent side}}$ let the adjacent side be 10 cm long. Thus a triangle with angles of 90°, 24° (and thus another of 66°) and an adjacent side of 10 cm length has to be constructed.

Let the length XY be drawn horizontally 10 cm long to represent the adjacent side. At X an angle of 24° is drawn to XY. A perpendicular to XY is drawn at Y and the point where the opposite side and hypotenuse intersect is the vertex Z. The resulting triangle XYZ is shown in Fig. 13.8

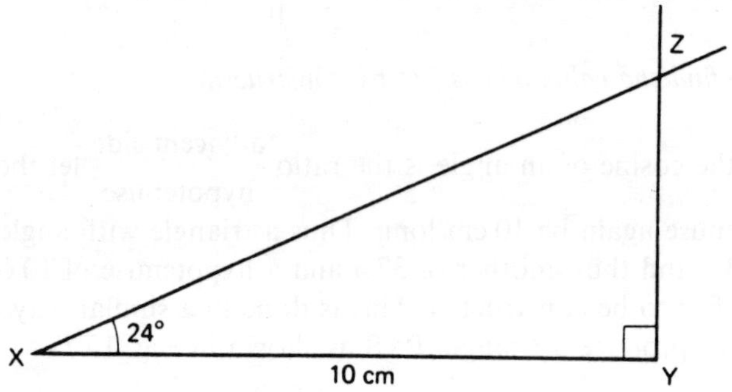

Fig. 13.8

$$\tan 24° = \frac{\text{opposite}}{\text{adjacent}} = \frac{YZ}{XY}$$

Therefore by measurement, $\mathbf{\tan 24°} = \frac{4.45}{10} = \mathbf{0.445}$

Further problems on trigonometric ratios of acute angles may be found in Section 13.9 (Problems 1–6), page 340.

13.3 Evaluating trigonometric ratios

It is possible to find the sine, cosine or tangent of any angle between 0° and 90° by scale drawing as in Problem 4 of the previous section. However, to produce a scale drawing every time a trigonometric ratio is required would be a very inconvenient and time-consuming exercise. 4-figure tables are available which give values of sines, cosines and tangents for angles between 0° and 90°. However, the easiest method of evaluating trigonometric functions of any angle is by using a calculator.

The following values, correct to 4 decimal places, may be checked:

$$\text{sine } 17° = 0.2924, \text{ sine } 62° = 0.8829,$$

$$\text{cosine } 23° = 0.9205, \text{ cosine } 79° = 0.1908,$$

$$\text{tangent } 34° = 0.6745 \text{ and tangent } 83° = 8.1443$$

To evaluate, say, sine 37° 43′, using a calculator means finding $37\frac{43°}{60}$ since there are 60 minutes in 1 degree.

$\frac{43}{60} = 0.716\dot{6}$, thus $37° 43' = 37.716\dot{6}$

and sine $37° 43' =$ sine $37.716\dot{6} = 0.6118$, correct to 4 decimal places

Similarly, cosine $68° 20' =$ cosine $68\frac{20°}{60} = 0.3692$

and tangent $52° 52' =$ tangent $52\frac{52°}{60} = 1.3206$, correct to 4 decimal places

To find, for example, the angle whose sine is 0.2648, enter into the calculator 0.2648 and then press the inverse sine (or $\sin^{-1}$) function. This gives 15.3550691...°. Subtracting 15 leaves 0.3550691...°. Multiplying by 60 gives 21′ to the nearest minute. (Many calculators have a function which enables the angle to be displayed in degrees, minutes and seconds.) 'Arcsin θ' is an abbreviation for 'the angle whose sine is equal to θ'. Thus from above:

$$\arcsin 0.2648 = 15.355° \text{ or } 15° 21'$$

Similarly, $\arccos 0.4921 = 60.52°$ or $60° 31'$

and $\arctan 1.2795 = 51.99°$ or $51° 59'$

Use your calculator to check the values of the trigonometric ratios in the following worked problems.

Worked problems on evaluating trigonometric ratios

Problem 1. Evaluate, correct to 4 decimal places:
(*a*) sine 67° (*b*) sine 11.49° (*c*) sine 36° 49′

(*a*) sine 67° = 0.92050485... = **0.9205**, correct to 4 decimal places (*b*) sine 11.49° = 0.199 196 90 ... = **0.199 2**, correct to 4 decimal places

(*c*) sine 36° 49′ = sine $36\frac{49^\circ}{60}$ = sine $36.81\dot{6}$

= 0.599 256 49 ...

= **0.599 3**, correct to 4 decimal places

Problem 2. Evaluate, correct to 4 significant figures:
(*a*) cosine 12° (*b*) cosine 75.41° (*c*) cosine 41° 51′

(*a*) cosine 12° = 0.978 147 6 = **0.978 1**, correct to 4 significant figures
(*b*) cosine 75.41° = 0.251 900 45 ... = **0.251 9**, correct to 4 significant figures

(*c*) cosine 41° 51′ = cosine $41\frac{51^\circ}{60}$ = cosine 41.85°

= 0.744 894 05 ...

= **0.744 9**, correct to 4 significant figures

Problem 3. Evaluate, correct to 5 significant figures:
(*a*) tangent 83° (*b*) tangent 25.784° (*c*) tangent 49° 19′

(*a*) tangent 83° = 8.144 346 42 ... = **8.144 3**, correct to 5 significant figures
(*b*) tangent 25.784° = 0.483 074 42 ... = **0.483 07**, correct to 5 significant figures

(*c*) tangent 49° 19′ = tangent $49\frac{19^\circ}{60}$ = tangent $49.31\dot{6}$

= 1.163 291 55 ...

= **1.163 3**, correct to 5 significant figures

Problem 4. Evaluate, correct to 4 significant figures:

(*a*) sine $\frac{\pi}{4}$ (*b*) cosine 1.324 (*c*) tangent 0.734 2

(*a*) 'sine $\frac{\pi}{4}$,' means 'the sine of $\frac{\pi}{4}$ radians.'

A calculator needs to be on the radian function.

$\frac{\pi}{4}$ = 0.785 398 16 ...

Hence $\sin\frac{\pi}{4} = \sin 0.785\,398\,16\ldots$

$= \mathbf{0.707\,1}$, correct to 4 significant figures

Actually, $\frac{\pi}{4}$ radians $\equiv 45°$ (see page 259) and sine $45° = 0.707\,1$, as above.

(*b*) cosine 1.324 means the cosine of 1.324 radians.
cosine $1.324 = 0.244\,298\,61\ldots = \mathbf{0.244\,3}$, correct to 4 significant figures

(*c*) tangent $0.734\,2 = 0.902\,509\,79\ldots = \mathbf{0.902\,5}$, correct to 4 significant figures
Note that if there is no degrees sign (i.e. °) then it is assumed that the function is in radians. Be careful; with some calculators, when you switch off, it automatically reverts to the 'degrees function'.

Problem 5. Determine the acute angles represented by:
(*a*) arcsin 0.732 9 (*b*) arccos 0.112 14 (*c*) arctan 3.469 2

(*a*) 'arcsin 0.732 9' means 'the angle whose sine is equal to 0.732 9'. 0.732 9 is entered into the calculator and then the inverse (or $\sin^{-1}$) function is pressed.
Hence arcsin $0.732\,9 = 47.130\,065\,4\ldots°$ or $47°\,8'$ if the calculator is on degrees mode,
or arcsin $0.732\,9 = 0.822\,574\,81\ldots = 0.822\,6$ radians, correct to 4 significant figures, if the calculator is on radian mode.
Hence arcsin $0.732\,9 = \mathbf{47.13°}$ or $\mathbf{47°\,8'}$ or **0.822 6 radians**

(Note $47.13° = 47.13 \times \frac{\pi}{180}$ radians $= 0.822\,6$ radians)

(*b*) arccos $0.112\,14 = 83.561\,308\,0\ldots°$ or $83°\,34'$ or arccos $0.112\,14 = 1.458\,419\,95\ldots$ radians
Hence arccos $0.112\,14 = \mathbf{83.56°}$ or $\mathbf{83°\,34'}$ or **1.458 4 radians**

(*c*) arctan $3.469\,2 = \mathbf{73.92°}$ or $\mathbf{73°\,55'}$ or **1.290 2 radians**

Problem 6. Evaluate $\dfrac{5.24 \tan 37°\,14' - 2.91 \sin 65°\,3'}{7.42 \cos 28°\,31'}$, correct to 3 significant figures.

$$\tan 37°\,14' = \tan 37\frac{14°}{60} = 0.759\,96$$

$$\sin 65°\,3' = \sin 65\frac{3°}{60} = 0.906\,68, \text{ and}$$

$$\cos 28°\,31' = \cos 28\frac{31°}{60} = 0.878\,68$$

$$\text{Hence } \frac{5.24 \tan 37^\circ 14' - 2.91 \sin 65^\circ 31'}{7.42 \cos 28^\circ 31'}$$

$$= \frac{5.24(0.759\,96) - 2.91(0.906\,68)}{7.42(0.878\,68)}$$

$$= \frac{1.343\,75}{6.519\,81} = \mathbf{0.206}, \text{ correct to 3 significant figures}$$

Further problems on evaluating trigonometric ratios may be found in Section 13.9 (Problems 7–16), page 342.

13.4 Graphs of sin θ, cos θ and tan θ from 0° to 90°

(*a*) If θ denotes any angle between 0° and 90°, a graph can be plotted with θ degrees horizontal and sin θ vertical. Using a calculator the following table of values is produced:

θ degrees	0	10	20	30	40	50	60	70	80	90
sin θ	0	0.174	0.342	0.500	0.643	0.766	0.866	0.940	0.985	1.000

The curve of sin θ is shown in Fig. 13.9

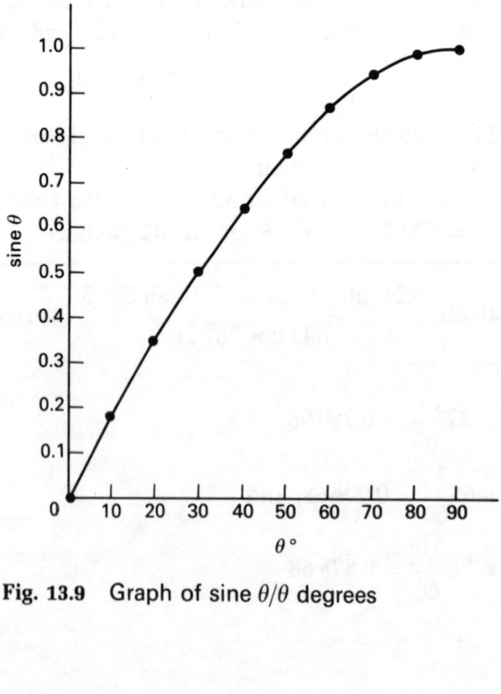

Fig. 13.9 Graph of sine θ/θ degrees

(*b*) If θ denotes any angle between 0° and 90°, a graph can be plotted with θ degrees horizontal and $\cos\theta$ vertical.
Using a calculator the following table of values is produced:

θ degrees	0	10	20	30	40	50	60	70	80	90
$\cos\theta$	1.000	0.985	0.940	0.866	0.766	0.643	0.500	0.342	0.174	0

The curve of $\cos\theta$ is shown in Fig. 13.10

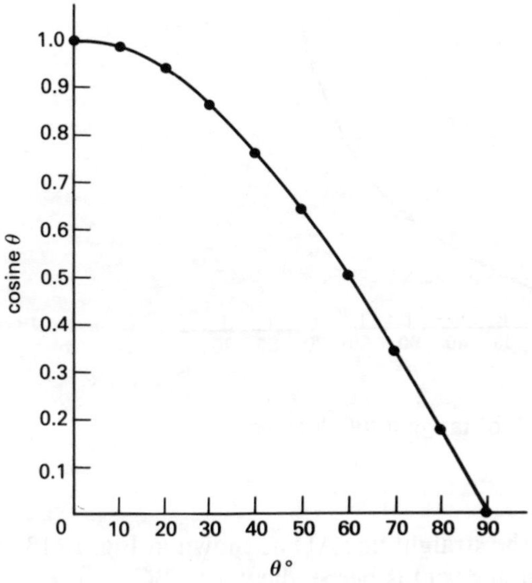

Fig. 13.10 Graph of cosine θ/θ degrees

(*c*) If θ denotes any angle between 0° and 90°, a graph can be plotted with θ degrees horizontal and $\tan\theta$ vertical.
Using a calculator the following table of values is produced:

θ degrees	0	10	20	30	40	50	60	70	80	90
$\tan\theta$	0	0.176	0.364	0.577	0.839	1.192	1.732	2.748	5.671	∞

The curve of $\tan\theta$ is shown in Fig. 13.11

13.5 The fractional and surd form of trigonometric ratios for 30°, 45° and 60°

Trigonometric ratios for 30° and 60°

Consider an equilateral triangle ABC with each side equal to 2 units and each angle equal to 60°, as shown in Fig. 13.12.

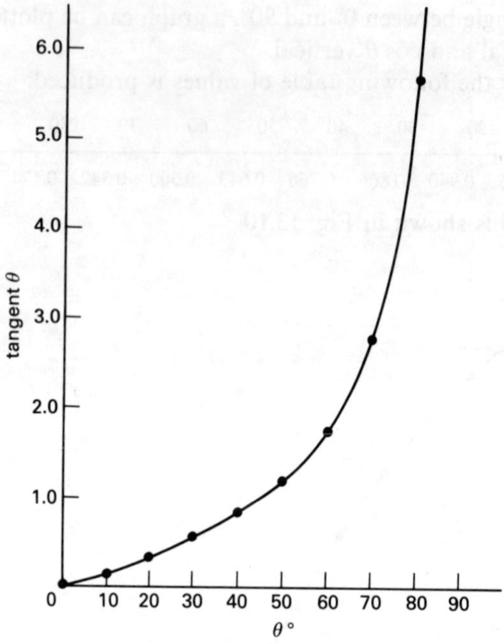

Fig. 13.11 Graph of tangent θ/θ degrees

Let $\angle$A be bisected by the straight line AD as shown in Fig. 13.13. The side BC will also be bisected and AD is perpendicular to BC.
Thus, $\angle BAD = \angle CAD = 30°$ and BD = DC = 1 unit.

If Pythagoras's theorem is applied to triangle ABD then:

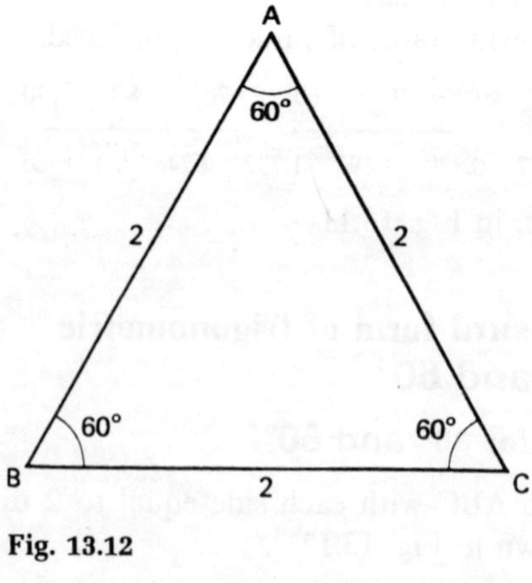

Fig. 13.12

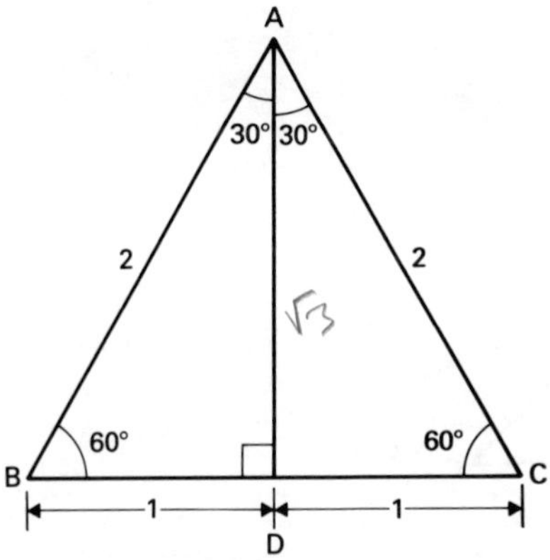

Fig. 13.13

$$(AB)^2 = (AD)^2 + (BD)^2$$

Therefore $(2)^2 = (AD)^2 + (1)^2$

$$AD^2 = (2)^2 - (1)^2$$

$$= 4 - 1$$

$$= 3$$

Therefore $\mathbf{AD = \sqrt{3}}$

Using the definitions of the trigonometric ratios:

$\sin 30^\circ = \frac{BD}{AB} = \frac{1}{2}$ $\quad \sin 60^\circ = \frac{AD}{AB} = \frac{\sqrt{3}}{2}$ $\quad \cos 30^\circ = \frac{AD}{AB} = \frac{\sqrt{3}}{2}$

$\cos 60^\circ = \frac{BD}{AB} = \frac{1}{2}$ $\quad \tan 30^\circ = \frac{BD}{AD} = \frac{1}{\sqrt{3}}$ $\quad \tan 60^\circ = \frac{AD}{BD} = \frac{\sqrt{3}}{1} = \sqrt{3}$

Trigonometric ratios for 45°

Consider a right-angled isosceles triangle ABC with the equal sides each 1 unit in length and $\angle B = 90^\circ$ as shown in Fig. 13.14
As the triangle ABC is isosceles, $\angle A = \angle C = 45^\circ$. By Pythagoras's theorem:

$$(AC)^2 = (AB)^2 + (BC)^2$$

Therefore $(AC)^2 = (1)^2 + (1)^2$

$$= 1 + 1$$

$$= 2$$

Therefore $\mathbf{AC = \sqrt{2}}$

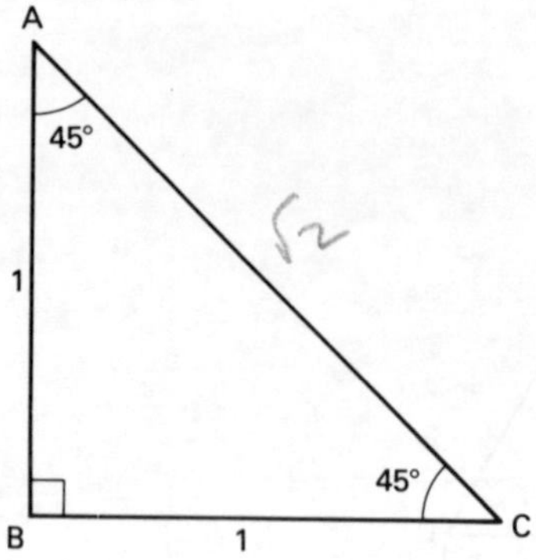

Fig. 13.14

Using the definitions of the trigonometric ratios and referring to $\angle$C:

$$\sin 45° = \frac{AB}{AC} = \frac{1}{\sqrt{2}}$$

$$\cos 45° = \frac{BC}{AC} = \frac{1}{\sqrt{2}}$$

$$\tan 45° = \frac{AB}{BC} = \frac{1}{1} = 1$$

A quantity which is not **exactly** expressible in figures is called a **surd**. For example, $\sqrt{3}$ or $\sqrt{2}$ are called surds because they cannot be expressed as a fraction and the decimal part may be continued indefinitely. Obviously an approximation can be made, such as letting $\sqrt{3} = 1.7321$ (instead of 1.732 050 8 . . .) and letting $\sqrt{2} = 1.414$ (instead of 1.414 213 6 . . .). Surd form is used often in trigonometry.

A summary of 30°, 45° and 60° trigonometric ratios is shown in Table 13.1 It will be noticed that:

$$\sin 30° = \cos 60°$$

$$\sin 60° = \cos 30°$$

and $\sin 45° = \cos 45°$

It can be checked by reference to tables that such results are true for any pair of **complementary angles**. (That is, angles which add up to 90°.) If θ is any angle then:

$$\sin \theta = \cos(90 - \theta)$$

and $\cos \theta = \sin(90 - \theta)$

Table 13.1

Trigonometric ratio	Exact value in surd form	Approximate value
Sine 30°	$\frac{1}{2}$	
Sine 45°	$\frac{1}{\sqrt{2}}$	0.707 1
Sine 60°	$\frac{\sqrt{3}}{2}$	0.866 0
Cosine 30°	$\frac{\sqrt{3}}{2}$	0.866 0
Cosine 45°	$\frac{1}{\sqrt{2}}$	0.707 1
Cosine 60°	$\frac{1}{2}$	
Tangent 30°	$\frac{1}{\sqrt{3}}$	0.577 4
Tangent 45°	1	
Tangent 60°	$\sqrt{3}$	1.732 1

For example $\sin 20° = \cos 70°$

$\sin 36° = \cos 54°$

$\cos 41° = \sin 49°$

$\cos 72° = \sin 18°$ and so on.

Problems on trigonometrical ratios in surd form and complementary angles may be found in Section 13.9 (Problems 17–19), page 345.

13.6 Solution of right-angled triangles

To '**solve** a right-angled triangle' means 'to find the unknown sides and angles'.

In order to solve a right-angled triangle it is sufficient to know the length of one side and the value of one angle other than the right angle. Alternatively, if the lengths of two sides of a right-angled triangle are known then the triangle can be solved by using the theorem of Pythagoras in addition to trigonometric ratios.

Worked problems on the solution of right-angled triangles

Problem 1. In the triangle ABC shown in Fig. 13.15, find the lengths AC and AB.

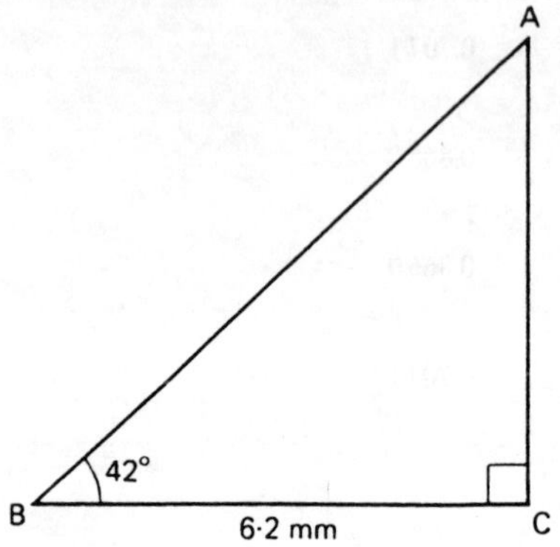

Fig. 13.15

There is usually more than one way in which such a problem can be solved.

$$\tan 42^\circ = \frac{AC}{BC} = \frac{AC}{6.2}$$

therefore $AC = 6.2 \tan 42^\circ$

$= 6.2(0.900\,4)$

$\mathbf{AC = 5.582\ mm}$

$$\cos 42^\circ = \frac{BC}{AB} = \frac{6.2}{AB}$$

therefore $AB = \dfrac{6.2}{\cos 42^\circ} = \dfrac{6.2}{0.743\,1}$

$\mathbf{AB = 8.343\ mm}$

Alternatively, if BC = 6.2 mm and AC = 5.582 mm, then Pythagoras's theorem can be used to find AB. Thus:

$(AB)^2 = (AC)^2 + (BC)^2$

$= (5.582)^2 + (6.2)^2$

$= 31.16 + 38.44$

$= 69.60$

$AB = \sqrt{69.60}$

$\mathbf{AB = 8.343\ mm}$

Thus, **AC is 5.582 mm and AB is 8.343 mm**

[Note: when alternative methods of solution are possible, it is advisable to use those methods which utilise the original data. If derived data are used in a calculation there is a possibility that they may be incorrect.]

Problem 2. Solve the triangle DEF shown in Fig. 13.16

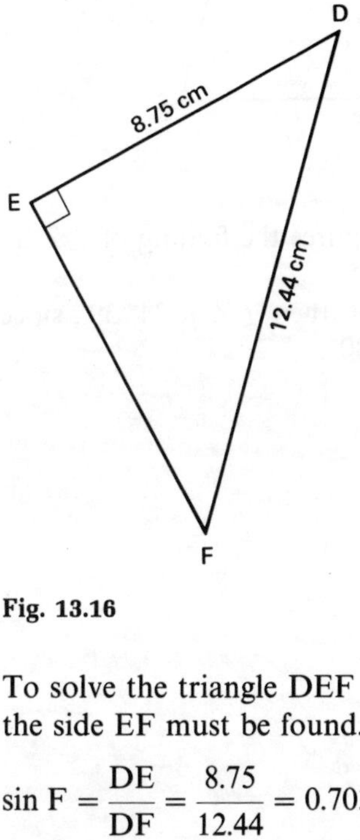

Fig. 13.16

To solve the triangle DEF the angles D and F and the length of the side EF must be found.

$$\sin F = \frac{DE}{DF} = \frac{8.75}{12.44} = 0.7034$$

Therefore $\angle$**F** = arcsin 0.703 4 = **44.70° or 44° 42′**

Hence $\angle D = 180° - 90° - 44° 42'$

Therefore $\angle$**D = 45° 18′**

$$\text{Now } \cos F = \frac{EF}{12.44}$$

Therefore EF = 12.44 cos 44° 42′

= 12.44(0.710 8)

Therefore **EF = 8.842 cm**

(EF can be checked using Pythagoras's theorem.)

Hence $\angle$**D is 45° 18′**, $\angle$**F = 44° 42′ and EF = 8.842 cm**

Problem 3. Solve triangle XYZ given $\angle Y = 90°$, $\angle X = 15° 21'$ and XZ = 17.68 mm. Find also the area of the triangle.

The triangle XYZ is shown in Fig. 13.17

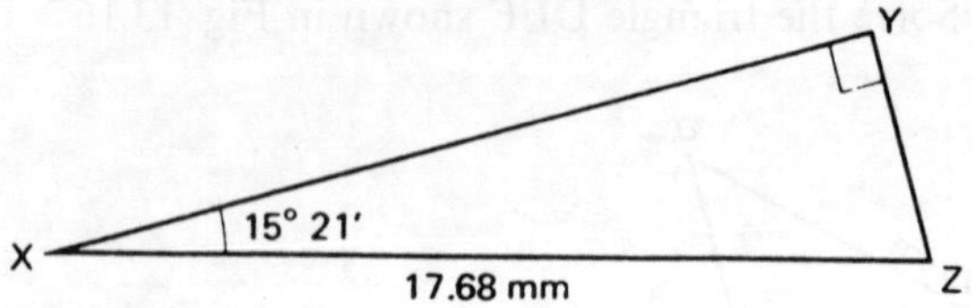

Fig. 13.17

Solution of the triangle XYZ requires the finding of $\angle Z$ and the lengths of sides XY and YZ.

If $\angle Y = 90°$ and $\angle X = 15° 21'$ then $\angle Z = 74° 39'$, since the angles in a triangle add up to 180°.

$$\sin 15° 21' = \frac{YZ}{XZ} = \frac{YZ}{17.68}$$

Therefore $YZ = 17.68 \sin 15° 21'$

$= 17.68(0.264\,7)$

Therefore **YZ = 4.680 mm**

$$\cos 15° 21' = \frac{XY}{XZ} = \frac{XY}{17.68}$$

Therefore $XY = 17.68(\cos 15° 21')$

$= 17.68(0.964\,4)$

Therefore **XY = 17.05 mm**

(YZ or XY can be checked using Pythagoras's theorem.)

Area of triangle XYZ $= \frac{1}{2}$ (base)(perpendicular height)

$= \frac{1}{2}(YZ)(XY)$

$= \frac{1}{2}(4.68)(17.05)$

$= \mathbf{39.90\ mm^2}$

Hence **$\angle$Z is 74° 39′, YZ is 4.68 mm, XY is 17.05 mm and the area of triangle XYZ is 39.90 mm²**

Further problems on the solution of right-angled triangles may be found in Section 13.9 (Problems 20–30), page 345.

13.7 Angles of elevation and depression

Angle of elevation

Consider a point A on horizontal ground 100 m away from a vertical flag-pole BC which is 30 m high, as shown in Fig. 13.18.

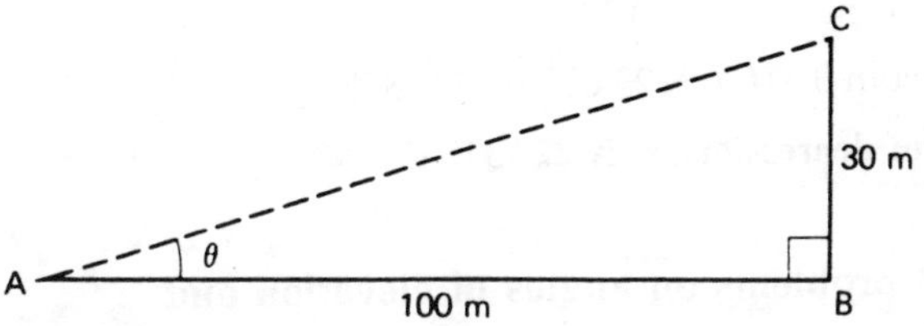

Fig. 13.18

The **angle of elevation** of the top of the flag-pole C is the angle that the imaginary straight line AC must be raised (or elevated) from the horizontal AB.

Hence the angle of elevation in Fig. 13.18 is θ

Now $\tan\theta = \dfrac{\text{opposite}}{\text{adjacent}} = \dfrac{BC}{AB} = \dfrac{30}{100} = 0.300\,0$

Hence θ is the angle whose tangent is 0.300 0

that is, $\theta = \arctan 0.300\,0$

$= 16.70°$ or $16'42'$

Hence **the angle of elevation, θ, is 16° 42′**

Angle of depression

Consider a point D on the edge of a 50 m high vertical cliff DE looking out to sea. A ship, F, is situated 120 m from the base of the cliff. This is shown in Fig. 13.19

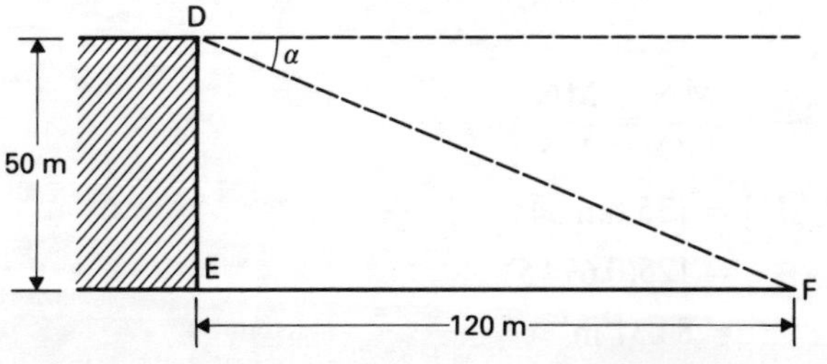

Fig. 13.19

The **angle of depression** of the ship from the point D is the angle through which the imaginary straight line DF must be lowered (or depressed) from the horizontal to the ship.

Hence the angle of depression in Fig. 13.19 is $\boldsymbol{\alpha}$

Angle DFE is also equal to α (alternate angles between parallel lines).

$$\text{Now } \tan\alpha = \frac{\text{opposite}}{\text{adjacent}} = \frac{\text{DE}}{\text{EF}} = \frac{50}{120} = 0.416\,7$$

Therefore $\alpha = \arctan 0.416\,7 = 22.62°$ or $22°\ 37'$

Hence **the angle of depression, α, is 22° 37′**

Worked problems on angles of elevation and depression

Problem 1. An electricity pylon stands on level ground. At a point 125 m from the foot of the pylon the angle of elevation of the top of the pylon is 34°. Find the height of the pylon and the angle of elevation of its mid point.

Figure 13.20 shows the pylon MN and the angle of elevation of M from the point O as 34°

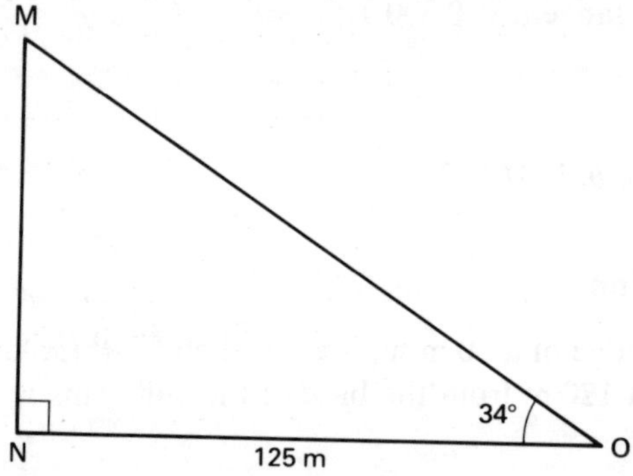

Fig. 13.20

$$\text{Now } \tan 34° = \frac{\text{MN}}{\text{NO}} = \frac{\text{MN}}{125}$$

Therefore MN = 125 tan 34°

$= 125(0.674\,5)$

$= 84.31$ m

Therefore **height of pylon = 84.31 m**

The mid point, X, shown in Fig. 13.21 is at a height of $\frac{84.31}{2}$ m, that is, 42.155 m, or 42.16 m to 4 significant figures.

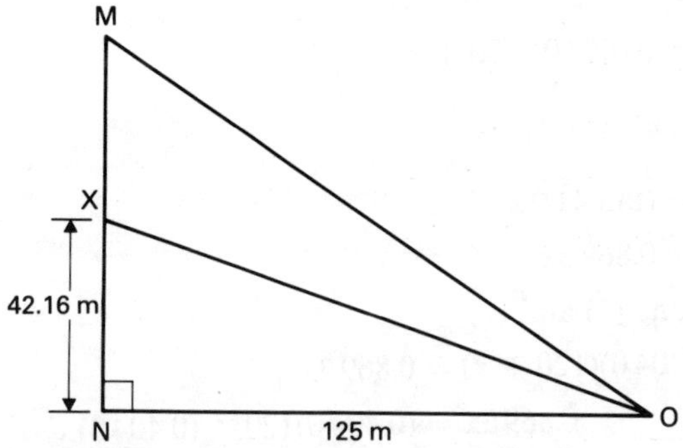

Fig. 13.21

Now $\tan XON = \frac{42.16}{125} = 0.337\,3$

There $\angle XON = \arctan 0.337\,3 = 18.64°$ or $18°\ 38'$

Hence **the angle of elevation of the mid point of the pylon is 18° 38′**

[Note the angle of elevation of the mid point is **not** half of the angle of elevation of the top of the pylon, that is, it is **not** 17°. This is because the graph of θ against $\tan\theta$ is not linear.]

Problem 2. A theodolite is an instrument by which a surveyor measures angles. When a surveyor sets up a theodolite on horizontal ground some distance from a vertical tower the angle of elevation is 22°. The surveyor moves 120 m nearer to the tower and finds the angle of elevation is now 41°. Find the height of the tower.

The tower BC and the angles of elevation are shown in Fig. 13.22

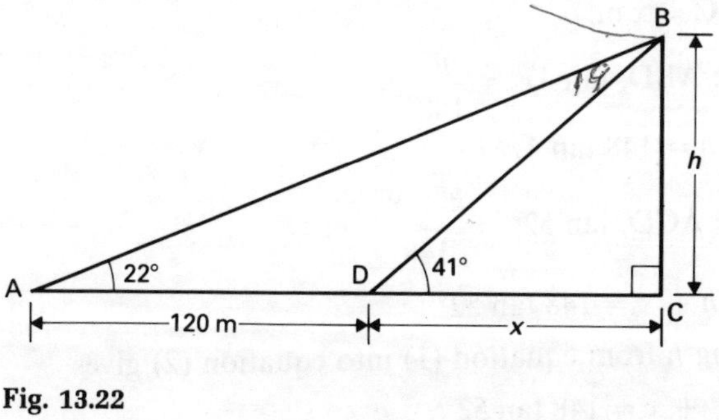

Fig. 13.22

Let DC = x m and BC = h m.

In triangle ABC, $\tan 22° = \frac{h}{120 + x}$

Therefore $h = (\tan 22°)(120 + x)$

$$h = (0.4040)(120 + x) \qquad (1)$$

In triangle BDC, $\tan 41° = \frac{h}{x}$

Therefore $h = (\tan 41°)x$

$$h = 0.8693x \qquad (2)$$

From equations (1) and (2)

$$(0.4040)(120 + x) = 0.8693x$$

Therefore

$$0.8693x = (0.4040)120 + (0.4040)x$$
$$0.8693x = 48.48 + 0.4040x$$
$$0.8693x - 0.4040x = 48.48$$
$$0.4653x = 48.48$$

Therefore $x = \frac{48.48}{0.4653} = \mathbf{104.2\ m}$

From equation (2), $h = (0.8693)(104.2)$

$= \mathbf{90.58\ m}$

Hence the height of the tower is **90.58 m**

Problem 3. A vertical flagstaff stands on the edge of the top of a vertical building. A surveyor 148 m from the building measures the angles of elevation of the top and bottom of the flagstaff as 52° and 47°. Calculate the height of the flagstaff.

A flagstaff BC is shown in Fig. 13.23 on top of a building AB, with the surveyor standing at D.

Let the height of the building, AB = h m and the height of the flagstaff BC = x m.

In triangle ABD, $\tan 47° = \frac{h}{148}$

Therefore $h = 148 \tan 47° \qquad (1)$

In triangle ACD, $\tan 52° = \frac{h + x}{148}$

Therefore $h + x = 148 \tan 52° \qquad (2)$

Substituting h from equation (1) into equation (2) gives

$148 \tan 47° + x = 148 \tan 52°$

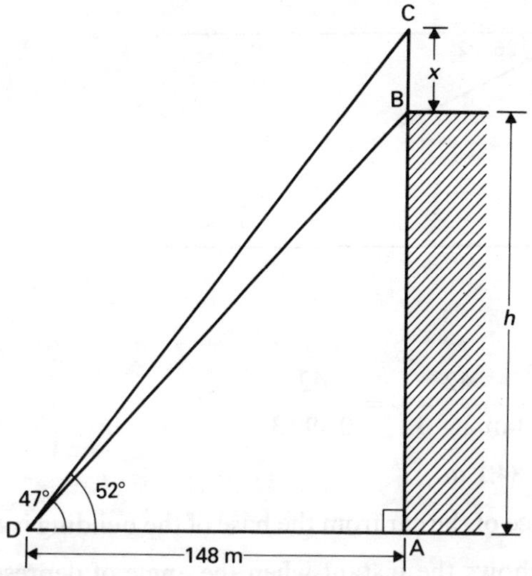

Fig. 13.23

$$\text{Therefore } x = 148 \tan 52° - 148 \tan 47°$$
$$= 148(\tan 52° - \tan 47°)$$
$$= 148(1.279\,94 - 1.072\,37)$$
$$= 148(0.207\,57)$$
$$\mathbf{x = 30.72\ m}$$

Hence the height of the flagstaff is **30.72 m**

Problem 4. The angle of depression of a car viewed at a particular instant from the top of a 42 m high vertical building is 26° 32′. What is the angle of elevation of the top of the building viewed from the car? Find the distance of the car from the building at this instant. If the car is travelling away from the building at $10\ \text{km h}^{-1}$, how long will it take for the angle of depression from the top of the building to the car to become 10°?

Figure 13.25 shows the building AB with an angle of depression of 26° 32′ to the car C.

The angle of elevation of the top of the building B from the car C is 26° 32′ (alternate angles between parallel lines). That is, **∠ACB is 26° 32′**.

To find AC, triangle ABC is used:

$$\tan 26° 32' = \frac{AB}{AC} = \frac{42}{AC}$$

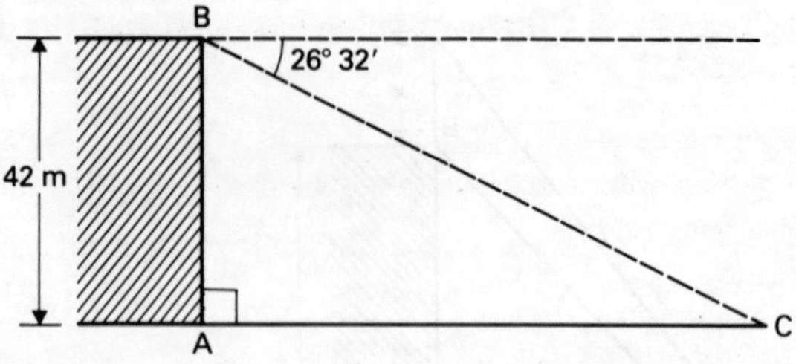

Fig. 13.24

$$\text{Therefore AC} = \frac{42}{\tan 26°\ 32'} = \frac{42}{0.499\,3}$$

$$\text{AC} = 84.1 \text{ m}$$

Hence the distance of the car from the base of the building is **84.1 m**

Figure 13.25 shows the instant when the angle of depression from B to point D is 10°

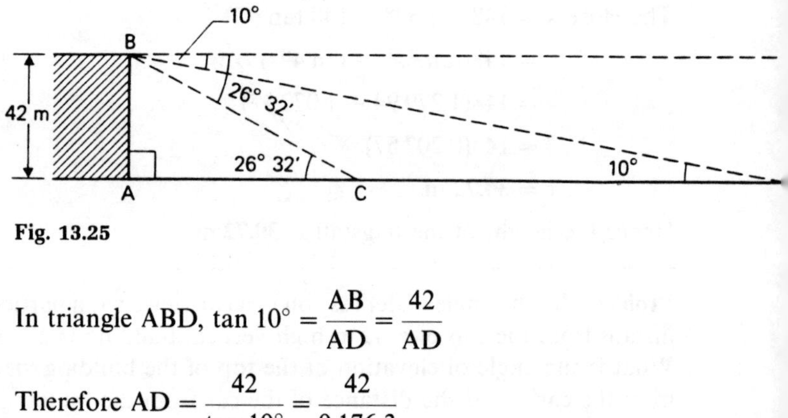

Fig. 13.25

$$\text{In triangle ABD, } \tan 10° = \frac{\text{AB}}{\text{AD}} = \frac{42}{\text{AD}}$$

$$\text{Therefore AD} = \frac{42}{\tan 10°} = \frac{42}{0.176\,3}$$

$$\text{AD} = 238.2 \text{ m}$$

Now CD = AD − AC

= 238.2 − 84.1

Therefore **CD = 154.1 m**

$$10 \text{ km h}^{-1} = \frac{10\,000}{3\,600} \text{ m s}^{-1} = 2.778 \text{ m s}^{-1}$$

Thus the car travels 2.778 m in 1 second, and so the time to travel 154.1 m is $\frac{154.1}{2.778}$ s, that is, 55.5 s

Thus the car takes 55.5 s for the angle of depression from B to change from 26° 32′ to 10°

Further problems on angles of elevation and depression may be found in Section 13.9 (*Problems 31–44*), page 349.

13.8 Sine and cosine curves

In Section 13.4 graphs of sin θ and cos θ were plotted for values of θ between 0° and 90°. However, the curves do not end at 90°.

Consider a circle, drawn on rectangular axes Ox and Oy, of centre O and radius 1 unit. The circle is divided into 360° (in 15° sectors) as shown in Fig. 13.26. The horizontal radius arm OA is assumed to represent 0° and 360° and angles are labelled anticlockwise from this radius arm.

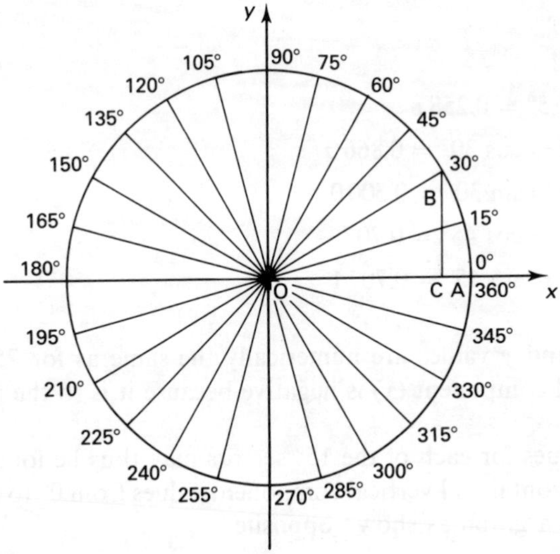

Fig. 13.26

Each radius arm can be considered to have a horizontal component x and a vertical component y. For example, for the 30° radius arm in Fig. 13.26 the horizontal component is shown by the length OC and the vertical component is shown by BC.

At 0°: the horizontal component, x = OA = 1 unit
the vertical component, $y = 0$

At 15°: the right-angled triangle shown in Fig. 13.27 shows the horizontal component x and the vertical component y.

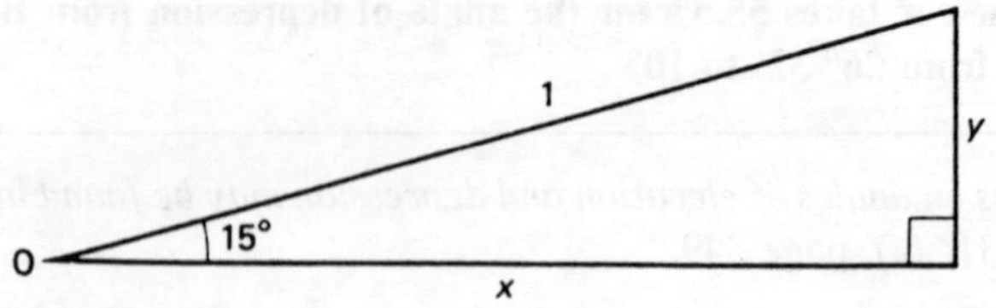

Fig. 13.27

The horizontal component x is calculated using the cosine:

$$\cos 15° = \frac{x}{1}$$

Therefore $x = \cos 15° = 0.9659$

The vertical component y is calculated using the sine:

$$\sin 15° = \frac{y}{1}$$

Therefore $y = \sin 15° = 0.2588$

Similarly, at 30°: $x = \cos 30° = 0.8660$

$y = \sin 30° = 0.5000$

and at 45°: $x = \cos 45° = 0.7071$

$y = \sin 45° = 0.7071$

and so on.

At 105°, the x and y values are numerically the same as for 75°, except that the horizontal component (x) is negative because it is to the left of the origin.

The x and y values for each of the 15° sectors may thus be found.

Each of the horizontal and vertical component values from 0° to 360° may be projected on to a graph as shown opposite.

1. Sine curve

The vertical components (that is, the y values from above) are each projected on to a graph of y against angle in degrees, as shown in Fig. 13.28.

The vertical component EF for 15° is projected across to E′F′, the corresponding value of 15° on the graph. Similar projections are made for each of the other 15° increments.

The graph produced shows a complete cycle of a **sine curve**. If values of angles greater than 360° are taken then the graph merely repeats itself indefinitely. If increments smaller than 15° are taken a more accurate sine curve is produced since more points are obtained.

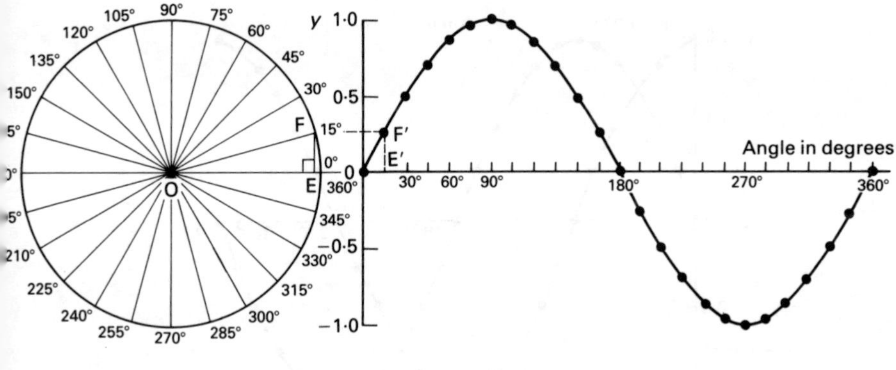

Fig. 13.28

2. Cosine curve

The circle is now redrawn with the radius arm OA in a vertical position and the 360° labelled once again in an anticlockwise direction.

The vertical components are each projected on to the graph of y against angle in degrees as shown in Fig. 13.29

The component GH, shown in Fig. 13.29, for 15° is projected across to G′H′, the corresponding value of 15° on the graph. Similar projections are made for each of the other 15° increments.

The graph produced shows a complete cycle of a **cosine curve**.

If a sine curve, $y = \sin \theta$, and a cosine curve, $y = \cos \theta$, were sketched on the same axis over one complete cycle the result would be as in Fig. 13.30

It may be seen that the cosine curve is of the same form as the sine curve, but displaced by 90°.

The sine curve representation has many important applications, including depicting the values obtained in simple harmonic motion, wave motion and in electrical currents and voltages.

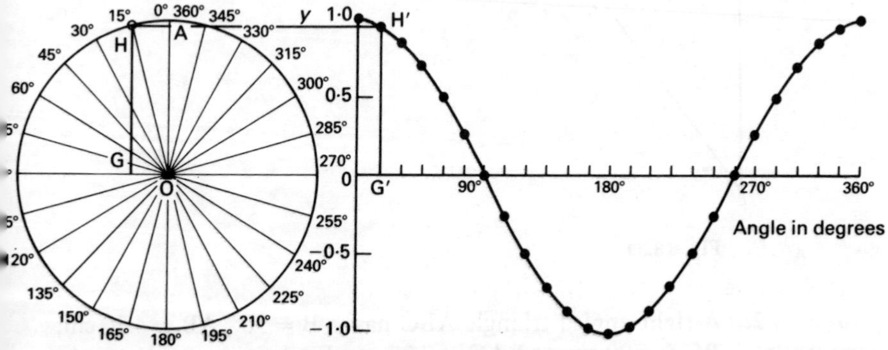

Fig. 13.29

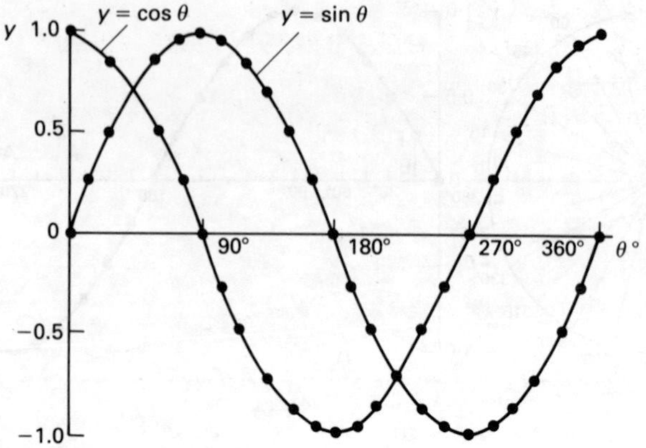

Fig. 13.30

13.9 Further problems

Trigonometric ratios

1. In Fig. 13.31 find sin E, cos E, tan E, sin F, cos F and tan F.

[Sin E = $\frac{4}{5}$ or 0.8, cos E = $\frac{3}{5}$ or 0.6, tan E = $\frac{4}{3}$ or 1.333 3, sin F = $\frac{3}{5}$ or 0.6, cos F = $\frac{4}{5}$ or 0.8, tan F = $\frac{3}{4}$ or 0.75]

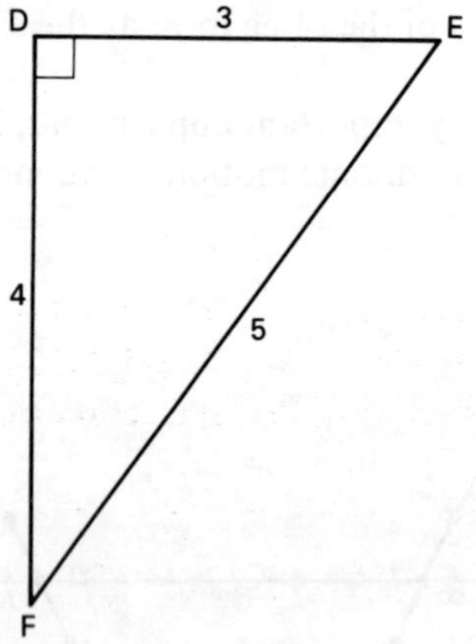

Fig. 13.31

2. A right-angled triangle ABC has ∠B = 90°, AB = 4.20 cm, BC = 5.90 cm and AC = 7.24 cm. Find sin A, cos A, tan A, sin C, cos C and tan C.

[sin A = 0.8149, cos A = 0.5801, tan A = 1.4048, sin C = 0.5801, cos C = 0.8149, tan C = 0.7119]

3. Find sin J and tan G in Fig. 13.32
[Sin J = 0.2641, tan G = 3.6526]

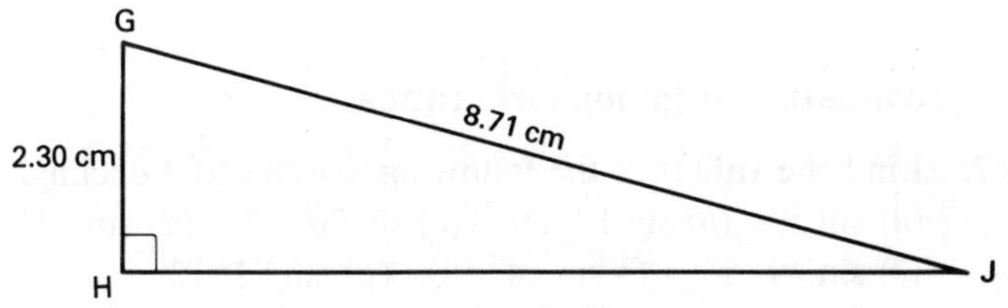

Fig. 13.32

4. Find sin K, cos K, tan K, sin M, cos M and tan M in Fig. 13.33

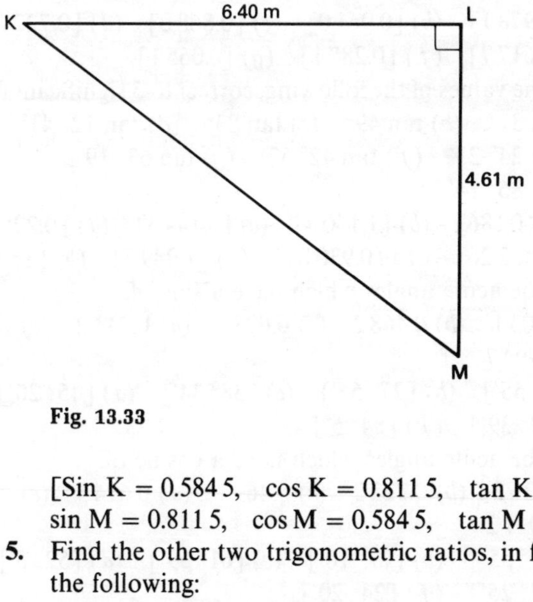

Fig. 13.33

[Sin K = 0.5845, cos K = 0.8115, tan K = 0.7203, sin M = 0.8115, cos M = 0.5845, tan M = 1.3883]

5. Find the other two trigonometric ratios, in fraction form, given the following:
(*a*) tan A = $\frac{7}{24}$, (*b*) cos C = $\frac{5}{13}$, (*c*) sin D = $\frac{9}{41}$, (*d*) tan E = $\frac{15}{112}$
(*e*) sin F = $\frac{13}{85}$, (*f*) cos G = $\frac{35}{37}$, (*g*) sin H = $\frac{19}{181}$
(*a*) [sin A = $\frac{7}{25}$, cos A = $\frac{24}{25}$] (*b*) [sin C = $\frac{12}{13}$, tan C = $\frac{12}{5}$]
(*c*) [cos D = $\frac{40}{41}$, tan D = $\frac{9}{40}$] (*d*) [sin E = $\frac{15}{113}$, cos E = $\frac{112}{113}$]
(*e*) [cos F = $\frac{84}{85}$, tan F = $\frac{13}{84}$] (*f*) [sin G = $\frac{12}{37}$, tan G = $\frac{12}{35}$]
(*g*) [cos H = $\frac{180}{181}$, tan H = $\frac{19}{180}$]

6. Find, by drawing suitable triangles, the values (to 3 significant

figures) of:
(*a*) sin 33° (*b*) sin 84° (*c*) cos 25° (*d*) cos 76° (*e*) tan 18°
(*f*) tan 66°
(*a*) [0.545] (*b*) [0.995] (*c*) [0.906] (*d*) [0.242] (*e*) [0.325]
(*f*) [2.246]

Evaluating trigonometric ratios

7. Find the values of the following, correct to 4 decimal places:
(*a*) sin 5° (*b*) sin 12° 30′ (*c*) sin 20° 52′ (*d*) sin 37° 19′
(*e*) sin 49° 5′ (*f*) sin 64° 58′ (*g*) sin 82° 10′
(*a*) [0.087 2] (*b*) [0.216 4] (*c*) [0.356 2] (*d*) [0.606 2]
(*e*) [0.755 6] (*f*) [0.906 1] (*g*) [0.990 7]
8. Find the values of the following, correct to 4 decimal places:
(*a*) cos 12° (*b*) cos 16° 4′ (*c*) cos 29° 46′ (*d*) cos 42° 53′
(*e*) cos 51° 51′ (*f*) cos 73° 19′ (*g*) cos 86° 16′
(*a*) [0.978 1] (*b*) [0.961 0] (*c*) [0.868 0] (*d*) [0.7327]
(*e*) [0.617 7] (*f*) [0.287 1] (*g*) [0.065 1]
9. Find the values of the following, correct to 5 significant figures:
(*a*) tan 31° (*b*) tan 49° (*c*) tan 83° (*d*) tan 12° 41′
(*e*) tan 24° 23′ (*f*) tan 42° 57′ (*g*) tan 63° 19′
(*h*) tan 86° 14′
(*a*) [0.600 86] (*b*) [1.150 4] (*c*) [8.144 3] (*d*) [0.225 05]
(*e*) [0.453 26] (*f*) [0.930 88] (*g*) [1.989 7] (*h*) [15.189]
10. Find the acute angles which have a sine of:
(*a*) 0.103 1 (*b*) 0.468 2 (*c*) 0.623 4 (*d*) 0.711 2 (*e*) 0.862 9
(*f*) 0.994 7
(*a*) [5° 55′] (*b*) [27° 55′] (*c*) [38° 34′] (*d*) [45° 20′]
(*e*) [59° 39′] (*f*) [84° 6′]
11. Find the acute angles which have a cosine of:
(*a*) 0.036 3 (*b*) 0.222 2 (*c*) 0.469 7 (*d*) 0.614 5 (*e*) 0.794 2
(*f*) 0.911 2
(*a*) [87° 55′] (*b*) [77° 10′] (*c*) [61° 59′] (*d*) [52° 5′]
(*e*) [37° 25′] (*f*) [24° 20′]
12. Find the acute angles which have a tangent of:
(*a*) 0.462 5 (*b*) 0.984 4 (*c*) 1.211 7 (*d*) 1.984 1 (*e*) 2.472 6
(*f*) 5.987 0
(*a*) [24° 49′] (*b*) [44° 33′] (*c*) [50° 28′] (*d*) [63° 15′]
(*e*) [67° 59′] (*f*) [80° 31′]
13. Find the dimension marked *x* in Fig. 13.34(*a*) to (*f*)
(*a*) [12.22] (*b*) [5.619] (*c*) [14.87] (*d*) [8.350] (*e*) [5.595]
(*f*) [5.275]

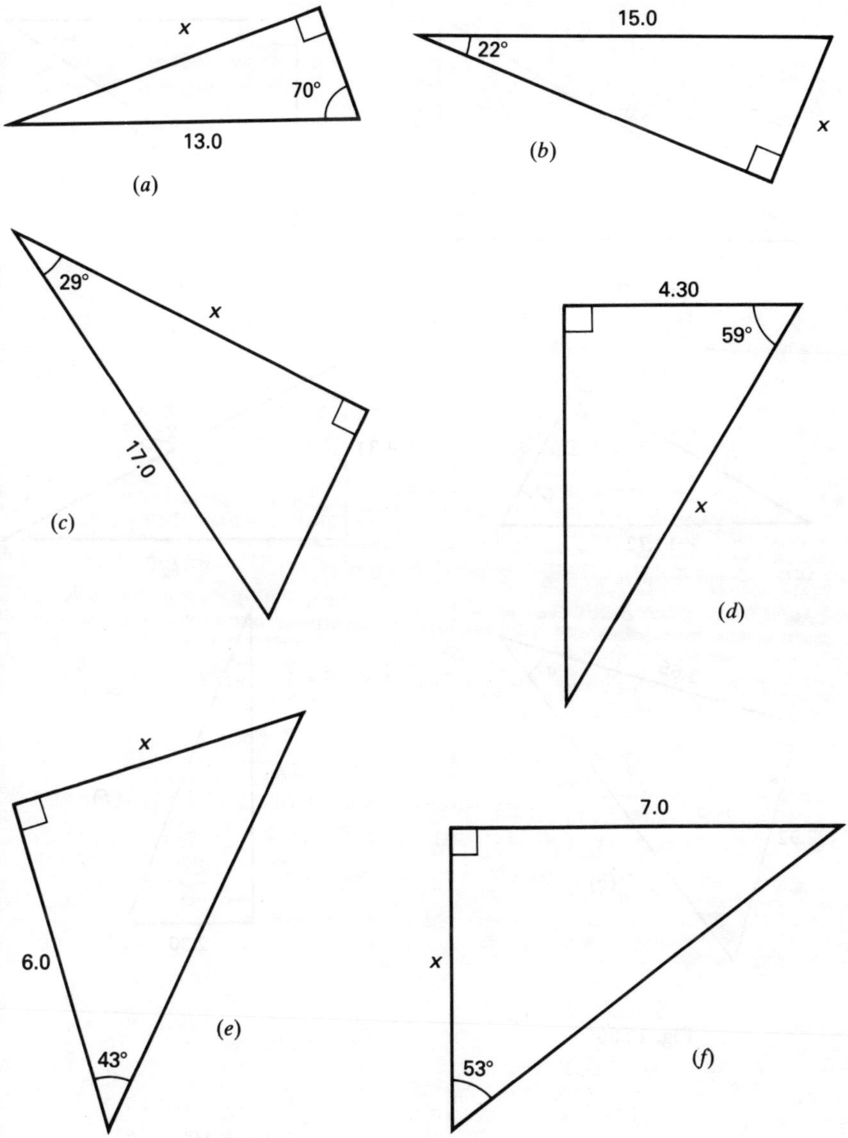

Fig. 13.34

14. Find the angles marked α in Fig. 13.35(*a*) to (*f*)
(*a*) [26° 23′] (*b*) [33° 49′] (*c*) [61° 21′] (*d*) [63° 58′]
(*e*) [38° 40′] (*f*) [22° 20′]

15. Evaluate the following:
(*a*) $4 \cos 63° + 2 \sin 27°$
(*b*) $2.6 \cos 22° - 4.1 \sin 67° + 6.2 \tan 55°$

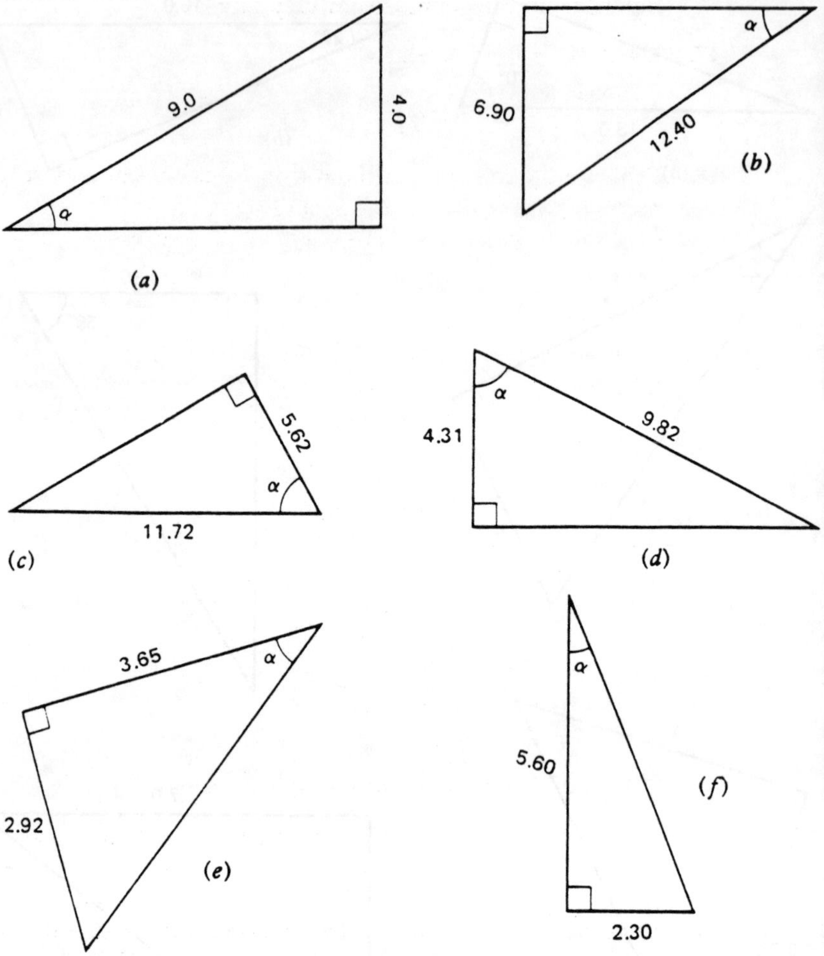

Fig. 13.35

(*c*) $\tan 82° \, 6' - 3 \cos 12°$ (*d*) $\dfrac{5.3 \tan 10°}{4 \cos 17° - 3 \sin 47°}$

(*a*) [2.724] (*b*) [7.491] (*c*) [4.272] (*d*) [0.5729]

16. If $\angle \mathrm{D} = 32° \, 17'$ and $\angle \mathrm{E} = 73° \, 52'$ evaluate the following (to 4 significant figures):

(*a*) $2 \sin \mathrm{D} \cos \mathrm{E}$ (*b*) $\dfrac{\tan \mathrm{D} + \tan \mathrm{E}}{1 - \tan \mathrm{D} \tan \mathrm{E}}$ (*c*) $\dfrac{3 \tan \mathrm{D}}{2 \sin \mathrm{D} \sin \mathrm{E}}$

(*a*) [0.2968] (*b*) [−3.453] (*c*) [1.847]

Surd forms and complementary angles

17. If $\cos 41° = 0.766$ evaluate:

$$2\left(\frac{5.4 \sin 49°}{3 \cos 41°} - \sin 90°\right) \qquad [1.6]$$

18. Evaluate the following without using a calculator and, if necessary, leave the answer in surd form:

(*a*) $2 \sin 30° + 3 \cos 60°$ (*b*) $3 \tan 60° - 2 \sin 60°$

(*c*) $\dfrac{3 \tan 30°}{\tan 60°}$

(*d*) $4 \cos 60° - 2 \sin 60°$

(*e*) $(\tan 45°)(2 \cos 60° - \sin 45° \cos 45°)$

(*a*) [2.5] (*b*) [$2\sqrt{3}$] (*c*) [1] (*d*) [$2 - \sqrt{3}$] (*e*) [0.5]

19. Show that:

(*a*) $\sin 30° \cos 60° + \cos 30° \sin 60° = 1$

(*b*) $\cos 60° \cos 30° - \sin 60° \sin 30° = 0$

(*c*) $(\cos 45°)^2 = 1 - (\sin 45°)^2$

(*d*) $\dfrac{1 - (\tan 45°)^2}{2 \tan 45°} = 0$ (*e*) $\dfrac{\sin 60°}{\cos 60°} = \tan 60°$

(*f*) $\dfrac{\tan 60° - \tan 30°}{1 + \tan 30° \tan 60°} = \dfrac{1}{\sqrt{3}}$

Solution of right-angled triangles

20. In an isosceles triangle ABC, AB = BC = 7 cm. If ∠ABC is 76° find the length of AC and the value of angles BAC and ACB. [8.169 cm, 52°, 52°]

21. Find the unknown sides and angles in the right-angled triangles shown in Fig. 13.36. The dimensions shown are centimetres.

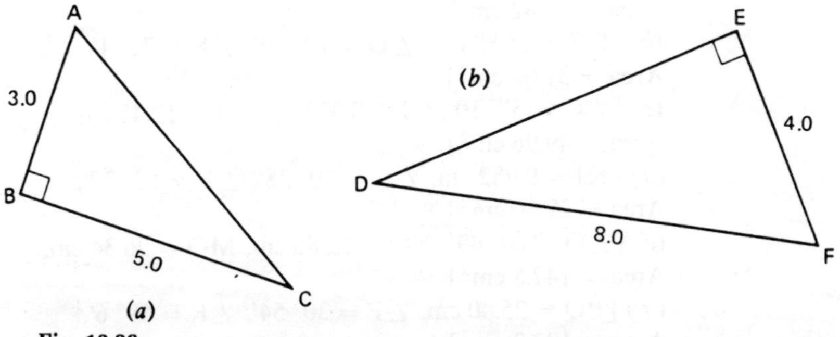

Fig. 13.36

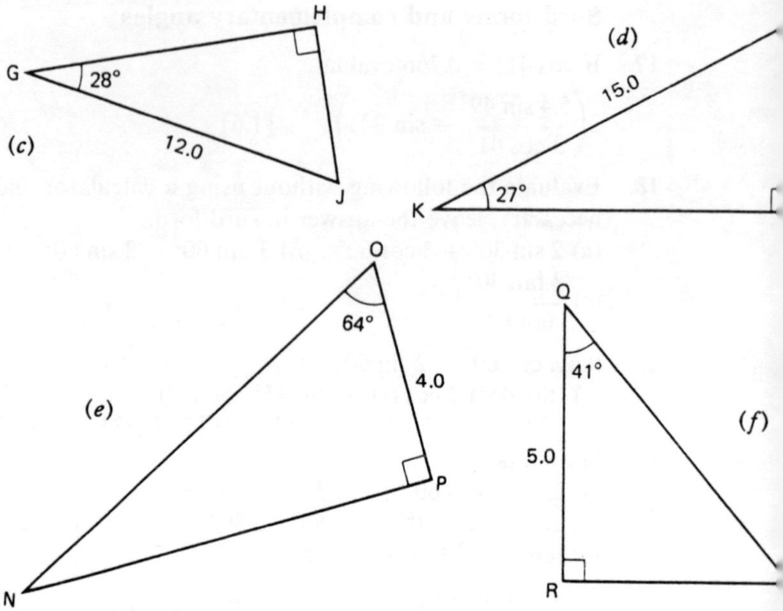

Fig. 13.36 (*continued*)

(*a*) [AC = 5.831 cm, ∠A = 59° 2′, ∠C = 30° 58′]
(*b*) [DE = 6.928 cm, ∠D = 30°, ∠F = 60°]
(*c*) [∠J = 62°, HJ = 5.634 cm, GH = 10.59 cm]
(*d*) [∠L = 63°, LM = 6.810 cm, KM = 13.37 cm]
(*e*) [∠N = 26°, ON = 9.125 cm, PM = 8.201 cm]
(*f*) [∠S = 49°, RS = 4.346 cm, QS = 6.625 cm]

22. In a triangle XYZ, ∠X = 90°, ∠Y = 15° 43′ and XZ = 4.23 cm. Find the lengths of XY and YZ.
[15.03 cm, 15.61 cm]

23. Solve the triangles shown in Fig. 13.37 and find their areas. The dimensions are in centimetres.
(*a*) [AB = 5.140 cm, AC = 7.900 cm, ∠A = 49° 25′, Area = 15.42 cm^2]
(*b*) [DF = 15.88 cm, ∠D = 17° 59′, ∠F = 72° 1′, Area = 37.00 cm^2]
(*c*) [∠G = 57° 10′, GI = 8.025 cm, IH = 12.45 cm, Area = 49.96 cm^2]
(*d*) [KJ = 9.982 cm, ∠K = 30° 38′, ∠L = 59° 22′, Area = 29.50 cm^2]
(*e*) [∠O = 60° 49′, NO = 12.85 cm, MO = 26.34 cm, Area = 147.8 cm^2]
(*f*) [PQ = 25.40 cm, ∠P = 30° 54′, ∠R = 59° 6′, Area = 193.0 cm^2]

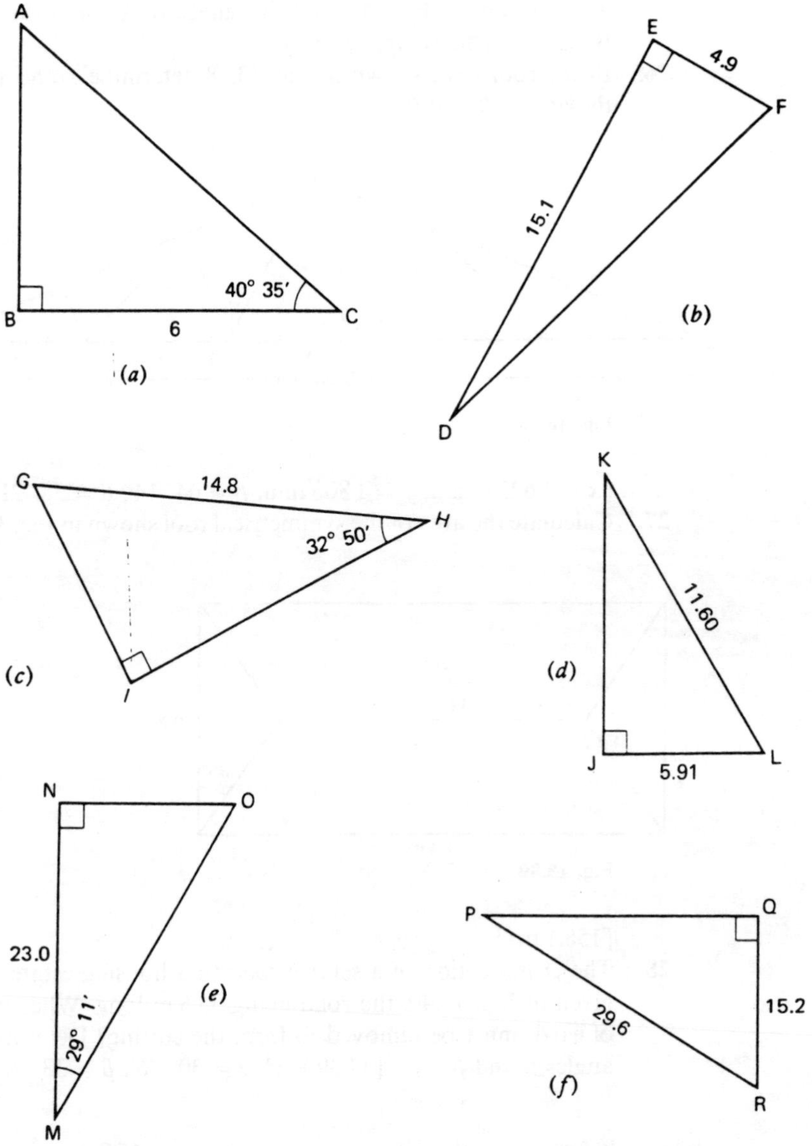

Fig. 13.37

24. A ladder rests against the top of the wall of a house and makes an angle of 77° with the ground. If the foot of the ladder is 8 m from the wall find the height of the house. [34.65 m]

25. A right-angled triangle ABC has an area of 16.8 cm^2. If BC is 5.2 cm and ∠B is 90° find the length of AC and the angle BAC. [8.294 cm, 38° 50′]

26. In the roof strut shown in Fig. 13.38 determine x and y and the angles β and θ.

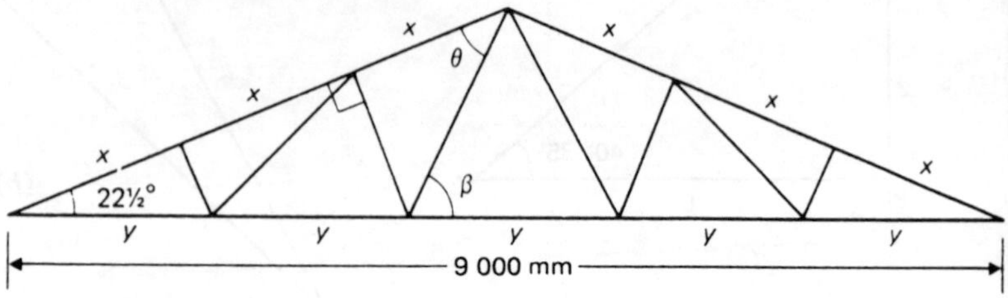

Fig. 13.38

[$x = 1\,623.6$ mm, $y = 1\,800$ mm, $\beta = 64°\ 14'$, $\theta = 38°\ 21'$]

27. Calculate the area of the symmetrical roof shown in Fig. 13.39.

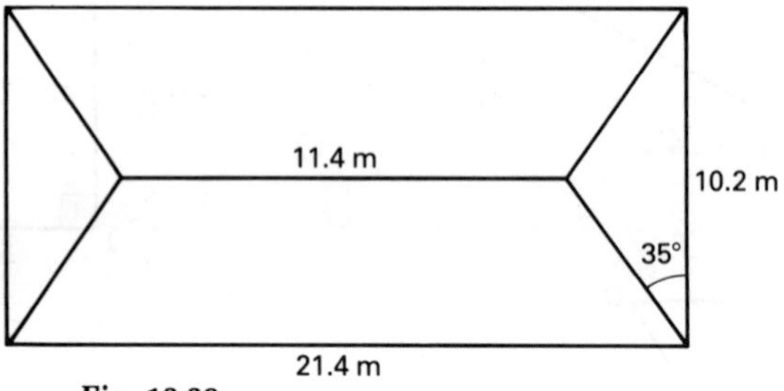

Fig. 13.39

[158.1 m^2]

28. The cross-section of a service road to a housing estate is given in Fig. 13.40, the road being 35.6 m long. What volume of earth must be removed to form the cutting? Determine the angles α and β. [11 390 m^3, $\alpha = 30°\ 46'$, $\beta = 33°\ 20'$]

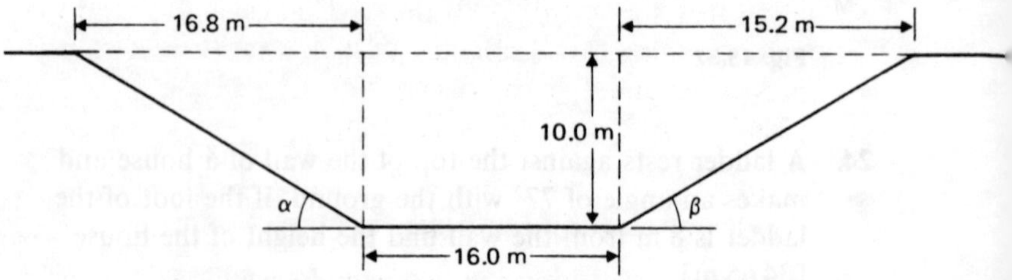

Fig. 13.40

29. Figure 13.41 shows a symmetrical scissors truss with BC = 9.0 m and ∠DBC = 30°. Determine the lengths of the struts and ties.

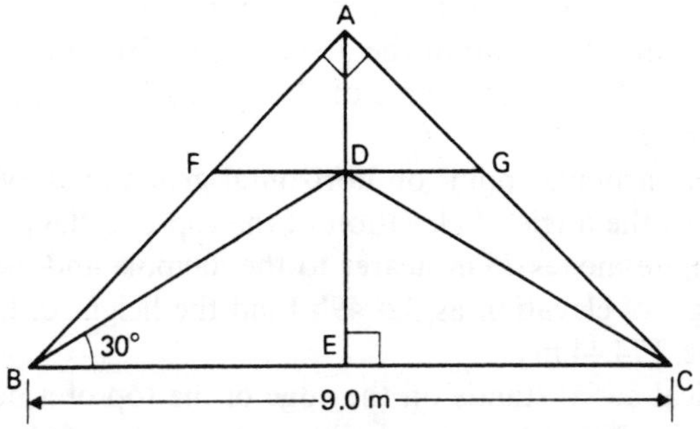

Fig. 13.41

[AF = AG = 2.69 m, FB = GC = 3.67 m,
FD = DG = 1.90 m, BE = EC = 4.50 m,
BD = DC = 5.20 m, AD = 1.90 m, DE = 2.60 m]

30. In the molecule $(CH_3)_3P$ the C—P—C bond is represented by Fig. 13.42

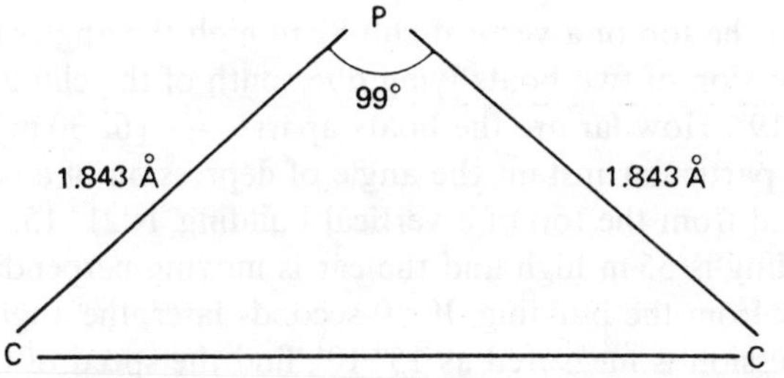

Fig. 13.42

What is the distance between the C atoms in centimetres given that 1 m = 10^{10} Å? If the CNC angle is 111° and the corresponding C—N distance 1.470 Å what is the distance between C and C in say $(CH_3)_3N$?
[2.803×10^{-8} cm, 2.423×10^{-8} cm]

Angles of elevation and depression

31. A vertical tower stands on horizontal ground. At a point 120 m from the foot of the tower the angle of elevation of

the top is 21°. Find the height of the tower.
[46.06 m]

32. If the angle of elevation of the top of a vertical 52 m high building is 31° 15′, how far is it to the building? [85. 69 m]
33. At a point 275 m from the foot of a vertical aerial, the angle of elevation of the top of the aerial is 16°. Find the height of the aerial. What is the angle of elevation of its mid point? [78.85 m, 8° 10′]
34. From a particular point on horizontal ground a surveyor measures the angle of elevation of the top of a flagpole as 21° 21′. He moves 75 m nearer to the flagpole and measures the angle of elevation as 32° 49′. Find the height of the flagpole. [74.44 m]
35. A vertical aerial stands on the edge of the top of a building. At a point 230 m from the building the angles of elevation of the top and bottom of the aerial are 37° and 34°. Calculate the height of the aerial. [18.18 m]
36. From the window of a house 110 m away from a tower, the angle of elevation of the top of the tower is 28° and the angle of depression of the bottom is 10°. Find the height of the tower. [77.88 m]
37. From the top of a vertical cliff 125 m high the angle of depression of a boat is 18° 50′. Find the distance of the boat from the cliff. [366.6 m]
38. From the top of a vertical cliff 82 m high the angles of depression of two boats lying due south of the cliff are 25° and 19°. How far are the boats apart? [62.30 m]
39. At a particular instant the angle of depression of a car, viewed from the top of a vertical building, is 21° 15′. The building is 55 m high and the car is moving perpendicularly away from the building. If, 10 seconds later, the angle of depression is measured as 15° 10′, find the speed of the car in km h^{-1}. [22.12 km h^{-1}]
40. A vertical flagstaff BE, 10 m high, stands at the corner B of a horizontal rectangular parade ground ABCD. The angles of elevation of the top of the flagstaff from A and D are 15° 11′ and 13° 16′, respectively. Calculate (*a*) the length of AB; (*b*) the length of BD; (*c*) the length of BC; and (*d*) the angle of elevation of E from C.
(*a*) [36.85 m] (*b*) [42.43 m] (*c*) [21.03 m] (*d*) [25° 26′]
41. From a boat at sea, the angles of elevation of the top and bottom of a vertical lighthouse standing on the edge of a cliff are 33° and 27° respectively. If the lighthouse is 23 m high, calculate the height of the cliff. [83.76 m]

42. From the top of a cliff 143 m high, the angles of depression of the top and bottom of a lighthouse which is at sea level are observed to be 35° and 42°. Find the height of the lighthouse. [31.80 m]
43. From a window 5 m above the ground a man notices that the angle of depression of the foot of a building across the road is 26°, and the angle of elevation of the top of the same building is 32°. Find the height of the building and the width of the road (assuming the road to be horizontal). [11.41 m, 10.25 m]
44. The elevation of a tower from two points, one due south of the tower and the other due north of it are 16° and 21° respectively and the two points of observation are 400 m apart. Find the height of the tower. [65.65 m]

Sine and cosine curves

45. Draw to scale a sine curve and a cosine curve on the same axis over one complete cycle. Make the maximum value 4 cm.

Chapter 14

An introduction to statistics

14.1 Introduction

Although statistics has been little more than a form of State Arithmetic in the past, as used today it is a scientific approach to collecting, organising, summarising and analysing numerical data and presenting them in a form such that reasonable predictions and decisions can be based on the analysis.

14.2 Data classification

Quantities which vary, such as the masses or heights of a group of people, are called **variables**. When variables can have any value at all within certain limits they are called **continuous** and since, for example, the mass of a child passes through all possible values between certain limits, then a set of data based on masses is continuous. Other variables can only have certain specific values within limits; for example, the number of children in a family can only be integer values, one or two or three and so on, and not, for example, two and a quarter. Data in a set of this sort are called **discontinuous** or **discrete**.

In statistics, it is often the method of analysis which is adopted which determines whether a variable is discrete or continuous. When data which are continuous are formed into a frequency distribution (see section 14.4) – for example, heights of a group of people measured correct to the nearest centimetre – then they become discontinuous. Similarly, discrete data with only a relatively small difference in values compared with the size of the variable are often treated as if they are continuous; for example, analysis of the population of a country. In general, measurements give rise to continuous data and counting gives rise to discrete data.

A **set** is a group of data and an individual value within the set is called a **member** of the set. Thus, if the heights of five objects are measured and are found to be 10 cm, 13 cm, 15 cm, 6 cm and 9 cm, correct to the nearest centimetre, then the set of the heights of the five objects is [10, 13, 15, 6, 9] and one of the members of the set is 13. A set containing all the members is called a **population**, thus, for example, **all** the telephone numbers for a given area form a population. However, if 30 telephone numbers are selected at random from a directory for the given area, the set of those 30 numbers is called a **sample**. The number of times that the value of a member occurs in a set is called the **frequency** of that member. Thus, for the set:

2, 3, 7, 4, 2, 5, 3, 4, 2, 5, 11, 7, 4

members 2 and 4 each have a frequency of three, since they each occur three times, members 3, 5 and 7 each have a frequency of two and member 11 has a frequency of one. The **relative frequency** with which a member of a set occurs is given by:

$$\text{relative frequency} = \frac{\text{frequency of member}}{\text{total number of members in the set}} \times 100\%.$$

Thus, the relative frequency of the member 2 is $\frac{3}{13} \times 100\%$, that is, approximately 23%.

Worked problems on data classification

Problem 1. State whether data obtained on the following topics are likely to be discrete or continuous:
(*a*) the number of premium bonds sold by a Post Office,
(*b*) temperature of air recorded every hour at a weather station,
(*c*) the time to failure of an electric light bulb, and
(*d*) the annual income of a group of workers.

(*a*) Discrete; since when counting the numbers sold, only integer values can result.
(*b*) Continuous; the air temperature is measured and can be any value within certain limits.
(*c*) Continuous; the amount of time is measured and can be any value within certain limits.
(*d*) Discrete; the annual income is likely to be to the nearest pound or even if an annual income is based on pounds and pence it will still be discrete.

Problem 2. For the set of data given, determine (*a*) the frequency of each member of the set and (*b*) the percentage relative frequency of each member.

$$\begin{Bmatrix} 19 & 14 & 15 & 17 & 15 \\ 17 & 15 & 17 & 15 & 14 \\ 15 & 19 & 15 & 14 & 15 \\ 12 & 15 & 17 & 14 & 17 \end{Bmatrix}$$

Table 14.1

Number	Tally	Frequency	Relative frequency %
12	1	1	5
14	1111	4	20
15	~~1111~~ 111	8	40
17	~~1111~~	5	25
19	11	2	10
	Total:	20	

One method of determining the frequencies is to use a 'tally'. An inspection of the set shows that it contains the numbers 12, 14, 15, 17 and 19 only. These are listed in the left-hand column of Table 14.1. Each member of the set is inspected in turn and allocated to a particular row, depending on its value, by placing a mark in the tally column. Each group of 4 marks is 'barred' when the fifth mark is allocated to a particular row, to assist in the counting process, The frequency values are as shown in Table 14.1, obtained by adding the appropriate tally marks. The relative frequency values are shown in the right-hand column of Table 14.1, and are obtained from

$$\text{relative frequency} = \frac{\text{frequency of member}}{\text{total frequency of all members}} \times 100\%$$

For the first row, relative frequency $= \dfrac{1 \times 100}{20} = 5\%$. For the second row, relative frequency $= \dfrac{4 \times 100}{20} = 20\%$, and so on.

Further problems on data classification may be found in Section 14.5 (Problems 1–4), page 370.

14.3 Presentation of data

A column of numbers can be uninspiring and does not really indicate trends at a glance. Graphs can be used to present data visually, but there are also many other ways of doing so. Some of these are dealt with in this section, for sets having a few members only.

Pictograms

A pictogram uses pictures to represent data, the usual method being to have a group of pictorial units, all the same size, the numerical value being represented by the number of pictures shown. For example, a survey was carried out in a firm to find out the mode of travel used by employees to get to work. It showed that 35% used a bus, 25% used a car or motor cycle, 20% came on a bicycle, 10% used a train and 10% walked. These data could be depicted as shown in the pictogram in Fig. 14.1.

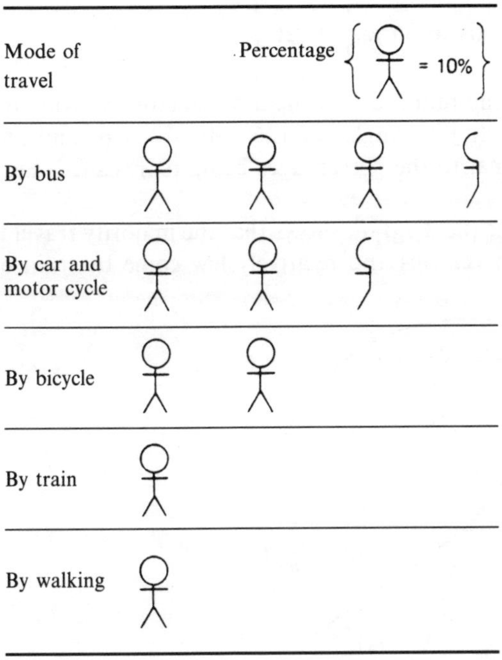

Fig. 14.1 Pictogram showing mode of travel to work for people employed by a firm

A glance at the pictogram shows that the majority of employees arrive at work either by bus or in their own vehicles. Relatively few come by train or live near enough to walk.

Pie diagrams

A pie diagram is constructed by drawing a circle of any radius and subdividing it into sectors, so that the area of each sector is proportional to the quantity it is representing. To achieve this, the angles at the centre of the circle must also be in proportion to the quantities that are being represented. The information about the mode of travel depicted in Fig. 14.1 by a pictogram could equally well have been shown on a pie diagram. Since

100% corresponds to 360°
10% corresponds to 36°
20% corresponds to $36 \times 2 = 72°$

25% corresponds to $36 \times \frac{5}{2} = 90°$

and 35% corresponds to $36 \times \frac{7}{2} = 126°$,

by drawing a circle of any radius and using a protractor to divide the angle at the centre of the circle into angles of 126°, 90°, 72°, 36° and 36°, these angles being proportional to the percentages being represented, Fig. 14.2 is produced.

An examination of the pie diagram shows that the majority travel to work by bus or in their own vehicles and relatively few come by train or walk.

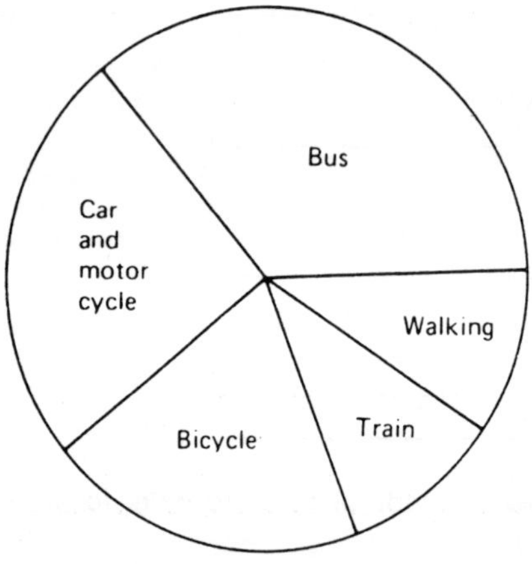

Fig. 14.2 Pie diagram showing mode of travel to work for employees of a firm

Bar charts

In a bar chart, data are represented by a series of bars or strips, the width of each bar being unimportant but the length of the bar being proportional to the quantity being represented. The bars can be drawn either horizontally or vertically. Figure 14.3 shows the data concerning mode of travel shown by means of a vertical bar chart, whereas Fig. 14.4 shows the same data presented as a horizontal bar chart.

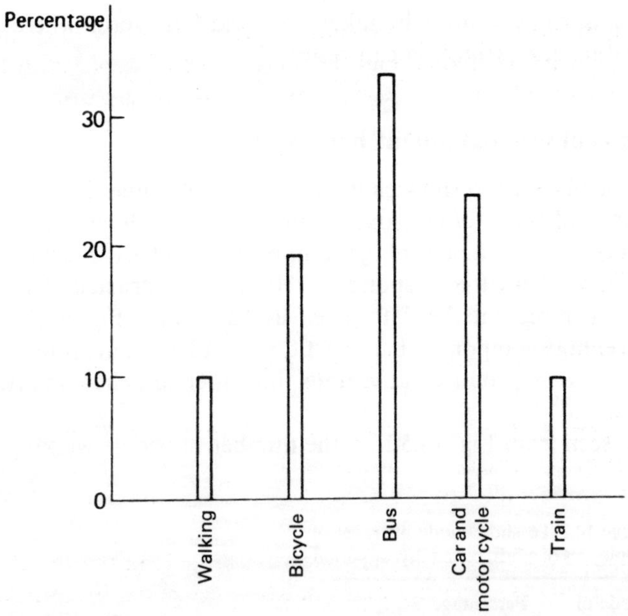

Fig. 14.3 Vertical bar chart showing mode of travel to work for employees of a firm

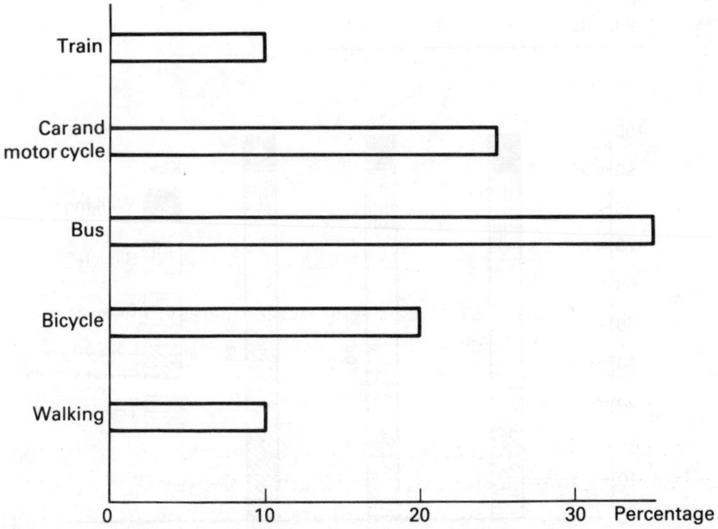

Fig. 14.4 Horizontal bar chart showing mode of travel to work for employees of a firm

As with graphs, care must be taken to avoid false zeros where possible, or a misleading representation can result.

Percentage component bar charts

These are usually used to show changing trends with time. Table 14.2 shows how the mode of travel of employees changed over a three-year period. To represent these data on a percentage component bar chart, a rectangle of any width but whose height corresponds to 100% is constructed. The height is sub-divided to represent the data given in Table 14.2. For just one set of data, a percentage component bar chart does not have a great deal of value, but if these bars are produced at various time intervals, changing trends can be seen readily.

It can be seen from Fig. 14.5 that the number of people walking to work

Table 14.2 To show mode of travel of employees of a firm over a 3-year period

Mode of travel	Percentage		
	Year 1	Year 2	Year 3
Train	10	5	5
Car and motor cycle	25	40	50
Bus	35	30	25
Bicycle	20	15	10
Walking	10	10	10

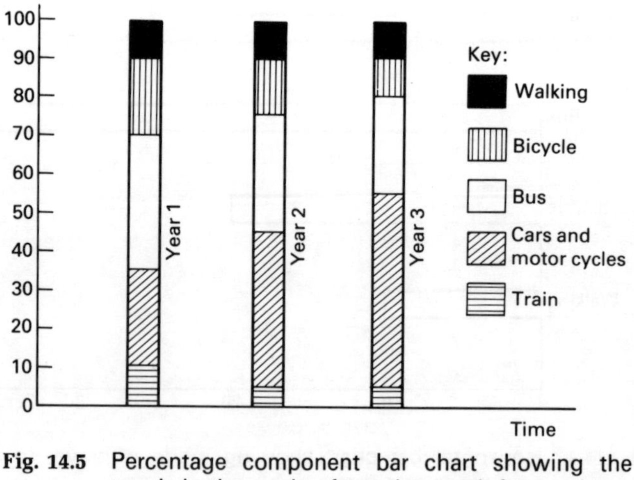

Fig. 14.5 Percentage component bar chart showing the changing trends in the mode of travel to work for employees of a firm

is remaining constant, but more employees are bringing cars and motorcycles at the expense of those cycling or travelling by public transport.

Problems on the presentation of data may be found in Section 14.5 (*Problems 5–11*), page 371.

14.4 Grouped data

When a set has a large number of members, it is not practical to use the methods introduced in Section 14.3 to depict the information given. Consider the set containing the heights of fifty children of the same age group, measured correct to the nearest centimetre, as shown.

171	170	172	170	171	166	169	169	169	168
175	168	167	168	168	172	168	173	168	170
169	173	169	173	170	170	177	167	167	172
173	171	174	168	174	171	166	169	167	165
167	174	170	168	163	169	173	174	170	168

These data are called **raw** or **ungrouped data**. The data are presented in a haphazard form and no conclusions can be drawn easily from such data. To present and analyse the information given in the set, a **frequency distribution** can be formed and the data are then said to be **grouped**. To form a frequency distribution, the **range** of the set is found by taking the smallest value from the largest; for this set, the range is 177–163, that is, 14 centimetres. It is normal to group the information together to give about six to eight **classes**.

Table 14.3 A frequency distribution showing the heights of children of the same age group

Class (centimetres)	Tally	Frequency
162–164	1	1
165–167	~~1111~~ 111	8
168–170	~~1111~~ ~~1111~~ ~~1111~~ ~~1111~~ 111	23
171–173	~~1111~~ ~~1111~~ 11	12
174–176	~~1111~~	5
177–179	1	1
	Total	50

The information shown in Table 14.3 has been selected to give six classes with no overlap between any two numbers forming them. For example, the first class ends at 164 centimetres and the second class starts at 165 centimetres. The class shown as 168–170 has **class limits** of 168 and 170 and 168–170 is called the **class interval**. Figure 14.6 shows that this class actually extends from 167.5 to 170.5 centimetres since the boundaries are halfway between 167 and 168, and 170 and 171. These two boundaries are called the **lower** and **upper class boundaries** respectively. The midpoint between these two boundaries is called the **class midpoint** and for the 168–170 class it is

$$\frac{167.5 + 170.5}{2} = 169 \text{ cm}.$$

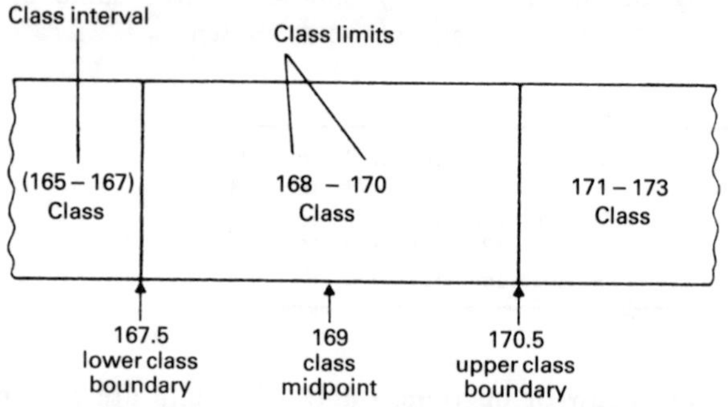

Fig. 14.6 A class of grouped data showing the terminology used

Selecting each height in turn from the ungrouped data, a mark is made in the appropriate row corresponding to the class required, thus for 171 centimetres, the class selected will be the 171–173 class. To assist with counting, a tally system is used, as introduced in section 14.2. By adding the tally, the frequency of each class is obtained. A table containing class intervals together with their frequencies is called a frequency distribution. Grouped data are usually presented pictorially either by a **histogram** or a **frequency polygon**.

Histograms

A histogram is probably one of the most widely used methods of depicting grouped data. Rectangles are drawn vertically, the width of the rectangles representing the class intervals and the height of the rectangles the class frequencies. Since the area is equal to class interval multiplied by frequency, it is the area which is proportional to the number of observations in the class concerned. When class intervals are equal in size, the heights of the rectangles also represent the frequencies.

A histogram depicting the frequency distribution given in Table 14.3 is shown in Fig. 14.7. Class midpoint values are used on the horizontal axis and

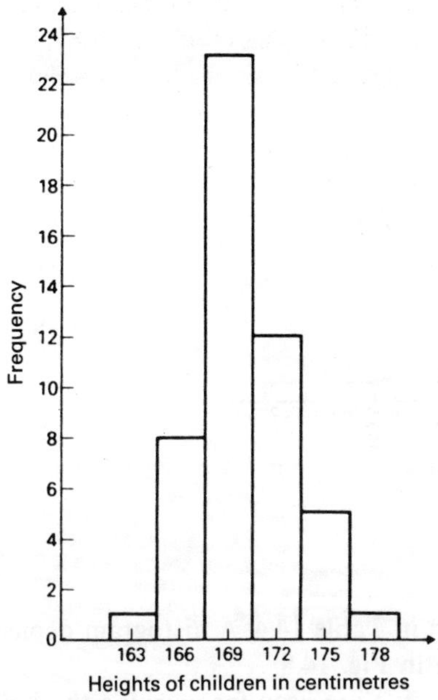

Fig. 14.7

Table 14.4 Determining the height of histogram rectangles for unequal class intervals

Class intervals	Frequency	Height of rectangle of histogram
0–9	15	$\frac{15}{10} = 1.5$ units
10–19	40	$\frac{40}{10} = 4.0$ units
20–39	70	$\frac{70}{20} = 3.5$ units
40–99	36	$\frac{36}{60} = 0.6$ units
100–199	20	$\frac{20}{100} = 0.2$ units

rectangles having heights directly proportional to the class frequencies are constructed vertically as shown.

For unequal class intervals, the heights of the rectangles must be adjusted so that the area is proportional to the frequency of the class. The height of a rectangle is proportional to the frequency divided by the class interval,

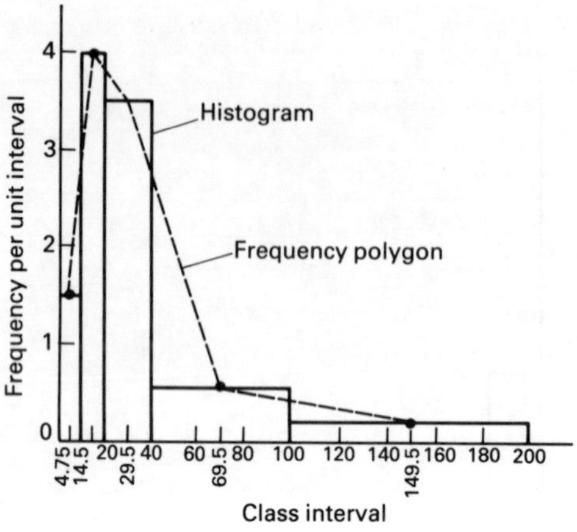

Fig. 14.8

an illustration of this being shown in Table 14.4. A histogram depicting the data given in Table 14.4 is shown in Fig. 14.8.

Class boundaries or class midpoints are used for marking the horizontal scale. It is usually easier to mark the class midpoints and then to draw the bases of the rectangles symmetrically about the midpoints. Frequency is used for the vertical scale for equal class intervals or frequency per unit class interval for unequal class intervals.

Frequency polygons

By connecting the midpoints of the tops of the rectangles of a histogram together by straight lines and joining the midpoint of the top of the first rectangle to the base at the centre of the preceding class interval and the midpoint of the top of the last rectangle to the base at the centre of the next class interval, a figure called a **frequency polygon** is formed. Frequency polygons for data having unequal class intervals cannot be 'anchored', because the class midpoint of the next class interval is not known. A frequency polygon for the data given in Table 14.4 is shown in Fig. 14.8 by a broken line.

Ogives

The curve obtained when a cumulative frequency is plotted as a graph is called an **ogive** (pronounced o-jive) or **cumulative frequency curve**. A frequency distribution can be presented as an ogive in two ways. By taking upper class boundaries, frequencies which are 'less than' based can be

Table 14.5 'Less Than' and 'Or More' based cumulative frequency distributions, based on the heights of 60 people

Frequency distribution		'Less Than' based cumulative frequency distribution		'Or More' based cumulative frequency distribution	
Class intervals	Frequency	Class	Cumulative frequency	Class	Cumulative frequency
		Less than 164.5	0	164.5 or more	60
165–166	6	Less than 166.5	6	166.5 or more	54
167–168	14	Less than 168.5	20	168.5 or more	40
169–170	20	Less than 170.5	40	170.5 or more	20
171–172	11	Less than 172.5	51	172.5 or more	9
173–174	6	Less than 174.5	57	174.5 or more	3
175–176	3	Less than 176.5	60	176.5 or more	0

determined. Alternatively, by taking the lower class boundaries, an 'or more' based cumulative frequency distribution can be determined, sometimes called a reverse cumulative frequency distribution. Both 'less than' and 'or more' based cumulative frequency distributions are shown in Table 14.5, the frequency distribution relating to the heights of 60 people measured at random.

A graph of the ogive of these values is shown in Fig. 14.9

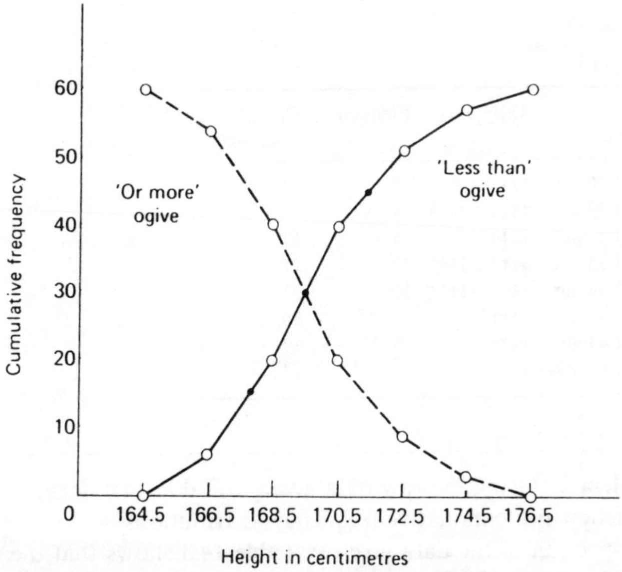

Fig. 14.9

Worked problems on grouped data

Problem 1. Form a frequency distribution and cumulative frequency distribution of the data given in Table 14.6, which refer to the diameter of components produced by a machine, measured correct to the nearest hundredth of a millimetre.

Table 14.6 Measurements of 40 components produced by a machine correct to the nearest hundredth of a millimetre

1.37	1.38	1.34	1.30	1.39	1.36	1.28	1.33
1.27	1.30	1.37	1.42	1.29	1.35	1.24	1.31
1.38	1.33	1.33	1.35	1.45	1.43	1.40	1.38
1.35	1.32	1.25	1.38	1.35	1.34	1.28	1.36
1.41	1.34	1.26	1.38	1.36	1.42	1.40	1.39

Each member of the set is examined to find the largest and smallest members. This gives the range and is (1.45 – 1.24) or 0.21 mm. Classes are selected to give, say, a total of eight classes. These are shown in Table 14.7. Each member of the set is placed in its appropriate class, using a tally system, and hence the frequency of each class is determined. By adding each frequency to the sum of the previous frequencies the cumulative frequencies are determined.

Table 14.7

Class intervals	Tally	Frequency	Cumulative frequency
1.24–1.26 mm	111	3	3
1.27–1.29 mm	1111	4	7
1.30–1.32 mm	1111	4	11
1.33–1.35 mm	~~1111~~ ~~1111~~	10	21
1.36–1.38 mm	~~1111~~ ~~1111~~	10	31
1.39–1.41 mm	~~1111~~	5	36
1.42–1.44 mm	111	3	39
1.45–1.47 mm	1	1	40

Problem 2. Using the raw data given in Table 14.8, form frequency and cumulative frequency distributions.

Inspection of the data given in Table 14.8 shows that the range is (182 – 2) or 180. However, it is immediately apparent that there are very few numbers larger than 100. A quick count, to give

Table 14.8

47	18	35	8	12	182	19	27	15	49
73	25	3	77	29	53	71	6	82	7
6	58	36	38	5	6	9	42	57	28
104	22	44	16	30	13	27	137	11	64
87	45	5	15	23	10	79	12	2	40
17	44	123	9	43	7	14	37	35	161
36	94	14	20	97	10	67	32	59	18

some idea of the distribution of the numbers, shows that:

Class intervals	Frequency
0–49	51
50–99	14
100–149	3
150–199	2

This shows that to overcome the disadvantage of having most of the data in very few classes, unequal class intervals should be selected. A second count in the range 0–100 shows that:

Class intervals	Frequency
0–19	27
20–39	16
40–59	12
60–79	6
80–99	4

The majority of numbers are less than 40. There is no hard and fast rule to help determine the final selection of class intervals and one possible selection is shown in Table 14.9. Many other arrangements of class intervals could have been selected and would have been equally correct.

Table 14.9

Class intervals	Tally	Frequency	Cumulative frequency
1–4	11	2	2
5–9	~~1111~~ ~~1111~~	10	12
10–19	~~1111~~ ~~1111~~ ~~1111~~	15	27
20–49	~~1111~~ ~~1111~~ ~~1111~~ ~~1111~~ 1111	24	51
50–99	~~1111~~ ~~1111~~ 1111	14	65
100–199	~~1111~~	5	70

Problem 3. Draw a histogram, frequency polygon and ogive for the data given in Table 14.10

Table 14.10

Class intervals	Frequency	Cumulative frequency
1.24–1.26 mm	2	2
1.27–1.29 mm	4	6
1.30–1.32 mm	4	10
1.33–1.35 mm	10	20
1.36–1.38 mm	11	31
1.39–1.41 mm	5	36
1.42–1.44 mm	3	39
1.45–1.47 mm	1	40

The class midpoints are selected for the horizontal scale of the histogram and frequency for the vertical scale. Rectangles are drawn as shown in Fig. 14.10 so that they symmetrically span the class midpoints and touch one another at the vertical common line. The frequency polygon is also shown in Fig. 14.10 by the broken line and is produced by joining the central points of the

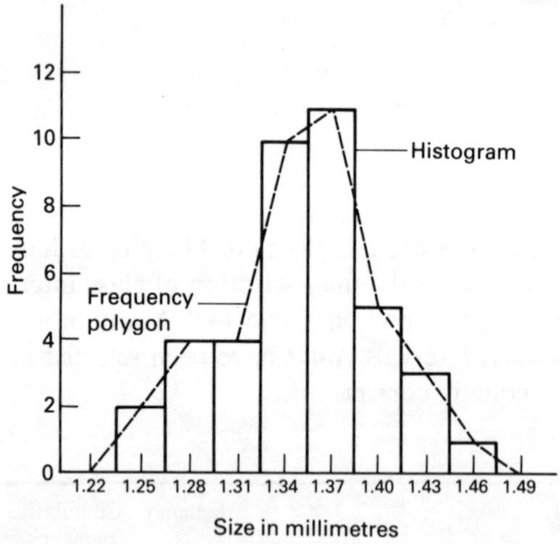

Fig. 14.10

tops of the rectangles together by straight lines. The polygon is 'anchored' at the start by drawing a straight line from the top centre of the first rectangle to the zero frequency of the midpoint of the class preceding it. It is 'anchored' at the other end by

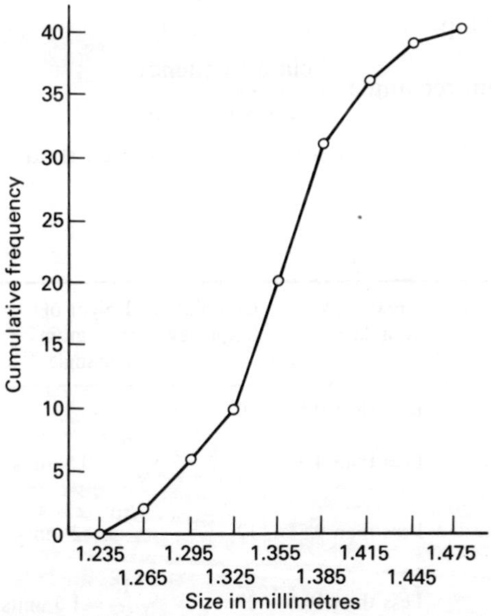

Fig. 14.11

drawing a straight line from the top centre of the last rectangle to the zero frequency of the next class after it.

Because the ogive is not specified as having a 'less than' or an 'or more' basis, the 'less than' basis is assumed. Upper class boundary values are selected for the horizontal scale and cumulative frequency as the vertical scale. The ogive is shown in Fig. 14.11.

Problem 4. Using the data given in Table 14.11, construct a histogram, frequency polygon and ogive.

Table 14.11

Class intervals	Frequency
1–4	2
5–9	10
10–19	15
20–49	24
50–99	14
100–199	5

The class intervals in this distribution are unequal and the heights of the rectangles of the histograms have to be calculated using the relationship:

$$\text{height of histogram rectangle} = \frac{\text{class frequency}}{\text{class interval}}$$

The data from Table 14.11 together with calculated data are given in Table 14.12

Table 14.12

Class intervals	Frequency	Upper class boundary	Cumulative frequency	Height of histogram rectangle
		Less than 0.5	0	
1–4	2	Less than 4.5	2	$\frac{2}{4} = 0.5$ units
5–9	10	Less than 9.5	12	$\frac{10}{5} = 2$ units
10–19	15	Less than 19.5	27	$\frac{15}{10} = 1.5$ units
20–49	24	Less than 49.5	51	$\frac{24}{30} = 0.8$ units
50–99	14	Less than 99.5	65	$\frac{14}{50} = 0.28$ units
100–199	5	Less than 199.5	70	$\frac{5}{100} = 0.05$ units

For the horizontal axis of the histogram, the class midpoints are $\frac{0.5 + 4.5}{2}$ or 2.5, $\frac{4.5 + 9.5}{2}$ or 7 and so on, and these are used as the central points of the rectangles. For histograms having unequal class intervals, the heights of the rectangles are not proportional to frequency. The units of the vertical scale are $\frac{\text{frequency}}{\text{class intervals}}$ and are shown as frequency per unit class interval. In this way, the area of the histogram is kept proportional to frequency. The histogram is shown in Fig. 14.12. By joining the central points of the tops of the rectangles together by straight lines, the frequency polygon is produced, shown by the broken line in Fig. 14.12. It is not possible to anchor the end points because the midclass points of the classes on either side of the polygon are not known.

The ogive is shown in Fig. 14.13 and is produced by plotting

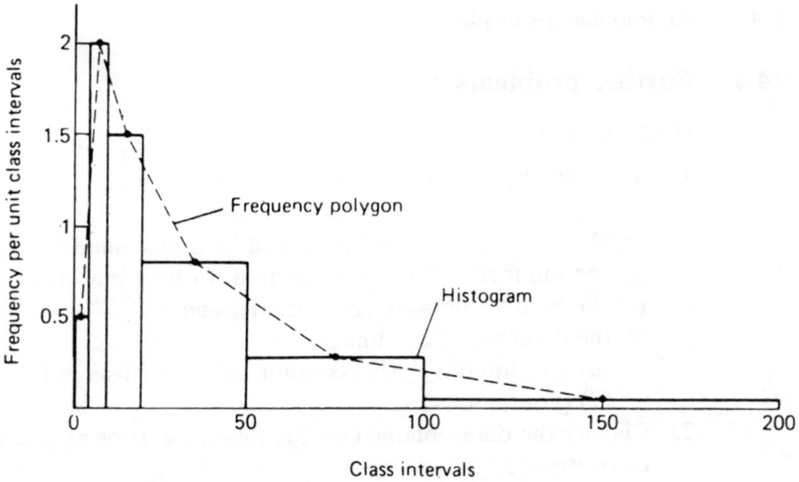

Fig. 14.12

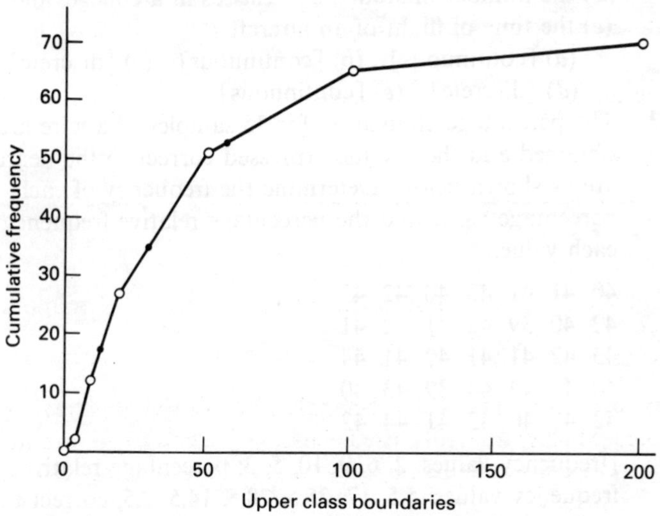

Fig. 14.13

the values of the cumulative frequencies against the values of the upper class boundaries.

Further problems on grouped data may be found in Section 14.5 following (Problems 12–21), page 372.

14.5 Further problems

Data classification

1. State whether the data obtained on the following topics are likely to be discrete or continuous:
(*a*) the diameter of a bolt produced by a machine tool,
(*b*) the number of litres of water in a washing machine,
(*c*) the number of books in a library, and
(*d*) the diameter of a cylinder.
(*a*) [continuous] (*b*) [continuous] (*c*) [discrete]
(*d*) [continuous]
2. Classify the data obtained on the following items as discrete or continuous:
(*a*) the rainfall measured daily in a weather station,
(*b*) the speed of a car measured in km h^{-1},
(*c*) the number of £1 coins in circulation,
(*d*) the number of students in classes in a college, and
(*e*) the time of flight of an aircraft.
(*a*) [continuous] (*b*) [continuous] (*c*) [discrete]
(*d*) [discrete] (*e*) [continuous]
3. The percentage elongation for 35 samples of a wire are obtained and the results, expressed correct to the nearest 1%, are as shown below. Determine the frequency of each percentage value and the percentage relative frequency for each value.

40	41	41	43	40	42	43
42	40	39	42	41	42	41
43	42	41	41	42	41	44
40	43	42	44	39	43	40
42	41	40	42	41	44	42

[frequency values: 2, 6, 9, 10, 5, 3; percentage relative frequency values: 5.5, 17, 25.5, 28.5, 14.5, 8.5, correct to the nearest 0.5%]
4. The capacitance values of a batch of 36 similar capacitors are as shown below, expressed correct to the nearest microfarad. Determine the frequency and percentage relative frequency for each value:

29	30	32	32	30	29
30	31	29	28	31	30
29	30	32	31	30	29
31	31	30	32	29	30
30	32	28	30	31	29
31	30	31	28	30	28

[frequency values: 4, 7, 12, 8, 5; percentage relative frequency values: 11, 19.5, 33.5, 22, 14, correct to the nearest 0.5%]

Presentation of data

5. The proportions of raw materials produced in certain areas are given below:

   Area A – 40 units
   B – 16 units
   C – 16 units
   D – 12 units
   E – 8 units
   F – 8 units

   Represent these data on (*a*) a pictogram, (*b*) a pie diagram, (*c*) a horizontal bar chart and (*d*) a vertical bar chart.
6. A company produces five products in the following proportions:

Product P	25
Product Q	17
Product R	15
Product S	10
Product T	5

   Draw (*a*) a pie diagram, (*b*) a horizontal bar chart and (*c*) a vertical bar chart to present these data visually.
7. An analysis of 100 items produced in a factory gave the following results:

Perfect	60
Defect due to poor finish	10
Defect due to being oversized	15
Defect due to being undersized	5
Defect due to mal-operation	10

   Represent these data visually in three different ways.
8. The water used by the chemical industry in the United Kingdom may be divided into classes as follows:

Pure uncontaminated water	77.8%
Contaminated water without pretreatment	3.5%
Contaminated water with pretreatment	18.7%

   Express these results as a pie chart.
9. The Gas Industry on assessing quality and reliability costs obtained the following values (Column 1) and considered that by involving management and workers in the quality control process this could be improved (Column 2).

	Column 1	Column 2
Costs to prevent defects	5%	10%
Cost of appraising output quality	30%	20%
Cost from defect or quality failure	65%	35%

Express these results as pie charts, showing clearly the savings involved in the second process.

10. The percentage sales in four departments of a company over a period of 5 years are as follows:

Year	1	2	3	4	5
Department A	10	15	17	16	8
Department B	27	22	20	21	29
Department C	34	38	42	47	50
Department D	29	25	21	16	13

Represent these data visually by means of a percentage component bar chart.

11. Over a three-month period, the six components produced by a factory varied as shown:

Month	1	2	3
Component P	9	10	8
Component Q	4	2	2
Component R	25	30	35
Component S	27	22	15
Component T	22	26	30
Component U	13	10	10

Represent these data by a visual diagram method.

Grouped data

12. Draw a histogram showing the haemoglobin levels in a number of children from the following data:

Level	60–64	65–69	70–74	75–79	80–84	85–89	90–94
Number	2	3	32	91	165	248	200
Level	95–99	100–104	105–109	110–114			
Number	156	98	23	9			

13. The lengths of components in centimetres cut off by an automatic guillotine machine are given below. Group these data into a frequency distribution.

5.38	5.36	5.26	5.37	5.35	5.36	5.39	5.45
5.29	5.41	5.39	5.28	5.24	5.33	5.35	5.36
5.43	5.35	5.34	5.37	5.33	5.42	5.40	5.28
5.40	5.31	5.27	5.36	5.42	5.38	5.25	5.33
5.34	5.32	5.33	5.30	5.32	5.30	5.39	5.34

[The distributions obtained will depend on the class intervals selected.]

14. The annual incomes of 50 employees of a company are listed below, the amount shown being correct to the nearest £10. Produce a frequency distribution for these data.

8 500	9 650	10 220	7 930	9 900	16 500	9 120	9 800
8 000	10 170	10 600	8 750	7 490	9 550	11 700	7 750
9 750	10 000	9 000	9 550	9 900	11 250	9 500	13 750
9 940	9 360	10 050	8 250	10 480	11 750	10 110	7 980
10 150	8 450	10 170	9 180	9 900	10 000	7 480	9 700
10 050	8 800	10 350	9 750	9 450	10 127	9 950	10 330
9 400	10 420						

[A distribution having unequal class intervals should be selected. The distribution obtained will depend on the class intervals selected.]

15. A batch of transistors was tested to destruction by operating them on overload conditions. The number of hours they functioned satisfactorily before failure occurred are shown below. Form frequency and cumulative frequency distributions for these data.

63	94	76	73	84	60	75	89	83	56
63	76	77	82	88	62	80	75	73	88
93	54	78	77	97	93	57	96	60	90
74	76	86	76	77	73	95	87	91	71
81	78	79	53	78	71	95	74	59	82
60	75	61	85	75	85	75	74	72	75

[The distributions obtained depend on the class intervals selected.]

16. The percentage elongation for samples of a wire are obtained and the results are as shown. Produce a frequency distribution and a cumulative frequency distribution for these data.

43	39	37	40	36	40	37	35	39	38
43	41	40	42	38	42	41	39	40	41
39	35	38	42	40	37	38	39	42	37
36	43	39	35	39	38	39	36	39	40
40	39	37	40	38	41	39	42	38	39
37	41	36	38	40	36	41	38	41	39

[The distributions obtained depend on the class intervals selected.]

17. The length in millimetres of a sample of bolts is shown below. Produce a frequency distribution, an 'or more' ogive and a 'less than' ogive for these data.

Height (cm)	165	166	167	168	169	170	171
Frequency	5	14	18	28	36	29	29
Height (cm)	172	173	174	175	176	177	
Frequency	24	19	15	6	3	2	

18. The annual incomes of a group of employees are given below. Draw a histogram, frequency polygon and ogive for these data.

Income (£)	7 500–7 999	8 000–8 499	8 500–8 999	9 000–9 499
Frequency	12	38	40	30
Income (£)	9 500–9 999	10 000–10 499	10 500–11 999	12 000–16 999
Frequency	19	12	8	4

(Note that the class intervals are unequal.)

19. Draw a histogram and ogive for the data given below, which refer to the number of hours a sample of electric light bulbs burned before failure.

Bulb life (hours)	Under 499	500–699	700–899	900–999
Frequency	14	23	36	40
Bulb life (hours)	1 000–1 099	1 100–1 299	1 300–1 499	Over 1 500
Frequency	41	34	25	12

(Note that the class intervals are unequal.)

20. The data given below refer to the intelligence quotient of a group of children. Draw a histogram, frequency polygon and ogive for these data.

I.Q.	60–79	80–89	90–94	95–99	100–104	105–114	115–129	130–149
Frequency	9	29	51	57	61	53	31	9

21. The percentage of chlorine in a chloro hydrocarbon was estimated 80 times with the following results:

21.83	21.84	21.85	21.86	21.87	21.88	21.89	21.90	21.91
1	1	2	6	8	10	28	13	6
21.92	21.93							
3	2							

Construct a histogram and frequency polygon of these results.

Index